The Mail Order Shopper for Parents

The Mail Order Shopper for Parents

Where to Get the Very Best for
Your Children—from Infants to Teens

Hal Morgan

Doubleday

New York   London   Toronto   Sydney   Auckland

Photograph credits:
Page 35 illustration by Maurice Sendak, copyright © 1971 by Scholastic Inc., all rights reserved, used with permission; page 55, © Lands' End, Inc., photo reprinted courtesy of Lands' End Catalog; page 163 left illustration copyright © 1987 by Jane Dyer from the HBJ book, *Three Bears Rhyme Book*; page 163 right illustration copyright © 1963 by Maurice Sendak, all rights reserved, used with permission.

Published by Doubleday, a division of Bantam Doubleday Dell Publishing Group, Inc., 666 Fifth Avenue, New York, New York 10103.

Doubleday and the portrayal of an anchor with a dolphin are trademarks of Doubleday, a division of Bantam Doubleday Dell Publishing Group, Inc.

A Steam Press Book

Library of Congress Cataloging-in-Publication Data
Morgan, Hal.
    The mail order shopper for parents : where to get the very best for your children—from infants to teens / Hal Morgan.—1st ed.
      p. cm.
   "A Steam Press book"—T.p. verso.
   Includes index.
   ISBN 0-385-26389-9
    1. Children's paraphernalia—Catalogs. I. Title.
HD9970.5.C482M67 1990
670'.29'4—dc20   89-35398
                                                            CIP

ISBN 0-385-26389-9
Copyright © 1990 by Hal Morgan
All Rights Reserved
Printed in the United States of America
January 1990
First Edition

Contents

Introduction 7
If a Problem Arises with an Order 9

General Catalogs 13
Art Supplies 14
Baby Needs 17
Bath Accessories 29
Bicycles, Tricycles, and Wagons 31
Birth Announcements and
 Greeting Cards 33
Books 36
Book Clubs 45
Childbirth 47
Children's Clothes 47
Children's Magazines
 For Young Children 62
 For School-Age Children 64
 For Older Children and Teens 71
Computers and Software 76
Cooking 78
Costumes 79
Dollhouses and Miniatures 81
Dolls 83
Educational Supplies 88
Furniture 95
Games and Puzzles 102
Gardening 107
Gifts 108
Health and Safety 113
Hobby Supplies 118
Kites, Boomerangs, and
 Juggling Supplies 121
Magic Tricks and Novelties 122
Maps 124
Maternity and Nursing Clothes 124
Model Trains, Planes, Cars, and
 Boats 128

Music 132
Outdoor Gear 138
Parents' Magazines
 General Parenting and Family 141
 For New Parents 145
 For Parents of Young Children 145
 Adoption 146
 Book and Media Reviews 146
 Childbirth 147
 Education and Child Care 147
 Health and Safety 150
 Regional and City Publications 151
 Special Needs 155
 Travel 156
 Twins and Multiples 156
Party Supplies 157
Pet Supplies 158
Photographs, Albums, Frames,
 and Picture Plates 159
Posters and Decorations 161
Puppets 165
Rocking Horses 166
Science and Nature 168
Sheets, Blankets, and Sheepskins 174
Shoes 178
Sports Equipment 180
Stickers and Rubber Stamps 181
Swings and Climbing Sets 184
Teddy Bears 189
Toys 191
Travel 206
Videos 207
Woodworking 209

Associations
- *For Parents and Families* 213
- *For Children and Youth* 215
- *Adoption* 216
- *Childbirth* 217
- *Divorce* 218
- *Education and Child Care* 218
- *Family Planning* 220
- *Health* 220
- *Hobby* 222
- *Special Needs* 223
- *Travel* 225
- *Twins and Multiples* 225

Index 227

Introduction

Few parents are unaware of the appeals of shopping by mail—the convenience of ordering from home by letter or telephone call and the amazing diversity of merchandise available. The benefits of mail order answer the demands of parenthood so nicely.

I can remember poring over catalogs of children's clothes while my newborn son lay sleeping in his crib. In sharp color photographs and in crude drawings, the catalogs presented dozens of possibilities that couldn't be matched in several hours of store shopping. Here were soft cotton outfits, booties, hats, and blankets. Some were expensive, some cheap; some fit my idea of what baby clothes should look like, others didn't. The pleasure lay in sorting through and screening the good from the bad in a comfortable chair at home, without bundling up the baby and ferrying him from store to shopping mall.

As he's grown (and as the birthdays of nieces and nephews have arrived with shocking regularity), the horizons of the mail-order world have opened further. From baby clothes, sheepskins, and teething toys, we've moved on to rubber stamps, magic tricks, glow-in-the-dark bracelets, gardening tools, model trains, boomerangs, dolls, magnets, and picture plates.

Somewhere along the line, this attraction to the world of mail order turned into a mission to seek out its boundaries and to map the entire territory. If a few nice catalogs came in the mail without my asking, I supposed many more must be waiting for me to find them. As it turned out, several hundred more were out there waiting, many of the best of them sent only on request or on payment of a catalog fee. Some of the more obscure catalogs have turned into special favorites. Soft as a Cloud, for example, sells the softest cotton playsuits we've found anywhere; a beautifully trimmed cardigan came from Petit Pizzazz (we wish they sold it in grown-up sizes); and the Over the Moon Handpainted Clothing Co. paints wonderful moons and ice-cream cones on cotton shirts and pants. Gil & Karen's Toy Box carves a child's name on a box of wooden blocks. Tryon Toymakers makes a treasure chest decorated with a skull and crossbones or a flowered border. And for birthday presents it's hard to beat Edmund Scientific's warehouse of science gadgets and toys, Hearthsong's catalog of beautiful and wholesome playthings, or Archie McPhee & Company's array of curiosities.

Two sections of this book deserve special mention. The school-supply catalogs, described here under *Educational Supplies*, are often overlooked as shopping sources for parents. Most are happy to sell to individuals (some, like Environments, make an effort to attract nonprofessionals), and their pages are loaded with sturdy furniture and with such timeless playthings as wooden blocks, Bill Ding stacking clowns, and Lincoln Logs. And parents who think that children's magazines begin and end with *National Geographic World* and *Highlights for Children* should take a good look at the listings under *Children's Magazines*. Young readers now have a choice of dozens of excellent magazines, including some wonderful publications that focus on history, science, art, and anthropology.

Mail order is terrific, but the smart shopper should observe some basic safeguards. When in doubt, test an unfamiliar company with a small order. If you're satisfied with what you receive, buy more confidently the second time around. Whenever you order by mail, look carefully at the promises the company makes to its customers. When the merchandise arrives, note any warrantees that come with it and save any necessary receipts.

If a Problem Arises with an Order

Those who shop by mail regularly know that certain consumer safeguards apply to long-distance shopping. Virtually anything you order can be returned for a refund or an exchange if it doesn't meet your expectations. And if you're charged for something you haven't received, be sure to let the merchant know. Send a letter explaining the problem, and you'll generally have the matter solved by return mail. It's a good idea to keep copies of all correspondence in case things don't work out smoothly. When ordering by telephone, be sure to keep a written copy of your order.

If a merchant does not answer a complaint to your satisfaction, several avenues are open. You can refer the matter to your local postmaster if you ordered by mail and if the problem seems to involve consumer fraud. You can take your complaint to the Better Business Bureau (for the address of the appropriate local bureau, contact the Council of Better Business Bureaus, 1515 Wilson Boulevard, Suite 300, Arlington, VA 22209; telephone 703-276-0100). Or you can contact the Direct Marketing Association, an organization of businesses that sell by mail. The association acts as an intermediary in consumer complaints and has had good success in straightening out problems. Explain your complaint in a letter, with copies of any pertinent documentation, and send it to the Mail Order Action Line, Direct Marketing Association, Inc., 6 E. 43rd Street, New York, NY 10017.

The Direct Marketing Association also offers a service for mail-order shoppers who don't like receiving unsolicited catalogs and mailings. Send your name (with all variant spellings), address, and telephone number to the Mail Preference Service, Direct Marketing Association, Inc., P.O. Box 3861, Grand Central Station, New York, NY 10163. The association will add you to a computer list of names to be automatically screened from mailing lists. In three or four months, the unwanted mail should begin to disappear from your mailbox. If you add your name to the Mail Preference Service, be sure to alert any companies whose catalogs you *want* to receive, so they don't drop you from their list.

The Mail Order Shopper for Parents

General Catalogs

Best Products Co., Inc.
P.O. Box 25031, Richmond, VA 23260-5031
Telephone: 800-950-2378

Catalog price: free

Gadgets and toys for the whole family, from juicers and flashlights to skateboards and wooden puzzles. The selection changes dramatically from catalog to catalog. One recent edition included Sony's children's tape recorder, Sneak-R-Stilts, a wooden dollhouse, a jogging stroller, and a child's tool set. The next edition was devoted almost exclusively to jewelry, its only offerings for children being some twin-size down comforters.

Cracker Barrel Old Country Store
P.O. Box 787, Hartmann Drive, Lebanon, TN 37088-0787
Telephone: 615-444-0040

Catalog price: free

Food, gifts, and toys with an old-time aura, presented in a color catalog that almost smells of penny candy and nail bins. For children the store sells a dozen different teddy bears (priced from $8 to $35), Slinkies, Silly Putty, slippers made to look like duck heads, and cowboy and Indian costumes (if you can bear the idea of a working slingshot in your house, or a bow with suction-cup arrows).

Orvis
Blue Hills Drive, P.O. Box 12000, Roanoke, VA 24022
Telephone: 800-541-3541

Catalog price: free

Orvis doesn't sell its clothing in children's sizes, but the catalog always throws a few things for younger ones into the mix. Recent catalogs have sold frog- and duck-face rubber boots, BobShorts sleds that are worn like a pair of pants, a pine toy chest with rocking horses in relief on the front, and a soft chair in the form of a big teddy bear. The family dog gets about as much attention, with personalized collars, a log doghouse, electric heating pads, and beanbag sleeping cushions in five sizes.

J. C. Penney Company, Inc.
Circulation Dept., Box 2056, Dept. PR22, Milwaukee, WI 53201-2056
Telephone: 800-222-6161

Catalog price: $5, with certificate redeemable on first order

J. C. Penney's 1,400-page catalog has just about everything in the way of clothes and home furnishings. For new parents and parents-to-be the company offers maternity clothes, cribs, high chairs, car seats, nursing bras, baby clothes, bibs, and loads of other nursery supplies. All are also shown in the more manageable "Baby & You" catalog, available separately. The children's clothing section runs to 130 pages, covering sleepwear, winter coats, shoes, dress and play clothes, scouting uniforms, dance wear, shoes, and rain gear in sizes from toddler to teen. Campers can buy tents and sleeping bags; photographers will find cameras; pet owners can order cages, carriers, and pet beds; and home decorators can browse through the selection of wallpaper, curtains, and lighting supplies.

Sears, Roebuck & Co.
Call 800 number below or contact a local Sears store
Telephone: 800-366-3000

Catalog price: $5 for main "Home" and "Style" catalogs, credited toward first order; free "Toys" catalog

The Sears catalog hardly needs our introduction. In the world of mail order it is at once a respected grandparent and an almost mythic giant. Sears no longer sells houses by mail (as it did in the 1920s and '30s), but the 1,200-page "Home" and "Style" catalogs have just about everything else. The smaller "Toys" catalog is easier to handle and if it matches your needs may be the best route into the Sears network.

The "Toys" catalog presents playthings from Fisher-Price and Playmates, character toys from Disney and Sesame Street, natural and painted hardwood blocks, playhouses, play kitchens, easels, art supplies, Barbie and Ken dolls, ventriloquist's dummies, teddy bears, baseball cards, stamp and coin sets, nap mats, sleeping bags, tents for indoors and out, computer chess sets, books, board games, table soccer sets, child-size pool tables, radios, telescopes, microscopes, drum sets, electronic keyboards, Lego and Duplo sets, Lincoln Logs, pogo sticks, balls, swing sets, riding toys, tricycles, wagons, scooters, battery-powered vehicles, radio-controlled models, slot car sets, model trains, and such television-related toys as Transformers and Supernaturals. Toy chests are sold to help store it all away, along with a big collection of tables, chairs, and desks.

The main "Home" and "Style" catalogs offer a vast array of toys, baby clothes and supplies, clothes for older children, maternity clothes, nursing bras, children's beds and bedding, and hundreds of pages of adult-size clothes, appliances, furniture, tools, and decorating accessories.

Art Supplies

ABC School Supply
See entry under *Educational Supplies*.

All But Grown-Ups
A wooden easel with markerboard and chalkboard surfaces. See entry under *Furniture*.

Bellerophon Books
Coloring and cut-out books. See entry under *Books*.

Dick Blick Art Materials
P.O. Box 1267, Galesburg, IL 61401
Telephone: 309-343-6181
Catalog price: free
Four hundred pages of art supplies for the serious artist or the young person who wants to move beyond crayons and cheap watercolor sets. This catalog is aimed at art teachers and professionals, but most of the items can be had in small quantities, and as long as parents send payment with their orders they won't be turned away. Catalog shoppers will find easels, paints, children's smocks, markers, crayons, pastels, chalks, lettering pens, hundreds of brushes for every purpose, a veritable warehouse of paper, printmaking supplies, wood-carving tools, ceramic equipment and supplies (from clay to kilns), craft supplies, and instructional books. A professional printmaker or graphic designer will be satisfied here. A child will be overwhelmed with the possibilities.

Chaselle, Inc.
9645 Gerwig Lane, Columbia, MD 21046
Telephone: 800-CHASELLE (800-492-7840 in MD) or 301-381-9611
Catalog price: free (specify "Art & Craft Materials" catalog)
A big inventory of art and craft supplies, from crayons to potter's wheels. See entry under *Educational Supplies* for a more detailed description.

Childcraft
Markers, easels, crayons, paints, modeling clay and wax, and glitter. See entry under *Toys*.

Constructive Playthings
See entry under *Educational Supplies*.

Crate & Barrel
Easels. See entry under *Toys*.

Cumberland Crafters/Kid's Art
Your child's artwork reproduced on T-shirts, sweatshirts, tote bags, calendars, and placemats. See entry under *Children's Clothes*.

Dover Publications, Inc.
Coloring books, cut-and-assemble books, and books of lettering and design. See entry under *Books*.

Early Learning Centre
Art supplies for young children. See entry under *Toys*.

Educational Materials Library
Coloring books and color-your-own posters. See entry under *Educational Supplies*.

Educational Teaching Aids
See entry under *Educational Supplies*.

Emily's Toy Box
Short fat crayons, Brunzeel colored markers and pencils, watercolor crayons, face paints, foam paint, and Dover coloring books. See entry under *Toys*.

Environments, Inc.
Art supplies for younger children. See entry under *Educational Supplies*.

Geode Educational Options
Coloring books. See entry under *Books*.

Ginny Graves
5328 W. 67th Street, Prairie Village, KS 66208
Telephone: 913-262-0691
Catalog price: long self-addressed stamped envelope

Sells a crayon melter which turns old crayons into soup for crayon painting on plastic, glass, paper, fabric, or practically anything else. The melter plugs into an outlet and liquifies the crayons in ten separate cups. It comes with a squeeze pen for executing hot-wax art. Ms. Graves also sells an art activity book, titled *Discovery Stuff*, with 52 weekly projects. It tells how to make gift wrap from colored tissue and household bleach, how to grow a crystal garden, make a monoprint, and carve a pumpkin. Children aged three to six will find simpler activities in her *What Can I Do Now Mommy?*

J. L. Hammett Co.
A huge inventory of art and craft supplies, from crayons to potter's wheels. See entry under *Educational Supplies*.

HearthSong
Birch drawing boards, oversize colored pencils, block crayons, and Stockmar watercolor paints and modeling beeswax. See entry under *Toys*.

John Holt's Book & Music Store
Modeling beeswax, individual chalkboards, and other art supplies. See entry under *Books*.

Hoover Brothers, Inc.
See entry under *Educational Supplies*.

Karen Studios
Ceramic plates made from children's drawings. See entry under *Photographs, Albums, Frames, and Picture Plates*.

KidsArt
P.O. Box 274, Mt. Shasta, CA 96067
Telephone: 916-926-5076
Catalog price: free

An unusual mix for young artists, including coloring books, art activity books, powdered tempera paint, rubber stamps, Tintin comic books, and storage boxes for art supplies and childhood treasures. (See description of *KidsArt News* under *Children's Magazines, for School-Age Children*.)

The Left-Handed Complement
P.O. Box 447, Port Jefferson Station, NY 11776
Catalog price: $1

Tools, toys, gadgets, and books for lefties. Dozens of scissors for every purpose are offered, along with kitchen knives serrated for left-handed use, can openers, fruit peelers, pie servers, corkscrews, playing cards, boomerangs, calligraphy pens, rulers, coffee mugs, and an address book with a left-handed thumb index. A book, *Teaching Left-Handed Children*, may be a help to parents whose left-handed children are learning to write.

Ginny Graves's crayon melter

Manzanita Publications
1731 Hendrix Avenue, P.O. Box 1366, Thousand Oaks, CA 91360
Telephone: 805-495-4484
Catalog price: free

Coloring books, paper-doll books, and cut-and-assemble books, many of them difficult to find in stores. A cowgirl coloring book looks like a treasure, as does one titled *Eagles, Hawks, Falcons, and Owls of America*. The *Create Your Own Pictures* coloring book leaves some additional room for a child's imagination. The paper-doll books include *Asian Festival Figures* and *Great Women Paper Dolls*.

Nasco
901 Janesville Avenue, Ft. Atkinson, WI 53538
Telephone: 414-563-2446
Catalog price: free (request "Arts & Crafts" catalog)

Nasco is one of the country's largest educational supply houses, and it offers several thick catalogs. (See additional entries under *Science and Nature* and *Educational Supplies*.) The 300-page "Arts & Crafts" catalog is bursting at the seams with supplies for the art classroom. Markers for all ages are sold, from fat crayons for beginning artists to pastels, calligraphy pens, and airbrush sets. Brushes, paper, paints, and adhesives are sold in a huge range of choices, along with equipment and supplies for printmaking, ceramics, drafting, carving, woodworking, metal enameling, weaving, leatherwork, and other crafts.

The Natural Baby Co.
Art supplies for younger children from Galt Toys, and wooden easels and drawing desks from Beka Wood Products. See entry under *Toys*.

PlayFair Toys
A wooden easel, a tabletop loom, and a few other art and craft supplies. See entry under *Toys*.

S & S Arts & Crafts
Dept. 101, Colchester, CT 06415
Telephone: 203-537-3451

Catalog price: free

Art supplies and craft projects in a catalog aimed at teachers. Its big inventory of paints, markers, crayons, scissors, brushes, and paper should make this a useful source for parents. The art projects will translate less easily to the home, as most must be purchased in quantity. Parents with a number of artists in their charge might want to consider the projects, which include blank masks for decorating, materials for making sock monkeys, and wooden tops ready for painting.

Sax Arts & Crafts
P.O. Box 51710, New Berlin, WI 53151
Telephone: 414-784-6880

Catalog price: $3, deductible from first order

A weighty 400-page catalog of art and craft supplies "for the classroom and studio." Twenty pages of pastels, markers, and crayons should be enough for the most avid young colorist. Thirty pages of paper supplies offer a big choice of surfaces to draw on. Easels, smocks, scissors, brushes, and paints are stocked in similar abundance, along with materials and equipment for calligraphy, printmaking, ceramics, stone carving, woodworking, jewelry making, weaving, candle making, leather craft, and stained glass.

Small Fry Originals
Plates and mugs made from your child's artwork. See entry under *Photographs, Albums, Frames, and Picture Plates*.

The Timberdoodle
Cray-pas and coloring books. See entry under *Educational Supplies*.

Toys to Grow On
Art supplies and easels. See entry under *Toys*.

Treasured Toys
A wooden easel and Italian modeling dough. See entry under *Toys*.

Young Rembrandts
Children's artwork reproduced on plates, tiles, and trinket boxes. See entry under *Photographs, Albums, Frames, and Picture Plates*.

Easel, smocks, and floor mat from Environments, Inc.

Baby Needs

ABC School Supply
Baby toys and cribs. See entry under *Educational Supplies*.

A-Plus Products, Inc.
P.O. Box 4057, Santa Monica, CA 90405
Telephone: 800-359-9955 or 213-399-1177
Catalog price: free

A slim color brochure with a dozen useful products for babies and young children. Contoured bath cushions aid in bathing babies before they can sit up; support rings keep them safely up when they're just learning to sit. Animal-face cushions protect mobile babies and toddlers from bathtub faucets and handles, and an inflatable potty makes toilet training convenient and travel a little easier. Several styles of rattles are offered, as well as a clip-on tray for changing table or crib and a storage shelf for baby food jars.

Abilities International
See entry under *Toys*.

After the Stork
Cotton baby clothes from Soupçon, cotton diapers, and Nikky diaper covers. See entry under *Children's Clothes*.

American Bronzing Co.
P.O. Box 6504, Bexley, OH 43209
Telephone: 800-345-8112
Catalog price: free

For parents who want to bronze baby's first shoes. The company offers plain bronzing and a two-tone antique finish, and will mount the finished shoes on real walnut or wood-look stands. Parents can even have the shoes made into bookends or a picture frame stand. Cowboy boots cost a little more.

AnCar Enterprises, Inc.
Dept. B, P.O. Box 340124, Boca Raton, FL 33434
Telephone: 407-482-0646
Free brochure

This firm offers just one product: a padded baby seat that sets into a shopping cart for added comfort, straps onto an adult-size chair to make a safe child seat, and fits into a high chair for added height and support. In a pinch it snaps apart to make a soft changing pad.

Artisans Cooperative
Appliquéd bibs. See entry under *Gifts*.

'BabieKnit' Industries, Inc.
4301 S. Pinemont Street, Houston, TX 77041
Telephone: 713-462-5999
Catalog price: free

Soft, fluffy diaper covers from Australia, baby booties, and a series of videos for new parents on childbirth, breastfeeding, and bathing a baby. The company is moving away from direct-mail sales as it places its products in stores. Write to see if the products are still available by mail.

Bronzed baby-shoe display from American Bronzing Co.

SecureView mirror from Baby & Company, Inc.

Baby & Company, Inc.
P.O. Box 906, New Monmouth, NJ 07748
Telephone: 201-671-7777

Catalog price: $1, deductible from first order

Supplies for infants and toddlers, and gifts for new parents and grandparents. The catalog offers a lambskin comforter, a lambskin liner for car seats and strollers, a vinyl mat to protect the floor under the high chair, wooden baby toys, a rain poncho for babies, an extra rearview mirror so drivers can keep an eye on backseat children, a fully-stocked diaper bag, and polo shirts embroidered with "grandma" and "grandpa" emblems.

Baby Biz
P.O. Box 404, Eldorado Springs, CO 80025
Telephone: 303-499-2469

Free brochure

A compact selection of useful items for new parents: Nikky diaper covers (they eliminate the need for pins), a sunshade for the car window, the Infant Stim-Mobile, the Gentle Expressions battery-powered breast pump, the Baby Bag snowsuit, overall stretchers (they extend shoulder straps for another three months of wear), and Klutz Press's *Kids Cooking* and *Kids Songs & Holler Along Handbook*.

Baby Bunz & Co.
P.O. Box 1717-SP, Sebastopol, CA 95473
Telephone: 707-829-5347

Catalog price: $1

As its name suggests, Baby Bunz is in the baby bottom business. The company sells five styles of Nikky diaper covers, cotton terry and combined flannel-terry diapers, Dovetails biodegradable disposable diapers made from wood pulp without plastic, and Weleda baby soaps and creams. The company also sells Benay's playsuits—tough-looking cotton suits in bright solid colors.

Baby Care
P.O. Box 5620, San Mateo, CA 94402
Telephone: 415-572-2689

Free flier

Makes a bean-shaped cushion that nestles around mother's stomach to hold a nursing baby.

Baby Dreams
P.O. Box 3338, Gaithersburg, MD 20878
Telephone: 800-638-5965

Catalog price: $1, deductible from first order

Quilted covers for car seats, strollers, and booster seats keep surfaces cool in the summer and warm in the winter. All are made from calico cotton and cotton blends in a choice of ten colors. The firm also sells changing pads, head-support pillows, and diaper bags in the same fabrics. The color brochure shows a patch of each of the fabric choices.

Baby Furs by Scandinavian Origins
Ingrid Edstrom, 39 Rolling Lane, Dover, MA 02030
Telephone: 508-785-2622

$2 for brochure and fur samples

Fur child-carrier liners and sheepskin baby buntings. The buntings are made with the fleece inside and either suede or heavy nylon on the outside. The carrier liners are covered with corduroy. The sheepskin fleeces come in a choice of gray, cream, or golden brown, the coverings in bright blue or red.

Baby Wrap, Inc.
P.O. Box 10584, Denver, CO 80210
Telephone: 303-757-5564

Free brochure

A strapless baby carrier that wraps around the parent's torso. The child is secured in a fabric seat with leg holes, then hoisted onto the adult's back and tucked into place with some deft folds and knots. The design is based on traditional carriers used by West Africans and Native Americans.

Baby's Comfort Products
1740 N. Old Pueblo Drive, Tucson, AZ 85745
Telephone: 602-624-1892

Free brochure

Car-seat covers with pieces for covering harnesses, buckles, and straps; matching car-seat sunshades; hooded towels, bibs, and baby-room wall hangings. All are available in a choice of several print or solid fabrics, which are briefly described but not pictured in the brochure.

Babysling, Inc.
1 Mason, Irvine, CA 92718
Telephone: 800-541-5711 or 714-770-5095
Free brochure

Sells a padded over-the-shoulder baby sling in a choice of several fabrics. Newborns ride in it cradle-style across the belly; toddlers sit up in back or on the parent's hip. Shift baby slightly and the sling serves as a support for nursing.

L. L. Bean, Inc.
Sells the Baby Bag snowsuit and backpack child carrier. See entry under *Outdoor Gear*.

Bear Feet
Baby shoes. See entry under *Shoes*.

Best Selection, Inc.
2626 Live Oak Highway, Yuba City, CA 95991
Telephone: 916-673-9798
Catalog price: $2

An array of baby products that covers the field from crib toys to car seats to breast pumps. Just a couple of choices are offered in each category, the selections chosen for safety, convenience, usefulness, and value, with an effort made to pick items not easily found in stores. Among the more unusual offerings are a hammock for newborns that hangs between the top rails of a crib, a portable bottle warmer, and a light switch for the baby room that is activated by the sound of footsteps. Equally useful are baby gates, home safety kits, baby carriers, swings, jumpers, walkers, strollers, and bath aids.

Biobottoms
Cotton diapers and breathable diaper covers. See entry under *Children's Clothes*.

Birth & Beginnings
6828 Route 108, Laytonsville Shopping Center,
Laytonsville, MD 20879
Telephone: 301-990-7975
Catalog price: $1 each for Nikky and Birth & Beginnings catalogs

The complete line of Nikky diaper covers and training pants is displayed in the 16-page Nikky catalog. The covers are made of cotton or wool, with Velcro closures. Some are ornamented with prints, some with colored trim. One for older children is made to look like a pair of running shorts. Matching T-shirt and diaper-cover sets are offered for fashion-conscious toddlers. Birth & Beginnings' full catalog includes much of the Nikky merchandise plus Baby Bag snowsuits, PolarPlus jackets from American Widgeon, deerskin baby shoes, the Medela breast pump, and cotton clothing for babies and young children from Jeanne Mac of Vermont, Fusen Usagi, Fix of Sweden, and Petit Bateau.

Bumkins International, Inc.
291 N. 700 E., Payson, UT 84651
Telephone: 801-465-3995 or 9330
Free information; $5.95 for sample

Cotton flannel diapers with elastic edges, Velcro closures, and waterproof outer shields. These are not cheap ($50.50 per dozen in 1989, $68.50 per dozen for toddler size), but Bumkins claims they are better.

C.H.I.L.D.
c/o Christina Waller, P.O. Box 127, Langlois, OR 97450
Telephone: 503-348-2203 or 9910
Free brochure

C.H.I.L.D. stands for Computerized Holistic Infant Learning Development. Its primary product is a series of activity programs designed to "aid in the development of five growth areas: language, cognition, fine motor, gross motor, and self-help (feeding, dressing, etc.)" for children from birth to age three. Payment of $16.50 (check current price), sent with your child's name and age, will get you the first of these packages, a collection of 15 to 30 suggestions for interactive parent-child play designed around such daily routines as feeding and bathing. The company also sells the Dr. Sears baby sling, an over-the-shoulder cloth carrier.

Cabin Creek Quilts
Bibs, clutch balls, and crib quilts. See entry under *Sheets, Blankets, and Sheepskins*.

Chaselle, Inc.
9645 Gerwig Lane, Columbia, MD 21046
Telephone: 800-CHASELLE (800-492-7840 in MD) or 301-381-9611
Catalog price: free (specify "Pre-School & Elementary School Materials" catalog)

Baby supplies for day-care centers, many of which are equally suited to the home. Among the listings are cribs, high chairs, a large shatterproof mirror (2 × 4 feet), rattles, teething toys, balls, and a baby-safe rocking horse. See main entry under *Educational Supplies* for a description of Chaselle's other offerings.

The Children's Corner
520 Monument Square, Racine, WI 53403
Telephone: 800-445-7033
Free brochure

Merchandise for parents of babies and young children, including a wipe warmer, a convertible child carrier/stroller, press-on sunshades for car windows, a baby monitor (with sound transmitter and receiver), a portable playpen, and an alarm that sounds when a child opens the door to go out. The brochure lists about a dozen items in all.

Chock Catalog Corporation
Discounted sleepwear, underwear, diapers, bath towels, crib sheets, and receiving blankets. See entry under *Children's Clothes*.

Claire's Bears & Collectibles
Kiddicraft toys and Baby Björn baby supplies. See entry under *Toys*.

The Comfey Carrier

Comfey Carrier
P.O. Box 447, Santa Cruz, CA 95061
Telephone: 408-338-2017

Free brochure

Makes a cotton-velour baby carrier that can be worn in a number of ways to accommodate passengers from birth to about 30 pounds. Newborns can be held face-in, older babies face-out. The harness can be adjusted for use as a baby backpack or as a nursing sling.

Compare and Save Premium Catalogue
P.O. Box 88828, Seattle, WA 98188
Telephone: 800-COMPARE

Catalog price: free

Discounted baby merchandise available with "Compare and Save" seals from store-brand disposable diapers. Car seats, Playskool and Johnson & Johnson toys, high chairs, and strollers are among the items to be found.

Courier Health Care, Inc.
P.O. Box 1210, Agona Hills, CA 91301
Telephone: 800-543-5387 or 818-991-1931

Catalog price: $1

Breast pumps, nursing bras and nightgowns, MagMag training cups, and such home medical care items as childproof medicine box, a lighted scope for ear exams, and a digital pediatric scale. An adult scale with a wall-mounted remote readout looks like the perfect thing for mothers-to-be tired of asking their husbands to read their weight. The company sells both one- and two-shoulder baby slings, cotton and Gore-Tex diaper covers, lambskins, car sunshades, and a child's sun hat with a back flap that protects the neck.

Cozy Baby Products, Inc.
P.O. Box 399, Mt. Vernon, NY 10552-0399
Telephone: 914-668-1686

Free brochure

Sells a beanbag cushion for infants. Covered with cotton flannel and filled with polystyrene beads, it cradles and supports babies in a soft nest. Use it as a stroller insert or as a portable baby seat around the house.

Cozy Carrier
18105 Valley View Road, Eden Prairie, MN 55344
Telephone: 612-934-0687 or 0382

Free brochure and fabric samples

Manufactures a cloth baby carrier that straps around the shoulder and buckles around the waist. The carrier can be worn either in front, when holding an infant, or in back, for an older baby. (It's not recommended for children over 25 pounds.) Choose one of seven bright solid colors from swatches sent with the brochure.

Creative Parenting Resources
Sells a baby sling. See entry under *Books*.

Cuddlers Cloth Diapers
3020 Cheyenne Drive, Woodward, OK 73801
Telephone: 405-254-3518

Long self-addressed stamped envelope for brochure and fabric sample

Makes a deluxe cotton flannel diaper with Velcro closures, a fitted hourglass shape, and elastic through the sides.

Deerskin Trading Post
119 Foster Street, Box 6008, Peabody, MA 01961-6008
Telephone: 617-532-4040

Catalog price: free

Sells a luxurious sheepskin baby bunting—soft leather on the outside and shearling fleece on the inside. To find it, shoppers will have to plow through many pages of adult-size leather pants and jackets, fleece slippers, fur coats, and the like.

Designer Diapers
3800 Wendell Drive, Suite 403, Atlanta, GA 30336
Telephone: 800-541-7604 or 404-691-4403

Catalog price: free

Disposable diapers for special occasions (or for parents with lots of disposable income). The backs are decorated with a colorful ornamental band in a choice of 25 designs; the fronts close with color-coordinated Velcro strips. The design options include sheep, trains, colored ribbons and ruffles, rocking horses, balloons, and teddy bears. A box of 12 diapers cost about $20 in 1989.

Diap-Air
P.O. Box 103, Upton, NY 11973

Free brochure

Sells the Wabby diaper cover, made of Gore-Tex layered between a cotton shell and a cotton/polyester liner. Diap-Air claims the combination, wrapped around an absorbent cotton diaper, will keep the diaper wetness in while letting air reach the baby's skin. The cover also eliminates the need for pins.

Diaperaps
P.O. Box 3050, Granada Hills, CA 91344
Telephone: 800-251-4321 or 818-886-7377

Free brochure

Breathable diaper covers in a choice of ten colors and patterns. The cover is made with a cotton outer shell, a thin foam core, and a nylon tricot liner, a combination that Diaperaps claims will keep clothes dry while allowing air circulation to the baby's bottom. Velcro closures allow for easy changes without pins.

Direct-To-You Baby Products
4599 Peardale Drive, Las Vegas, NV 89117
Telephone: 702-364-1979

Catalog price: 50¢

Niceties and necessities for babies, from a soft bath sponge to a toilet lid lock. The firm sells the Infant Stim-Mobile (a hanging mobile of bold black-and-white patterns), a natural lambskin, an inflatable car-seat insert for infants, and a set of knee pads for little crawlers. The Baby's Cloud cushion is a cloth support pillow filled with styrofoam pellets that can be used as a shopping-cart bed, an around-the-house infant seat, or a nursing aid. The firm sells instructional videos for new mothers, and music and activity videos for babies and young children.

Walt Disney World
Mickey Mouse baby dishes and bibs. See entry under *Gifts*.

Dy-Dee Service
834 Emerson Street, Rochester, NY 14613
Telephone: 716-458-5770

Long self-addressed stamped envelope for information

Prefolded cotton diapers of "diaper service quality." The diapers are two-ply, measure 14 × 21 inches, and have a six-ply absorbency panel.

E-Z Enterprises, Inc.
5920 E. Central, Wichita, KS 67208
Telephone: 316-685-7766

Free flier

Makes the E-Z Baby Tote, a carrier that holds a baby snugly against its parent's front. The straps are adjustable, the fabric washable, and the price reasonable—about $20 postpaid in 1989, with discounts if you combine an order with friends.

The Enchanted Child
County Route 5, Canaan, NY 12029
Telephone: 518-392-6985 or 5400

Catalog price: $1

Sells a baby bouncing seat made of cotton mesh over a steel frame (the Baby Bouncinette); wooden name, number, and alphabet puzzles; name wall plaques; personalized children's stools; and baby blankets embroidered with a child's name and birth date. The catalog comes in the form of color cards—one for each item, with a photograph on the front and detailed description on the back.

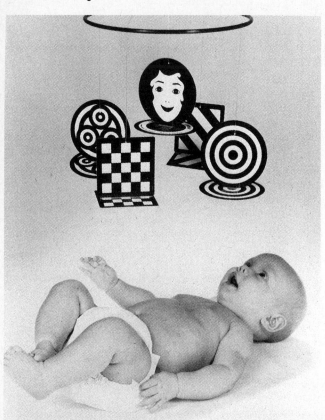

The Infant Stim-Mobile, sold by Direct-To-You Baby Products and Wimmer-Ferguson Child Products

F & H Child Safety Co.
Sells the Diaper Duck, a device for rinsing cloth diapers. See entry under *Health and Safety*.

Family Clubhouse
6 Chiles Avenue, Asheville, NC 28803
Telephone: 704-254-9236

Catalog price: $1

Quality cotton baby clothes from Kawamura and Fusen Usagi, Nikky diaper covers (and night pants for older children), terry and flannel diapers, Dovetails biodegradable disposable diapers, Weleda and Earthchild baby bath products, and such things as bibs, baby dishes, no-skid socks, and a diaper duck for rinsing soiled diapers. The color catalog shows all the fabrics and colors nicely.

Frostline Kits
Kits and ready-made buntings, baby carriers, comforters, and diaper bags. See entry under *Outdoor Gear*.

Garnet Hill
Cotton crib sheets, baby clothes, and nursing wear. See entry under *Children's Clothes*.

Good Gear for Little People
See entry under *Outdoor Gear*.

Hawk Meadow of New England
Sterling silver baby cups. See entry under *Gifts*.

Hazelwood
P.O. Box 4455, Panorama City, CA 91412-4455
Telephone: 818-891-1693

Catalog price: free

Makes a diaper cover with the Velcro closure in the back so babies can't take it off. It's constructed of cotton with a waterproof polyurethane liner. The catalog also offers airbrushed T-shirts with pictures of such things as a panda, a dinosaur, a clown, and a teddy bear. Sizes from 6 months to 14-16.

The Huggies Shopper
P.O. Box 90257, St. Paul, MN 55190
Telephone: 800-544-1847

Catalog price: free

Toys, books, and baby furnishings sold at a discount with the "points" saved from packages of Huggies disposable diapers. A Fisher-Price high chair or a Cosco changing table requires the labels from a lot of diaper packages, but *The Kids' Bathtub Songbook* can be had at half its bookstore price with the tags from just two large packages. Parents can outfit themselves with playpens, strollers, backpacks, car seats, baby gates, stacking toys, Golden Books, and other necessities, often at considerable savings. Anyone who writes for the catalog, whether they shop from it or not, receives an added benefit: periodic mailings containing discount coupons for Huggies diapers.

Infant Wonders
14431 Chase Street, Suite E-181, Panorama City, CA 91402
Telephone: 818-892-6407

Long self-addressed stamped envelope for brochure

Mobiles, soft toys, and crib sheets in bold black-and-white or colored designs to stimulate the senses of infants. The largest, simplest patterns are recommended for newborns, more complex patterns for babies over six months.

John Ingalls Designs
P.O. Box 707, Juneau, AK 99802
Telephone: 907-586-6293

Catalog price: free

John Ingalls manufactures a two-wheeled stroller of his own design that converts to a child-carrier backpack. It even collapses down for storage in the overhead compartments of airplanes. The frame is made of aluminum tubing, the seat of canvas with suede trim. Once the children are old enough to walk themselves, the stroller-pack can serve as a luggage cart. For those in his neighborhood, Mr. Ingalls also recommends it as a cart for hauling gold concentrate from the sluice.

The Initial Place
Personalized bibs and baby blankets. See entry under *Children's Clothes*.

Initials+ Collection
Personalized baby blankets and engraved baby cups. See entry under *Gifts*.

Iris Inc.
2525 Edgehill Road, Cleveland Heights, OH 44106
Telephone: 216-371-8765

Free brochure

Sells a baby feeding dish that straps onto a high-chair tray and comes off for microwave heating and dishwasher cleaning.

Irish World Imports
P.O. Box 98093, Lubbock, TX 79499
Telephone: 800-634-8554 or 806-793-0169

Catalog price: long self-addressed stamped envelope

Fleece-lined suede booties in two sizes for babies 6 to 18 months old. In 1989 they sold for $13 a pair, postpaid. Check current price before ordering.

Johnson & Johnson Child Development Toys
Baby toys. See entry under *Toys*.

Julia & Brandon
Toddler dinnerware and hook-on high chairs. See entry under *Toys*.

Kaplan School Supply Corp.
See entry under *Educational Supplies*.

Livonia
10 Main Street, P.O. Box 495, Chester, NY 10918-9989
Telephone: 914-469-2449
Catalog price: free

Livonia has made NameDate baby shoes for more than 30 years—white leather shoes with a child's name embossed in gold on one sole, her (or his) birthdate on the other. The current catalog also carries cotton terry bibs and bathrobes, child-size aprons, children's suede slippers, and wooden rattles.

Lovely Essentials/Snuggleups Diapers
P.O. Box 7, St. Francis, KY 40062
Telephone: 502-865-5501
Free brochure

Pinless cloth diapers made of cotton flannel. Elastic around the legs cuts down on leaking. Velcro closures facilitate quick changes. Tabs over the Velcro strips can be folded back during laundering to keep the diapers from gripping together. If you want to test one, send $4 for a sample. Specify small (up to 15 pounds) or large.

Maine Baby
P.O. Box 910, Morrill, ME 04952
Telephone: 207-342-5055
Free brochure

Makes a baby carrier, a diaper bag, and a child's apron under the Maine Baby label. The carrier slings over both shoulders for toting baby in front or in back. In front, the child can face either in or out. The diaper bag is made of the same soft, heavy cotton, lined with a vinyl changing pad. The apron is made of denim in two sizes: for toddlers to age two, and for growing chefs to age six. Other items in the brochure include two wooden rattles, a toddler block set in a bright red carpenter's box, the Baby Bag snowsuit (newborn and regular sizes), a crib-size futon, and a child's first bed made to fit a crib mattress.

J. Mavec & Company, Ltd.
625 Madison Avenue, New York, NY 10022
Telephone: 212-517-8822
No catalog

Antique and contemporary luxuries for babies, such as 19th-century rattles made of silver and coral, silver baby cups, children's hairbrushes. An elegant spoon with a duck head on the handle sells for about $120, and prices go up from there.

Metrobaby
P.O. Box 1572, New York, NY 10013-0869
Telephone: 212-966-2075
Free brochure

Cotton diapers, clothing, and bedding. The diapers are made of flannel or terry cloth, with layers of thick cotton terry sewn inside. They're sold trimmed to either the standard rectangle for old-fashioned pinning or an hour-glass shape to fit without pins in Nikky or Biobottom covers. The same soft cotton flannel is used for making crib sheets, bumpers, nursing pads, and such infant clothing as a gown, a bunting, and a footed creeper. A hooded robe and matching hooded towel are sewn of cotton terry cloth. The robe comes in sizes up to 3T; the creeper to 2T. To warm up the crib even more, Metrobaby sells flannel blankets and a down-filled crib quilt.

The Metropolitan Museum of Art's Baby's Journal *and* My Baby and Me

The Metropolitan Museum of Art
Publishes two baby record books. See entry under *Gifts*.

Milkduds
Infant coveralls and baby turtlenecks. See entry under *Children's Clothes*.

Mommy's Helper, Inc.
912 Army Trail Road, P.O. Box 1266, Addison, IL 60101
Telephone: 312-628-1142
Free information

Plastic lids for 32- and 13-ounce cans of infant formula.

The Montgomery Schoolhouse, Inc.
Wooden baby rattles. See entry under *Toys*.

Moonflower Birthing Supply
P.O. Box 128, Louisville, CO 80027
Telephone: 303-665-2120

Catalog price: two first-class stamps

Medical products for midwives, supplies for new parents, books on birth and parenting, and homeopathic remedies. Most parents will skip right over the several pages of birthing instruments to get to the listings of cotton diapers, Nikky diaper covers, breast pumps, and baby bounce seats. Older babies are not forgotten: the catalog offers a food grinder, the Sassy clip-on high chair, and a backpack child carrier. The book selection includes titles on pregnancy, natural childbirth, homeopathic medicine, and parenting, with an emphasis on alternative philosophies.

Motherwear
Diaper covers, flannel diapers, cotton newborn hats, and infant shoes. See entry under *Maternity and Nursing Clothes*.

Museum of the American Indian
Broadway at 155th Street, New York, NY 10032
Telephone: 212-283-2420

Catalog price: free

A small gift catalog from a fascinating museum. A beautiful silver baby rattle is handmade by Navaho silversmiths; infant and toddler moccasins (sizes 1 to 6) are made by Cherokee craftspeople. A stuffed buffalo toy is a cute alternative to the more common bears and bunnies.

Nasco
Baby toys, potties, gates, and other supplies. See entry under *Educational Supplies*.

The Natural Baby Co.
Rt. 1, Box 160, Titusville, NJ 08560
Telephone: 609-737-2895

Free brochure

Nikky diaper covers; flannel and terry diapers; a nursing bra made of pima cotton; and a line of baby creepers, gowns, and hooded towels made of cotton flannel and terry cloth. A more extensive packet of toy catalogs can be had for a $10 deposit (see separate entry under *Toys*).

Natural Elements
145 Lee Street, Santa Cruz, CA 95060
Telephone: 408-425-5448

Free brochure

Natural baby products and homeopathic remedies for "the new-age family." The company offers Nikky diaper covers, Rubber Duckie rain gear, Comfey baby carriers, Weleda and Country Comforts soaps and creams, a baby brush set, Steven Bergman's lullaby tapes, and Sara's Prints long johns. Among the homeopathic remedies are Hyland's teething and colic tablets and cough syrup.

Naturepath
Rt. 1, Box 99C, Hawthorne, FL 32640
Telephone: 904-481-2821

Catalog price: free

A birthing supply catalog that carries a good selection of pregnancy, birth, and parenting books, as well as parenting supplies like food grinders, diaper covers, diaper bags, and breast pumps. The medical equipment section of the catalog is extensive, much of it off-limits to those not certified to use it. A home ear scope is pitched at parents as a backup to doctor's visits, and a digital thermometer is almost a necessity. Vitamins, herbal remedies, and Weleda's natural soaps and skin care products back up the medical hardware. If you send your due date when you write for the catalog, Naturepath will enter your name in a monthly drawing for a gift certificate.

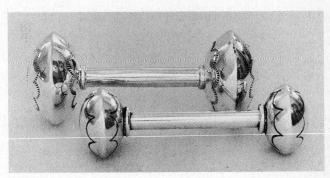

Silver Navajo rattles from the Museum of the American Indian

Nature's Little Shoes
Soft leather baby shoes. See entry under *Shoes*.

One Step Ahead
P.O. Box 46, Deerfield, IL 60015
Telephone: 800-274-8440

Catalog price: $2

Baby and toddler supplies, with an emphasis on the latest innovations. Lambskins offer age-old comfort, but almost everything else here has been invented or improved in the past few years. The catalog peddles Aprica's latest carriage/stroller, a portable bottle warmer from Snugli, a dishwasher basket for nipples and pacifiers, MagMag's rechargeable battery-operated breast pump, a novel circular bottle that's easy for little hands to grab, a dishwasher-safe microwaveable feeding tray, a video nursery monitor, and a device that vibrates a crib to simulate a car ride. For toddlers and young children the company offers a fold-out foam sofa and a couple of booster seats. Tendercare biodegradable disposable diapers are offered in six sizes.

Over the Moon Handpainted Clothing Co.
Bibs, diaper covers, and baby clothes. See entry under *Children's Clothes*.

Over the Shoulder Baby Holder
California Diversified Manufacturing, P.O. Box 635, San Clemente, CA 92672
Telephone: 714-361-1089
Free flier

A cloth over-the-shoulder baby carrier that can be worn in several different positions to hold an infant or a larger child, or to support a baby for nursing.

Pampers Baby Care Catalog
P.O. Box 8634, Clinton, IA 52736
Telephone: 800-543-7310
Catalog price: free

Toys, books, safety devices, furniture, and other childhood needs offered at discounted prices with the "Teddy Bear points" clipped from packages of Pampers disposable diapers. The catalog includes such childhood standards as the Radio Flyer wagon, the Fisher-Price Corn Popper push toy, and the book *Pat the Bunny*.

Papa Don's Toys
Hardwood rattles and crib toys. See entry under *Toys*.

J. C. Penney Company, Inc.
Cribs, high chairs, swings, carriers, strollers, car seats, baby clothes, christening sets, bedding, bottles, and other baby needs. See entry under *General Catalogs*.

Perfectly Safe
Baby products like a mat to spread under the high chair, a portable playpen, and a sheepskin stroller cover. See entry under *Health and Safety*.

Claudia Pesek Designs
P.O. Box 1184, Grants Pass, OR 97526
Catalog price: free

Patterns for making a baby carrier, a matching doll carrier, and a set of clothes for preemies and newborns.

Petit Pizzazz
Baby Björn's potty, bouncing seat, and baby toys. See entry under *Children's Clothes*.

Placenta Music, Inc.
2675 Acorn Avenue, N.E., Atlanta, GA 30305
Telephone: 404-262-1559
Free brochure

A relaxation tape for newborns that can also be played to calm mothers during labor and childbirth. Titled *Transitions*, it combines womb sounds with female vocals and soft synthesizer music.

Portland Soaker
P.O. Box 19827, Rochester, NY 14619
Catalog price: long self-addressed stamped envelope

Wool diaper covers in a choice of plaids and solid colors. The two-layer cover is said to draw moisture away from the diaper while permitting air to circulate back to the baby's skin. The covers can be purchased finished or in kits.

Prince Lionheart
2301 Cape Cod Way, Santa Ana, CA 92703
Telephone: 714-835-2626
Free brochure

Products to aid in the care of infants and toddlers. A shampoo shield keeps soap from dribbling down a child's face; insulated bottle cups keep baby bottles warm or cool; and the NAP container keeps nipples and pacifiers from flying around inside the dishwasher. For the car, Prince Lionheart makes window shades, a rubber seat cover, and a child-view mirror.

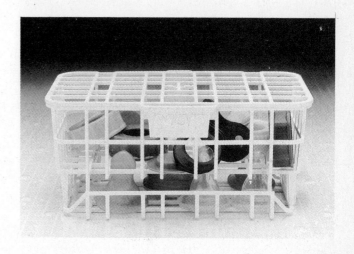

The NAP dishwasher container, sold by Prince Lionheart and Baby & Company, Inc.

Pro-Moms
161 Henry Street, Brooklyn, NY 11201
Telephone: 718-596-0846
Long self-addressed stamped envelope for brochure

Makes the MomSmock, a cover-up tunic that protects good clothes from baby messes. The smock allows mothers to feed and cuddle their babies after dressing for work in the morning.

Quiet Tymes, Inc.
2121 S. Oneida Street, #521, Denver, CO 80224
Telephone: 800-552-7360 or 303-757-5545
Free information

Quiet Tymes sells The Baby Soother, a tape of rhythmic sounds similar to those heard in the womb. To us the

The MomSmock by Pro-Moms
Photo: Tina Mucci

tape sounds a little like the inside of a factory, but we realize it's meant for infants, not for harried adults. Studies have shown that it gets the attention of and calms crying babies.

The R. Duck Company
Rubber Duckies diaper covers. See entry under *Children's Clothes*.

Racing Strollers, Inc.
The Baby Jogger stroller, for parents who run. See entry under *Sports Equipment*.

Rainbows & Lollipops, Inc.
13276 Paxton Street, Pacoima, CA 91331
Telephone: 818-897-7330

Catalog price: $1

A color catalog with a full range of baby merchandise and some helpful products for young children. New parents will find baby carriers, a bounce seat, a diaper bag that doubles as a nap bed, headrest cushions for car seats, the Gerry intercom nursery monitor, a crib mirror, the MagMag breast pump, and a wipe warmer. Baby gates, cabinet locks, stove-knob covers, and outlet seals keep toddlers out of trouble in the house. Car window sunshades, a shopping-cart seat, and a baby-view mirror keep them safe on the go. Knee pads, toilet-training aids, feeding dishes, and bath helps are among the other offerings.

Recreational Equipment, Inc.
PolarPlus infant suits, a diaper bag, and a frame child carrier. See entry under *Outdoor Gear*.

Richman Cotton Company
Baby clothes, diaper covers, wooden toys, and baby carriers. See entry under *Children's Clothes*.

The Right Start Catalog
5334 Sterling Center Drive, Westlake Village, CA 91361
Telephone: 800-548-8531

Catalog price: $2, refunded with order

The Right Start company somehow manages to find the latest inventions and innovations in child-rearing equipment, and to blend them in its big color catalog with the best of the old reliables. A familiar-looking hardwood high chair finds a place here right next to Aprica's ultra-modern collapsible model. A lambswool baby bunting loaded with old-fashioned charm shares the page with a state-of-the-art nursery monitor that not only broadcasts baby-room sounds but plays soothing music when baby cries. Other welcome novelties include a trough-shaped changing pad that keeps baby in place, an adjustable booster seat, and a car-seat sun hood. The catalog sells strollers, carriers, portable cribs and playpens, breast pumps, bedding, sheepskins, ear scopes, toddler toothbrushes, first-aid and child-safety kits, and

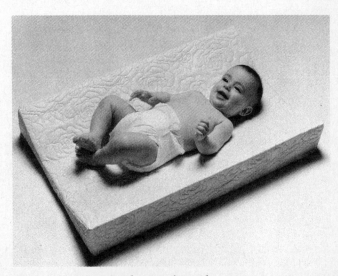

The Right Start Catalog's Safety Changer

toys like the Arcobaleno blocks and a rocking cow. Look for new improvements and even better ideas in the next catalog.

Ryder Products
20861 Collins Street, Woodland Hills, CA 91367
Telephone: 800-446-4416 or 818-999-9554

Catalog price: free

"Products invented *by* parents *for* parents," announces the cover of this booklet. The Ryders themselves invented the Little Shirt Anchor that leads off the catalog, a sort of upside-down garter belt that keeps baby's shirt from hiking up. Bunny- and bear-shaped overall stretchers, Flap Hat sun hats, the Sara's Ride child carrier, a bathtub spout cover, a car-seat sunshade, and an electric wipe warmer arrive to help out on the following pages.

Sara's Ride, Inc.
2448 Blake Street, Denver, CO 80205
Telephone: 303-292-2224

Free brochure

An over-the-shoulder baby carrier that doesn't look like an import from India. Made of a canvas seat and polypropylene straps, it clips with steel buckles and heavy plastic snap hooks. It can be worn in front, in back, or on the side to accommodate the riding styles of growing children.

Sears, Roebuck & Co.
A warehouse of baby needs, from cribs to clothes. See entry under *General Catalogs*.

Sensational Beginnings
Toys and books for babies and young children. See entry under *Toys*.

Sleeping Fawn Cradleboards
P.O. Box 9832, Santa Fe, NM 87504

Free brochure and photograph

Cradleboards made from a traditional American Indian design. With its rigid wooden back and laced-fabric front, a cradleboard can be hung as a swinging cradle, strapped to a parent as a backpack, or propped up so the baby can watch her surroundings. The makers have raised four girls in the cradleboards and claim the old concept still has an important place in child rearing.

Small Treasures
1951 Kelton Avenue, Los Angeles, CA 90025
Telephone: 213-477-6900

Catalog price: long self-addressed stamped envelope

Sells the Tot Tenders See & Snooze baby carrier, a two-strap fabric carrier that can be worn in several positions to accommodate children of different ages. Face-in, it snuggles a newborn to the parent's chest; face-out, it allows an older baby to see the world. It can also be worn as a backpack, used as a restraining seat attached to a ladder-back chair when no high chair is available, and inserted as a secure cushioned seat in a grocery cart.

Snapi Toy Tethers
4230 Progressive Avenue, Lincoln, NE 68504
Telephone: 402-464-1651

Free brochure

Makes a ribbon tether that keeps pacifiers and toys from flying out of high chairs and strollers.

Soft as a Cloud
Cotton baby clothes, booties, and bedding. See entry under *Children's Clothes*.

Soft Shoes
Elk-hide and wool slippers. See entry under *Shoes*.

Soft Star Shoes
Soft leather baby shoes. See entry under *Shoes*.

Soft Steps
Shoes for toddlers, sizes 4 to 8. See entry under *Shoes*.

Sound Sleep Products
P.O. Box 178, Mancos, CO 81328
Telephone: 303-533-7308

Free brochure

Sound Sleep Products makes an infant bed that attaches to the parents' bed. Made of either pine or oak, it is freestanding, with restraining walls on three sides. The fourth side opens onto the parents' bed for easy access during the night.

Cradleboards from Sleeping Fawn Cradleboards

Sweet Baby Dreams
Relaxation tapes for babies. See entry under *Music*.

Sweet Dreems, Inc.
130 E. Wilson Bridge Road, Suite 205, Worthington, OH 43085
Telephone: 800-662-6542 or 614-431-0496

Free brochure

Many parents have noticed that crying babies relax and fall asleep during car rides. The inventor behind Sweet Dreems translated the effect into a device to calm babies in their cribs. His two-piece SleepTight baby soother makes a crib vibrate and sound like a car traveling on a smooth highway. One piece attaches to the crib's springs to produce a gentle vibration. The other clips onto the railing to broadcast the sound of wind. According to the brochure, the machine reduced colicky behavior in 85 percent of infants tested in a recent independent study.

Tendercare Diapers
R-Med International, 5555 71st Street, Suite 8300, Tulsa, OK 74136
Telephone: 800-34 IM DRY or 918-491-9140

Free price list

Disposable diapers made without the super-absorbent gels found in other premium brands. The company points out in its literature that the gels haven't been subjected to long-term study, and that their contact over a period of years with a child's skin should be a cause for concern. Tendercare diapers are made with cellulose fibers and a plastic outer shell in a configuration that wicks moisture away from the skin.

Tiffany & Co.
Silver cups, rattles, and teething rings. See entry under *Gifts*.

Tot Tenders, Inc.
30441 S. Highway 34, Suite 7-AB, Albany, OR 97321
Telephone: 800-634-6870 or 503-967-9133

Catalog price: self-addressed stamped envelope

Manufactures the See and Snooze baby carrier, a six-position seat for toting babies, securing them in grocery carts, and holding them in adult-size chairs when there's no high chair at hand. The carrier is equipped with padded shoulder straps, plastic buckles and hooks, a sun hood, and a cushioned head support.

U-Bild
Patterns for building cradles, cribs, and changing tables. See entry under *Furniture*.

V.I.P. Bronzing Service
8126 Karen Drive, Tyler, TX 75703
Telephone: 214-561-4593

Catalog price: $2

V.I.P. Bronzing Service stands ready and willing to bronze your baby's shoes, but take a look at the other options first. Gold, silver, and pewter finishes present a more glamorous approach, and a Chinakote finish makes shoes look like pieces of glazed porcelain. For a little more money the finished shoes can be mounted on marble bases to make picture frames, bookends, ashtrays, lamps, or pen holders. The company encourages customers to think beyond baby shoes to a child's first toe shoes, the home-run ball from a Little League game, or a favorite hat.

Vermont Bird Company
A cashmere baby hat. See entry under *Outdoor Gear*.

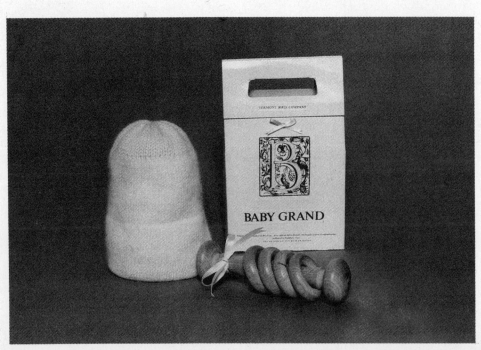

A rattle and cashmere hat gift set from Vermont Bird Company

The Vermont Country Store
Mail Order Office, P.O. Box 3000, Manchester Center, VT 05255-3000
Telephone: 802-362-2400

Catalog price: free

The "Voice of the Mountains" catalog brings the sensible approach of an old-time country store to bear on 20th-century merchandise, offering the best of the old with the most reliable of the new. The catalog sells reconditioned Electrolux vacuum cleaners, Handyaid rubber jar openers, hardwood clothes hangers, Timex watches, cotton nightgowns and nightshirts, and goose-down pillows. Parents of young children can order diaper cloth by the yard, Nikky diaper covers, flannel-covered rubber sheeting, and wooden baby gates. Cotton terry robes come in sizes for older children (2 to 14).

Lillian Vernon
Personalized diaper covers and other baby gifts. See entry under *Gifts*.

Vida Health Communications
6 Bigelow Street, Cambridge, MA 02139
Telephone: 617-864-4334 (800-526-4773 to order)

Free brochure

Sells a 110-minute videotape on baby care, titled *Baby Basics*, that has earned praise from such experts as T. Berry Brazelton and Ruth A. Lawrence. (To order directly by telephone you'll need to give the product number: NA2000.)

Wimmer-Ferguson Child Products
P.O. Box 10427, Denver, CO 80210
Telephone: 303-733-0848

Catalog price: free

Wimmer-Ferguson makes the Infant Stim-Mobile, the popular black-and-white pattern mobile that babies find so fascinating. The company's mail-order catalog features the mobile prominently, along with a mount for hanging it over a crib and vinyl Pattern Play cards that present the same geometric designs in a format for grabbing and chewing. To give the catalog some substance, baby products from other makers have been drawn into service: rubber squeaky toys, a ring rattle, the *Lullaby Magic* tapes, a simple busy box, a crib mirror, and the best freestanding play bar we've seen (for dangling toys over a reclining baby).

Wisconsin Wagon Co.
A baby cradle built of solid walnut or cherry. See entry under *Bicycles, Tricycles, and Wagons*.

Bath Accessories

The Body Shop
1341 Seventh Street, Berkeley, CA 94710
Telephone: 415-524-0216

Catalog price: free

Skin moisturizers, natural sea sponges, bubble bath, hairbrushes, and other pampering products for skin, hair, and bath. A lightly scented soapless soap is offered for babies.

Caswell-Massey Co. Ltd.
111 Eighth Avenue, New York, NY 10011
Telephone: 212-620-0900

Catalog price: $1

A wonderful collection of soaps, fragrances, and luxuries for the bath from a company that's been in the business since 1752. Caswell-Massey's floral soaps bring the scents of calendula, lily-of-the-valley, and wisteria into the bathroom. Fruit and vegetable soaps are made to smell of pears, apples, cucumbers, and tomatoes. Almond cold-cream soap pampers delicate skin; an array of colognes delights the nose (one was a favorite of George Washington). Swansdown powder puffs, skin creams, and bath oils may not be geared to young children, but parents deserve special treats too.

Delby System
450 Seventh Avenue, New York, NY 10123
Telephone: 212-594-5036

Free brochure

Natural Mediterranean sea sponges and non-natural cosmetic sponges in a variety of sizes, including some for babies and children.

Family Clubhouse
Weleda and Earthchild baby bath products. See entry under *Baby Needs*.

Hand in hand
No-slip bath and shower mats, a shampoo visor, and a floating bathtub carousel puzzle. See entry under *Toys*.

Metrobaby
Hooded towels, bath mitts, washcloths, and a hooded robe made of thick cotton terry. See entry under *Baby Needs*.

Naturepath
Weleda's natural soaps and skin care products. See entry under *Baby Needs*.

Nenuco
370 MacArthur Avenue, Long Branch, NJ 07740
Telephone: 201-222-4750
Free brochure
Hypoallergenic baby bath products imported from Spain. Nenuco makes baby shampoo, skin lotion, bath soap, talcum powder, and an after-bath splash. A brush and comb set comes in one combination package, a natural sponge in another. California customers should write to 10232 Mossy Rock Circle, Los Angeles, CA 90077. In Canada contact Nenuco at 160 George Street, Place Street George Mall, Ottawa, ON K1N 9M2, or telephone 613-563-3070.

Prince Lionheart
Makes a sun visor that doubles as a shampoo shield. See entry under *Baby Needs*.

The Right Start Catalog
Bath aids for young children. See entry under *Baby Needs*.

Towards Life Catalogue
P.O. Box 2243, Yountville, CA 94599
Telephone: 707-944-0713
Catalog price: $2
Herbal cosmetics, natural bath products, and homeopathic remedies. The company sells fragrances and skin conditioners from Kiehl's Pharmacy in New York City, Bare Escentuals makeup, Bach flower remedies, and several natural toothpastes. Sea wool bath sponges, recommended for babies, are described as "soft like silk."

Prince Lionheart's Eyes R' Dry shampoo shield

Bicycles, Tricycles, and Wagons

Action BMX Cycle Co.
255 Wolfner Drive, Fenton, MO 63026
Telephone: 314-343-9466
Catalog price: $2

BMX and freestyle bicycles, scooters, and skateboards in a big catalog full of parts and accessories. Complete bicycles are sold from Haro, GT, Dyno, Hutch, Kuwahara, CW, and several other makers. Skateboard manufacturers include Powell-Peralta, Santa Cruz, Zorlac, and Decks Vision. Videos, clothes, lubricants, and protective attire supplement the huge list of parts.

Blue Sky Cycle Carts
P.O. Box 704, Redmond, OR 97756
Telephone: 503-548-7753
$1 for brochure

A trailer for pulling children behind a parent's bicycle. The frame is made of welded tubular steel, the walls and seat of nylon. Outfitted with seats and restraining harnesses, the cart will cost about $300. A top canopy, available as an additional option, protects passengers from sun and rain.

California Hot Products
2511-P W. La Palma Avenue, Anaheim, CA 92801
Telephone: 800-HOT-3233 (800-4CALHOT in CA) or 714-995-2036
Catalog price: $1

More than 100 skateboard models in a bare-bones price list that assumes shoppers know what they want. Decks and wheels can be ordered separately. Shoes, knee and elbow pads, helmets, T-shirts, shorts, pants, hats, jackets, copers, nose guards, tail skids, lappers, and stickers are offered to complete the outfit. Be sure to ask about closeout specials.

Chaselle, Inc.
Tough-looking trikes and scooters from Lakeshore, Radio Flyer's Row Cart and wagons, and a pump-action riding horse. See main entry under *Educational Supplies*.

Crate & Barrel
Radio Flyer's Row Cart. See entry under *Toys*.

John Deere Catalog
A John Deere pedal tractor. See entry under *Toys*.

1st Class B*M*X
P.O. Box 66290, Portland, OR 97266
Telephone: 503-253-8688 (800-325-1048 to order)
Catalog price: $2

BMX bicycles and parts, with brand names like CW, GT, Haro, Mongoose, and Redline. Most of the bicycles are sold in kit form and will require assembly by an experienced mechanic. Those who know what they're doing can custom-build a bike from the list of components or soup up an ordinary bike with special handlebars, wheels, and number plates. Haro clothes, Dyno shoes, Vision street wear, and body guards may be necessary to make the rider look as great as the bike.

The Blue Sky bicycle trailer

The Radio Flyer Wagon— A Childhood Classic

For Citizen Kane, the idyllic image of childhood revolved around a sled called Rosebud. For most of the rest of us, a little wagon comes to mind—in bright red, the signature color of Radio Flyer.

The company now known as Radio Flyer set up shop in Chicago in 1919, making phonograph cabinets and furniture. At some point in its formative period—no one really remembers how it happened or even exactly when—the firm made a wooden wagon called the Liberty Coaster. The wagon turned out to be so popular that, over the next few years, the manufacturer phased out the furniture trade to concentrate on rolling stock. The company's breakthrough came in 1933, with the introduction of the first steel-body wagon, dubbed the Radio Flyer. The announcement was made in appropriate style for the vehicle that was destined to changed the face of childhood transportation: the company built a display at Chicago's Century of Progress Exposition that year in the form of a giant red wagon, manned by salespeople touting its obvious stylistic and functional advantages.

Radio Flyer still makes that first red wagon, now known as the No. 18. Its 36-inch body bears the same futuristic white lettering used in 1933. To it the company has added 12 other wagon models, most smaller and more compact in keeping with today's fuel-economy standards, all of them painted red. The firm also makes bicycles, toy wheelbarrows, a scooter, a Row Cart, a toddler tricycle, a rocking horse, and a rolling pony.

Chaselle, Inc. (see entry under *Educational Supplies*), and Crate & Barrel (see entry under *Toys*) sell a selection from the Radio Flyer inventory. The Natural Baby Co. (see entry under *Toys*) sends out the complete Radio Flyer catalog.

The Fordham-Scope Catalog
260 Motor Parkway, Happauge, NY 11788
Telephone: 516-435-8080

Catalog price: free

High-tech gifts for grown-ups and a few riding toys for children—scooters, trikes, sleds, and the pedal-powered Kettcar. A German-made tricycle with a tilting dump-box on the back should keep hardworking riders busy.

Gloucester Classics, Ltd.
A wooden wagon, a riding fire truck, and a riding school bus. See entry under *Rocking Horses*.

Hoover Brothers, Inc.
Tricycles. See entry under *Educational Supplies*.

Nasco
Wagons, scooters, and tricycles. See entry under *Educational Supplies*.

The Natural Baby Co.
Radio Flyer wagons, riding toys, and bicycles. See entry under *Toys*.

Perfectly Safe
A stable tricycle, a safe bicycle seat, and helmets. See entry under *Health and Safety*.

Sears, Roebuck & Co.
Tricycles, wagons, scooters, pogo sticks, and other riding toys are presented in the "Toys" catalog. Bicycles are sold in the main "Home" catalog. See entry under *General Catalogs*.

Stick-Em Up
P.O. Box 9108, Pleasanton, CA 94566
Telephone: 415-426-1040

Catalog price: free

Vinyl and Mylar stickers for decorating skateboards, bicycles, motorcycles, cars, and other treasured possessions. Many feature manufacturers' names and logos such as Haro, Shimano, and Harley-Davidson. Others display lines like "Get rad," "My other car is a bicycle," and "Boneless."

Wisconsin Wagon Co.
507 Laurel Avenue, Janesville, WI 53545
Telephone: 608-754-0026

Catalog price: free

Janesville wagons, scooters, and wheelbarrows made of solid oak with big red steel wheels and black rubber tires. These are reproductions of toys that were in production between 1900 and 1940, and their tough construction should give the designs a few more decades of use. A toddler tricycle, a teeter-totter, and an oak sled with steel runners are also patterned after older models. Of newer design is a baby cradle built of solid walnut or cherry.

Birth Announcements and Greeting Cards

Associated Photo Co.
Photographic birth announcements. See entry under *Photographs, Albums, and Frames*.

Baby Name-a-Grams™ Designer Birth Announcements
P.O. Box 8465, Dept. DD90, St. Louis, MO 63132
Telephone: 314-966-BABY

Free brochure and sample

Birth announcements that combine pictures and lettering in a novel way. Each announcement is drawn by hand with the baby's name repeated in calligraphy to form the contours of a drawing. Your child's name can be turned into a stork, a sailboat, a teddy bear, a baby carriage, a rocking horse, a kangaroo, or a sheep. A two-bear design is available for twins, a more conventional layout for baptismal and christening invitations.

Babygram Service Center
301 Commerce, Suite 1010, Fort Worth, TX 76102
Telephone: 800-345-BABY or 817-334-0069

Free brochure

Produces a photographic birth announcement that simulates the look of a real telegram, but with a color photograph (negative supplied by the parent) pasted into the center of the message.

Birth-O-Gram Company
1825 Ponce de Leon Boulevard, Coral Gables, FL 33134
Telephone: 305-446-6015

Free samples; 50¢ for catalog

Birth announcements and thank-you cards in dozens of different designs. Most show friendly cartoon babies

Birth Announcements and Greeting Cards

Custom birth announcement from Baby Name-a-Grams™ Designer Birth Announcements

with amusing messages. Some are die-cut in the shape of airplanes, milk bottles, telephones, ocean liners, and houses. Several are offered for adopted babies, one for twins, and one for families who live in mobile homes. Many are designed to reflect parents' work or hobby interests—they announce the arrival of new dog fanciers, tennis players, truck drivers, and salespeople.

BirthWrites
5 E. Gwynns Mill Court, P.O. Box 684, Owings Mills, MD 21117
Telephone: 301-363-0872
Free brochure and samples

BirthWrites offers three basic styles of birth announcements and thank-you cards: formal announcements typeset and printed in raised lettering, cards with realistic drawings of babies and brief phrases like "So tiny, so tender" and "A miracle of life," and funny cards with cartoon drawings and amusing inscriptions. The preprinted cards are shipped within 24 hours. Those that require typesetting take three days.

Cradle Gram
P.O. Box 16-4135, Miami, FL 33116-4135
Telephone: 305-595-6050
Long self-addressed stamped envelope for brochure

Birth and baptismal announcements printed on pink or blue parchment paper in a design suggestive of a telegram. Name and vital statistics are set at the top with a picture of a stork; a six-line rhyming message is run in scripty type below. Choose from eight messages, among them an announcement in Spanish, one to be sent by grandparents, and one by a single parent. Mention *The Mail-Order Shopper for Parents* on your order and Cradle Gram will add three announcements free of charge. The company also sells chocolate lollipops that say "It's a boy" or "It's a girl."

Current
Birthday, holiday, and anniversary cards. See entry under *Stickers and Rubber Stamps*.

Custom Cards
R.D. 2, Box 127, Dept. SP, Montgomery, NY 12549
Catalog price: $3.50 (includes samples)

Birth announcements, thank-you cards, gift cards, and more. Black-and-white drawings of such things as rocking horses, wildflowers, teddy bears, skaters, and Christmas wreaths adorn the fronts. The insides can be left blank or custom-printed with birth information, holiday greetings, or whatever else you may want. The cards can be printed in brown or black ink on a choice of color papers.

A Custom Cards birth announcement

Green Tiger Press
Note cards featuring the art of Beatrix Potter, Jessie Wilcox Smith, and Cicely Barker. See entry under *Books*.

H & F Products, Inc.
3734 W. 95, Leawood, KS 66206
Telephone: 800-338-4001 or 913-649-1444
Free brochure

Custom-printed birth announcements shipped within 24 hours of a telephoned order. The cards are decorated with pastel or primary-color borders of balloons, teddy bears, trains, alphabet blocks, flowers, and other appropriate designs. The type is printed in pale blue or pink, or in bright blue or red. The company offers 20 different designs in all.

A Maurice Sendak birthday card, published by The Peaceable Kingdom Press

Heart Thoughts, Inc.
6200 E. Central, Suite 100, Wichita, KS 67208
Telephone: 316-688-5781

Free brochure and samples

Birth announcements, thank-you cards, and other greetings, illustrated with soft pencil drawings. The cards are not personalized, so they can be ordered in advance. Insert cards with blanks for the baby's name, birth date, and weight can be filled in when the baby arrives. Special designs are offered for the parents of twins and for grandparents.

Joy Bee Designs
3650 Greenfield Avenue, #7, Los Angeles, CA 90034
Telephone: 213-473-8123

Catalog price: $3 (includes samples)

Birth announcements, invitations, thank-you cards, birthday cards, and other greetings custom-lettered in calligraphy. Choose from the catalog of illustrations and messages, or create your own message. Cards are lettered in black on a choice of colored papers.

The Metropolitan Museum of Art
Note cards and Christmas cards. See entry under *Gifts*.

New Moons
530 Rhodora Heights Road, Lake Stevens, WA 98258
Telephone: 206-334-6403

Long self-addressed stamped envelope for brochure

Birth announcements and thank-you cards drawn in a comfortable homey style. Four seasonal cards herald births at different times of the year. Special cards send word of twins, adoptions, and pregnancy. These cards leave blanks for the pertinent data, so they can be ordered in advance and kept on hand. New Moons will be moving to Texas in early 1990.

The New-York Historical Society
Note cards. See entry under *Posters and Decorations*.

The Peaceable Kingdom Press
Cards featuring the work of great children's-book artists. See entry under *Posters and Decorations*.

Printed Personals
138 Magnolia Street, Westbury, NY 11590
Telephone: 516-997-6906

Free brochure and sample

Hand-lettered birth announcements decorated with drawings of teddy bears, baby booties, rocking horses, and other childhood motifs. The cards blend lettering and pictures in integrated designs that are both simple and attractive. One places the baby's name and birth statistics inside a lace-edged bib, another in ribbons streaming from an old-fashioned rattle. The company suggests that customers order a month ahead of the due date, then call with the final wording and design choice as soon as the baby is born.

A Star Is Born
6462 Montgomery Avenue, Van Nuys, CA 91406
Telephone: 818-785-5656

Free brochure

Birth announcements printed to look like the chalk clapboards used to start each take in the filming of a Hollywood movie. Those serious about their movies may want to order a personalized clapboard or a living-room version of a classic stage light.

The Victorian Papers
P.O. Box 411352, Kansas City, MO 64141
Telephone: 816-561-5651

Catalog price: $1

Birth announcements, thank-you notes, birthday cards, and other greetings illustrated with color reproductions of Victorian paintings. Fifty different birthday cards are offered, two birth announcements, and several congratulations cards for weddings, pregnancies, and new babies.

The Writewell Co.
887 Transit Building, Torrington, CT 06790-1989
Telephone: 203-496-9655

Catalog price: free

Personalized stationery, return-address labels and rubber stamps, and such other personalized items as pencils, doormats, and mailboxes.

Books

ABC School Supply
Books for young readers. See entry under *Educational Supplies*.

Aims International Books, Inc.
3216 Montana Avenue, P.O. Box 11496, Cincinnati, OH 45211
Telephone: 513-661-9200
Catalog price: free

Books in Spanish for children and adults. Younger readers will find illustrated storybooks, books about animals, fairy tales, science books, adventure stories, and books about trains and trucks. The company also offers a few children's titles in French.

The Ark Catalog
See entry under *Toys*.

Baby-Go-To-Sleep Center
P.O. Box 1332, Florence, AL 35631
Telephone: 800-537-7748
Catalog price: free

A color catalog of children's books, ranging from old chestnuts like *The Secret Garden* (in a choice of three editions) and *Little Women* to recent classics like *Knots on a Counting Rope* and Chris Van Allsburg's *The Z Was Zapped*. About 100 books rate shelf space in this store. To balance the pricey new illustrated books, the company offers some inexpensive collections of fairy tales, myths, and Bible stories.

Bellerophon Books
36 Anapaca Street, Santa Barbara, CA 93101
Telephone: 805-965-7034
Catalog price: free

An amazing world of coloring and cut-out books. Bellerophon takes these booklets out of the realm of cute animals and licensed characters and uses the medium to explore history and culture. Your children's crayons can delve into ancient Greece, the art of Japan, the ships of the American Revolution, great dancers and composers, cowgirls, dinosaurs, and dragons. Little scissors can work on paper-doll books with such titles as *Great Women*, *Henry VIII & His Wives*, and *Infamous Women*. Or they can cut and assemble books of castles, trains, airplanes, and ancient masks.

Better Beginnings Catalog
345 N. Main Street, W. Hartford, CT 06117
Telephone: 800-274-0068 or 203-236-4907
Catalog price: $1

Books and tapes for babies and young children (to about age eight), with a few handbooks for parents. The books include William Steig's *Amazing Bone*, Syd Hoff's *Danny and the Dinosaur*, Margaret Wise Brown's *Runaway Bunny* and *Goodnight Moon*, Chris Van Allsburg's *The Polar Express*, and Janet and Allen Ahlberg's *The Jolly Postman*. Among the tapes are Pete Seeger's *American Folk Songs*, Rick Charette's *Alligator in the Elevator*, Raffi's *Baby Beluga*, and several by Sharon, Lois & Bram.

Blacklion Books
9 E. Oxford Avenue, Alexandria, VA 22301
Catalog price: free

Favorite children's books, many of them offered in a choice of hardcover and inexpensive paperback editions. *Goodnight Moon* can be ordered in the traditional cloth binding or as a paperback at one third the price. Helen Oxenbury board books are sold for toddlers; illustrated stories like *Blueberries for Sal*, *Harry the Dirty Dog*, and *Make Way for Ducklings* for young listeners. For children learning to read, Blacklion sells the *Little Bear* books, *Tales of Amanda Pig*, and *Frog and Toad are Friends*. Fairy tale and nursery rhyme collections should appeal to young and old; stories like *Charlotte's Web*, *Owls in the Family*, and *The Reluctant Dragon* to more advanced readers.

Bluestocking Press/Educational Spectrums
Guides to buying children's books. See entry under *Educational Supplies*.

Book Hunters
P.O. Box 7519, N. Bergen, NJ 07047
Telephone: 201-869-8786

No catalog

A search service that tracks down out-of-print books. If you're interested in sharing a lost childhood story with your family, Book Hunters may be able to ferret out a copy for you.

Books of Wonder
132 Seventh Avenue, Dept. AD, New York, NY 10011
Telephone: 212-989-3270

Catalog price: $3 for new book lists, $3 for list of old and rare books

New York City's largest children's bookstore, Books of Wonder sends out packets of catalogs to interested mail-order customers. The new book lists feature *all* of the Dr. Seuss books, *all* of the Beatrix Potter books, and virtually every other classic of children's literature. The entire Chris Van Allsburg library is here, Else Holmelund Minarik's *Little Bear* books, the first eight Paddington books, and other greats like Lynd Ward's *The Biggest Bear* and Carol Carrick's *Patrick's Dinosaur*. Books of Wonder is also big on facsimile editions of important older works, such as the *Rip Van Winkle* illustrated by N. C. Wyeth and the original *The Wonderful Wizard of Oz* illustrated by W. W. Denslow. Once you're on the mailing list, you'll receive a monthly newsletter with notices of new releases and opportunities to buy signed copies.

The old and rare book list may be a temptation for parents with an interest in book collecting. It's filled with first editions and signed copies, many affordable, some priced in the hundreds of dollars.

Cahill & Company
950 North Shore Drive, Lake Bluff, IL 60044
Telephone: 800-448-8311 or 312-295-8088

Catalog price: free

A catalog for people who love to read. The selection is weighted toward adult titles, but some treasures are dangled for children as well. *The Scottish Chiefs* by Jane Porter, with illustrations by N. C. Wyeth, is sure to provide many hours of enchantment. Sesyle Joslin's *What Do You Do, Dear?* and *What Do You Say, Dear?*, both illustrated by Maurice Sendak, teach manners with an unforgettable whimsy. *The Chronicles of Narnia*, *Ring of Bright Water*, and *Wild Animals I Have Known* plumped out the 1988 list, along with about 20 other pleasant choices for young readers.

Chaselle, Inc.
A big school-supply catalog well stocked with children's books. See entry under *Educational Supplies*.

Children's Book & Music Center
2500 Santa Monica Boulevard, Santa Monica, CA 90404
Telephone: 800-443-1856 or 213-829-0215

Catalog price: free

A delightful and comprehensive 80-page catalog of children's books, records, tapes, videos, and musical instruments. The mail-order bookshelf holds hundreds of titles, from board books for babies to storybooks for school-age children (up to about age nine). The catalog is nicely organized, so it's easy to flip to such sections as dinosaurs, Japanese culture, divorce, science, and beginning readers, and find books that match the interests of your children. For a description of the musical and

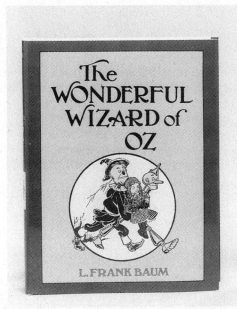 

Two facsimile editions from Books of Wonder

video offerings, see separate entries under *Music* and *Videos*. If you're in the area, we recommend a visit to the store. It's stocked with more than 20,000 titles.

Children's Recordings
Stories on record and cassette. See entry under *Music*.

The Children's Small Press Collection
719 N. 4th Avenue, Ann Arbor, MI 48104
Telephone: 800-221-8056 or 313-668-8056
Catalog price: free

Children's books from more than 100 small publishers. The selection includes picture stories, cookbooks, informational books on history and science, and books to help children deal with special situations like divorce, new siblings, and a death in the family.

A Child's Collection
155 Avenue of the Americas, 14th Floor, New York, NY 10013
Telephone: 212-691-7266
Catalog price: free

A carefully chosen collection of books presented in a 50-page color catalog. There's a temptation on every page, from beautiful illustrated books like *Crows*, *The Weaving of a Dream*, and *Where the Wild Geese Go* to captivating texts like *The Wind in the Willows*, *The Secret Garden*, and Laura Ingalls Wilder's *Little House* books. The catalog offers board books for little readers, pop-up books for those who like action in their libraries, and reprints of N. C. Wyeth's illustrated editions of *Treasure Island*, *Kidnapped*, *The Yearling*, and *The Last of The Mohicans*. Customers receive spring and fall catalogs, and the inventory changes radically from season to season. This company does a wonderful job of choosing the best of the new books and finding the best of the old.

Chinaberry Book Service
2830 Via Orange Way, Suite B, Spring Valley, CA 92078-1521
Telephone: 800-777-5205 or 619-670-5200
Catalog price: free

A fat 100-page catalog that's like having a bookstore with a knowledgeable clerk brought to your home. No pictures help shoppers make their choices, but a detailed and very personal description of each title does the job in a different way. We realize from the text that all of these books have been thoroughly tested on the owners' children, and that the comments of customers have been given careful attention through the years. The catalog progresses from wordless and board books for the very young (including Tana Hoban's and Fiona Pragoff's board books and Eric Hill's *Spot* books) to longer books for older children, like *The Adventures of Tintin* series, *Juggling for the Complete Klutz*, and C. S. Lewis's *The Chronicles of Narnia* series. Books for parents, the *Anti-Coloring Book* series, glow-in-the-dark ceiling stars, and a few pages of children's music and story cassettes are among the additional bonuses. An index helps shoppers find particular titles.

A Common Reader
175 Tompkins Avenue, Pleasantville, NY 10570
Telephone: 914-747-3388
Catalog price: free

Customers receive a monthly 48-page catalog stuffed with wonderful and hard-to-find titles. Most of the booklet is taken up with grown-up reading, but a few pages are always set aside for children's books. The *Madeline* library is a favorite here, as is *The Secret Garden* and *Harold and the Purple Crayon* (five of Harold's adventures are listed). Books on history and art complement the stories.

Create-A-Book
207 Leton Drive, Columbia, SC 29210
Long self-addressed stamped envelope with 25¢ for brochure

Personalized books with your child as the main character. The order form has blanks for the child's name, age, and hometown, and the names of three friends. The information is woven in computerized type into one of 11 stories with color illustrations.

Creative Parenting Resources
P.O. Box 7238, Capistrano Beach, CA 92624
Telephone: 714-240-7746
Free flier

Books of guidance and advice for parents of young children, written by William Sears. Titles deal with such subjects as sleep problems, fussy babies, breastfeeding, and fathering. Most of the same subjects are also covered on audiocassettes. As a sideline, the firm sells a cotton over-the-shoulder baby sling. If you're interested in the sling, ask for the brochure of available fabrics when you write.

Cumberland General Store
McGuffey's Readers. See entry under *Toys*.

The Disney Catalog
Disney books. See entry under *Gifts*.

Doubleday
666 Fifth Avenue, New York, NY 10031
Telephone: 800-223-5780 (416-977-7891 in Canada)
Catalog price: free

Among the stars on Doubleday's children's book list are *The Brave Little Toaster*, Marguerite de Angeli's *The Door in the Wall*, an edition of Rudyard Kipling's *Just So Stories* illustrated by Nicolas, Peter Spier's *Noah's Ark*, and the D'Aulaires' illustrated biographies of Abraham Lincoln, Benjamin Franklin, and Pocahontas. A number of drawing handbooks by Lee J. Ames guide young artists; songbooks by Tom Glazer and Ruth Crawford Seeger provide fodder for young musicians. Doubleday combines its sales efforts with Delacorte Press in a single backlist catalog. Together the publishers offer such parenting titles as John Holt's *How Children Learn*, T. Berry Brazelton's *Toddlers and Parents*, Marguerite Kelly's *Mother's Almanac*, and Cuthbertson and Scheville's *Helping Your Child Sleep Through the Night*.

Dover Publications, Inc.
31 E. 2nd Street, Mineola, NY 11501-3582
Telephone: 212-255-3755 or 516-294-7000

Catalog price: free (request juvenile book catalog)

Hundreds of inexpensive activity books and reprints of old editions. Dover publishes dozens of paper-doll books and coloring books in the $2 to $4 price range, cut-and-assemble castles and toy theaters, sticker books, collections of magic tricks and science activities, and unusual how-to books like *Hand Shadows to Be Thrown Upon the Wall*. One page of the catalog presents books of mazes, another logic puzzles, and another the works of Beatrix Potter in $1.75 paperback editions. One extra not to overlook here is the collection of low-priced posters. Two reproduction circus posters will liven up a bedroom wall. Dinosaur, wildflower, and bird identification posters are both attractive and educational. Dover also puts out separate "Needlecraft," "Chess," and "Pictorial Archive" catalogs.

Durkin Hayes Publishing Ltd.
1 Columbia Drive, Niagara Falls, NY 14304
Telephone: 716-298-5150 (800-962-5200 to order)

Catalog price: free

Books on cassette. Readers like Douglas Fairbanks, Jr., Susannah York, James Mason, and John Le Carré apply their talents to readings of *Ring of Bright Water*, *Kim*, *The Adventures of Robin Hood*, *Chitty Chitty Bang Bang*, *Pinocchio*, and 80 other popular works.

EDC Publishing
P.O. Box 470663, Tulsa, OK 74147
Telephone: 918-622-4522

Catalog price: $2, refunded with first order

This firm publishes more than 300 reference and information books for children, most of them bearing the Usborne imprint. Colorful illustrations help explain such subjects as evolution, archaeology, electricity, and how to draw monsters. Guidebooks help identify butterflies, trees, and flowers. A "World of the Unknown" series deals with UFOs, ghosts, and mysterious creatures; a "Mysteries and Marvels" series shows oddities of plant and animal life. Board books and simple readers are geared to young children, handbooks of technical drawing, photography, and fashion design to older readers. The *Book of Car Travel Games* and *Book of Air Travel Games* will have special appeal for parents.

Early Learning Centre
Books for young children. See entry under *Toys*.

Fille
5300 Santa Monica Boulevard, Suite 309, Los Angeles, CA 90036
Telephone: 213-462-3665

Free information

Publishes a series of cloth books that come in the form of felt "briefcases" (they have handles and Velcro closures) stuffed with cloth characters and objects that stick to the pages with Velcro backs. *Play and Learn Numbers* has a big number on each page, and children are encouraged to fasten a number of objects to the page to match the numeral. *Play and Learn Colors* and *Play and Learn Shapes* take the same approach with colors and shapes. If the activity isn't forced, children can have a great time inventing stories around the characters and objects while they play. Each book costs about $40, plus $3 shipping (check current prices before ordering).

Fille's Play and Learn Numbers

Geode Educational Options
P.O. Box 106, West Chester, PA 19381
Telephone: 215-692-0413

Catalog price: free

A catalog of alternative books and games for children, focusing on creative thought and cooperative play. Books range from math and logic puzzles to art books and science activities. A book about law for junior-high students looks like a thought-provoking volume. Pages of fascinating coloring books explore history, architecture, mythology, and many other subjects. The company also offers about 20 cooperative board games that discourage competitive play.

Gleanings
60 Priorway Drive, Novelty, OH 44072
Telephone: 216-321-0214

Catalog price: free

A new children's book catalog, launched in early 1989. Old favorites like *The Story About Ping, Harry the Dirty Dog, The Wind in the Willows,* and Howard Pyle's tales of King Arthur are grouped in a section titled "Classics." The rest of the catalog is loosely arranged by reading ability, starting with picture books like *Our Animal Friends at Maple Hill Farm* and moving through such early readers as *Fox on the Box* (written with only nine different words) to more mature tales like Martin Handford's *Where's Waldo?* and Alice and Martin Provensen's *Shaker Lane.*

David R. Godine, Publisher, Inc.
300 Massachusetts Avenue, Boston, MA 02115
Telephone: 617-536-0761

Catalog price: free

Godine books are noted for their high-quality production—the sewn bindings don't snap shut while you're trying to read, and the paper and printing are generally first-rate. The children's book list includes Mary Azarian's *A Farmer's Alphabet,* Lawrence Treat's solve-them-yourself picture mysteries, a beautiful photographic alphabet titled *The Ark in the Attic,* and an edition of *The Secret Garden* with color illustrations by Graham Rust.

Ginny Graves
Activity books. See entry under *Art Supplies.*

Green Tiger Press
1061 India Street, San Diego, CA 92101
Telephone: 800-424-2443 or 619-238-1001

Catalog price: free

Gorgeous illustrated books, some of them new creations and some rediscoveries from the past. *Good Dog Carl* tops the press's sales charts. Other treats include *The Story of a Little Mouse Trapped in a Book,* an illustrated edition of *The Teddy Bear's Picnic,* Jasper Tomkins's *The Catalog,* and a collection of Carl Larsson's paintings. Older gems include *The Teddy Bear That Prowled at Night* and *The Teenie Weenies Book.* A wonderful collection of note cards draws on the art of Beatrix Potter, Jessie Wilcox Smith, and Cicely Barker. Perhaps because the company deals primarily with stores rather than individuals, its response to catalog requests is something less than lightning quick. Our catalog arrived after six months, two letters, and three telephone calls.

Grey Owl Indian Craft Co.
Books on American Indian culture and crafts. See entry under *Hobby Supplies.*

Growing Child
See entry under *Toys.*

Gryphon House
P.O. Box 275, Mt. Rainier, MD 20712
Telephone: 800-638-0928 or 301-779-6200

Catalog price: free (request "Early Childhood Catalog")

More than 200 carefully selected books for preschoolers in an illustrated catalog. The old chestnuts are fairly represented—*The Story of Ferdinand, Goodnight Moon,* and the Berenstain Bears are here—but the strength of the catalog lies in its exceptional array of newer titles. Gryphon House sells poetry collections, books that deal with troubling experiences (such as *I Have Asthma* and *Sometimes a Family Has to Split Up*), books for special occasions and holidays, counting and alphabet books, and lots of wonderful story and picture books. The company offers a separate catalog for teachers of young children (request the "Learn By Doing" catalog if you want a copy). Books can be ordered straight from the catalog, or by joining the Gryphon House Preschool Book Club. When parents order through the book club, their child's school receives free bonus books.

Harper & Row, Publishers, Inc.
Trade Sales Department, 10 E. 53rd Street, New York, NY 10022
Telephone: 800-242-7737 (800-982-4377 in Pennsylvania)

Catalog price: free (request junior books backlist)

One of the bigger children's book publishers, Harper & Row will send a complete list of its titles to anyone who asks and is happy to fill orders to individuals by mail. No descriptions or pictures aid the shopper, but a careful scan turns up such favorites as *Goodnight Moon, Harold and the Purple Crayon,* Else Holmelund Minarik's *Little Bear* books, the Laura Ingalls Wilder stories, and the Maurice Sendak tales.

Herron's Books for Children
P.O. Box 1389, Oak Ridge, TN 37830

Catalog price: free

Books for young children and beginning readers, and beginner's puzzles from Lauri. Madeline, Paddington, Babar, Corduroy, Ferdinand, Peter Rabbit, and George and Martha all make appearances on the book list. Some less commonly found titles include *The Country Bunny and the Little Gold Shoes, The Very Hungry Caterpillar,* and *The Bear's Toothache.* Discounts are offered on large orders at certain times of the year.

John Holt's Book & Music Store
2269 Massachusetts Avenue, Cambridge, MA 02140
Telephone: 617-864-3100

Catalog price: long self-addressed envelope with two first-class stamps

Books, art supplies, and musical instruments that are in keeping with the educational writings of John Holt. Books on learning theory and home schooling offer insight for parents. Children's bookshelves can be stocked with poetry collections, science activity books, reader-friendly math books, writing aids, and illustrated storybooks. The catalog carries David Macaulay's explanatory books of drawings (*Castle*, *Underground*, and others), the Tintin cartoon books, and titles like *Four Arguments for the Elimination of Television* and *Square Foot Gardening*. Young musicians can buy violins in a range of sizes down to $^1/_{16}$ scale, as well as recorders, pitch pipes, and pianicas. Songbooks and music lesson books are offered, including Suzuki music books for piano and violin. Artists will find modeling beeswax, watercolor pens, Cray-pas pastels, individual chalkboards, a rubber-stamp printing set, and a number of activity books. The same organization publishes the newsletter *Growing Without Schooling* (see entry under *Parents' Magazines, Education and Child Care*).

Hoover Brothers, Inc.
See entry under *Educational Supplies*.

Houghton Mifflin Company
Trade Sales Office, 2 Park Street, Boston, MA 02108
Telephone: 617-725-5959
Catalog price: free (request backlist catalog)
Houghton Mifflin publishes a number of favorite children's books, including all the Paddington Bear, George and Martha, and Curious George stories; Lynd Ward's *The Biggest Bear*; Virginia Lee Burton's *Mike Mulligan and His Steam Shovel*; and the books of David Macaulay and Chris Van Allsburg. The catalog lists these and some 400 others.

Kaplan School Supply Corp.
See entry under *Educational Supplies*.

La Leche League International
Books and pamphlets on breastfeeding and parenting. See entry under *Maternity and Nursing Clothes*.

Learn Me Bookstore
175 Ash Street, St. Paul, MN 55126
Telephone: 612-490-1805
Catalog price: $1
A catalog of books, records, and games chosen for their positive social messages and for their educational value. The store looks for parenting books that include fathers and people of other cultures, and books that show active grandparents and women in nontraditional roles. That selection process brings together well-known works like *Goodnight Moon* and *The Very Hungry Caterpillar* with titles like *Daddy Makes the Best Spaghetti* and *Just Us Women*. Several hundred titles are listed in all, each with a brief description. The slant of the catalog is perhaps most noticeable in the fairy tale section, where stories like *Ruby the Red Knight*, *The Woman in the Moon*, and *Paper Bag Princess* feature active heroines. A number of noncompetitive games are offered, along with a good-size collection of songs and recorded stories.

Little Ears
Stories on cassette. See entry under *Music*.

Living Skills Press
P.O. Box 88, Sebastopol, CA 95473
Telephone: 707-823-5483
Catalog price: free
Dozens of self-help books for children, written by Joy Berry. Entertaining illustrations help convey the messages in volumes like *Understanding Parents*, *Handling Feelings*, *Family Rules and Responsibility*, and *What to Do When Your Mom or Dad Says "We Can't Afford It."*

Los Angeles Birthing Institute
4529 Angeles Crest Highway, Suite 209, La Canada, CA 91011
Telephone: 818-952-6310
Catalog price: free
Pregnancy, birth, and parenting books, along with some classic children's books. The collection ranges from standard texts like *What to Expect When You're Expecting* and Penelope Leach's *Babyhood* to alternative views like Suzanne Arms's *Immaculate Deception*. Those who live in Southern California may want to find out more about the Institute's networking services, seminars, and childbirth education programs.

MB Enterprises
P.O. Box 2117, Myrtle Beach, SC 29578
Telephone: 803-626-8317
Catalog price: long self-addressed stamped envelope
Personalized storybooks starring the child of your choice. Send the child's name, birthday, and hometown, and the names of a friend and a pet, and you'll receive back a color-illustrated book with the names and information worked into the story in computer type. Choose from seven stories in paperback or hardcover editions.

Manzanita Publications
Coloring books, paper-doll books, and cut-and-assemble books. See entry under *Art Supplies*.

The Metropolitan Museum of Art
Illustrated children's books. See entry under *Gifts*.

The Mind's Eye
P.O. Box 6727, San Francisco, CA 94101
Telephone: 800-227-2020 or 415-883-7701
Catalog price: $2
Books and stories recorded on cassette, from *Cinderella* to *Out of Africa*. Gwen Watford reads *The Secret Garden*; full casts perform *Treasure Island*, *Little Women*, *Huckleberry Finn*, and *Alice in Wonderland*. In addition to well-known tales, the catalog offers audio biographies of such great Americans as Benjamin Franklin and Harriet Tubman, radio performances from the 1930s and '40s, foreign-language instruction tapes for children and adults, and a selection of musical recordings.

William Morrow & Company
105 Madison Avenue, New York, NY 10016
Telephone: 212-889-3050
Catalog price: free (request backlist catalog)

William Morrow & Company distributes the titles of Greenwillow Books and Lothrop, Lee & Shepard Books, as well as those from its own extensive inventory. The backlist catalog combines children's and adult books in a long descriptive list. If you're willing to take the time to search through, you may turn up some volumes that will be hard to find elsewhere. The stars of the children's list are the many books by Beverly Cleary, sold by Morrow in hardcover. Simon Seymour's books on such scientific subjects as optical illusions, the planets, stars, and volcanoes are popular with young readers, as are Helen Roney Satler's books on dinosaurs and her *Recipes for Art and Craft Materials*.

Museum Books Mail-Order
Illustrated children's books. See entry under *Gifts*.

Museum of Fine Arts, Boston
Illustrated books and books on the fine arts. See entry under *Gifts*.

Music for Little People
Videos and recorded stories. See entry under *Music*.

The Natural Baby Co.
Barron's wooden board books and a list of more than 100 favorite children's books offered at a discount. See entry under *Toys*.

Naturepath
Pregnancy, birth, and parenting books. See entry under *Baby Needs*.

The New-York Historical Society
Old-fashioned children's books like *The Slant Book* and a collection of rebuses. See entry under *Posters and Decorations*.

Orange Cat
442 Church Street, Garberville, CA 95440
Telephone: 707-923-9960
Catalog price: free

The catalog should be used as a starting point in dealing with Orange Cat. The owners are willing to suggest books based on a child's past likes, and they're happy to find other books that aren't on the list. That said, the catalog ropes in many wonderful works. The George and Martha books are here, as are Else Minarik's *Little Bear* stories, Lyle the Crocodile's tales, Arnold Lobel's *Frog and Toad*, Susan Cooper's five-volume *The Dark Is Rising* series, and Howard Zinn's *People's History of the United States*. On the parenting side are several beautiful baby journals, Penelope Leach's *Your Baby & Child*, and Tine Thevenin's *The Family Bed*.

The Peaceable Kingdom Press
Posters and cards featuring the art of popular children's book illustrators. See entry under *Posters and Decorations*.

Picture Book Studio
2515 E. 43rd Street, P.O. Box 182208, Chattanooga, TN 37422-2208
Telephone: 800-462-1252
Catalog price: free

Children's books with exceptional illustrations. Eric Carle's *Papa Please Get the Moon for Me* unfolds dramatically to make room for papa's *very* long ladder and the huge bulk of the moon. Lisbeth Zwerger often spends a month on each of her soft and luminous illustrations. She's added her remarkable vision to such favorite fairy tales as *Thumbeline*, *The Nightingale*, *Hansel and Gretel*, and Oscar Wilde's *The Canterville Ghost*. Other artists include Yoshi, Ivan Gantschev, Chihiro Iwasaki, and Eve Tharlet. The color catalog shows both covers and inside illustrations.

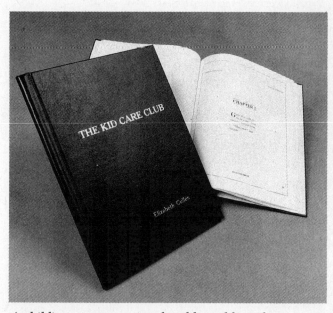

A child's own story, printed and bound by Polywog Press

Polywog Press
318 Massachusetts Avenue, N.E., Washington, DC 20002
Telephone: 202-546-1706
Catalog price: free

Polywog Press takes personalized children's books to a level more ambitious, more satisfying, and considerably more expensive than the fill-in-the-blank computerized stories. The press makes finished books from children's own stories, bound in simulated leather with the title and author's name stamped in gold. Customers are sent a writer's kit, which includes forms for the biography page and dedication and, if needed, some help in getting started (leading first lines and nudging questions). Children unable to write can dictate their stories onto an audiocassette. One long story or several short ones will make a book; the company will even handle poetry. When the manuscript or cassette is returned, Polywog Press sets the work in type, prints it, and binds it like a real book. A single volume costs about $50, additional copies a little less.

Practical Parenting
18326 Minnetonka Boulevard, Deephaven, MN 55391
Telephone: 612-475-3527
Catalog price: long self-addressed stamped envelope
Vicki Lansky is the author of *Feed Me I'm Yours*, a popular guide to baby nutrition. That book and a number of others she has written are offered here—on subjects like toilet training, travel with a baby, and getting a baby to sleep. Ms. Lansky has also assembled a good collection of books on divorce by other writers, for both parents and children.

Pro Libris
Gertrude Toll, 88 Ossipee Road, Somerville, MA 02144-1633
Telephone: 617-628-7487
Long self-addressed stamped envelope for search form
Looking for a long-lost children's book for your own child? Or an out-of-print book that can't be found in stores? For $10, Gertrude Toll will advertise and search for any title in the used-book trade for six months. She makes no guarantees of success, but she's been able to turn up some treasures. One family had read a favorite story every Christmas for 30 years, ad libbing when they came to a missing page. Pro Libris found them a fresh copy, complete from cover to cover.

Publishers Central Bureau
1 Champion Avenue, Avenel, NJ 07001-2301
Telephone: 201-382-7960
Catalog price: free
Discounted and remaindered books and a fair-size list of videos. Children's books are relegated to a single page, so don't expect a huge selection. All of the videos are available in both VHS and Beta formats.

Read It Again
32 Winter Street, Goffstown, NH 03045
Catalog price: $1, deductible from first order
An excellent collection of books for children, each tested as the kind children will say "Read it again" to when you finish. Board books for babies lead off the reading list, some with clear color photographs, a few with Helen Oxenbury's appealing drawings. Children old enough for genuine read-alouds can stock up on oldies like *Millions of Cats* and *Caps for Sale* or newer favorites like *Doctor De Soto*, *Tales of Amanda Pig*, and *The Jolly Postman or Other People's Letters*. About 150 books are offered in all, and the selection is so solid it doesn't look possible to make a mistake.

Recorded Books
270 Skipjack Road, Prince Frederick, MD 20678
Telephone: 800-638-1304 or 301-535-5590
Catalog price: free
Recorded books for sale or rent. The children's page lists such tales as *Black Beauty*, *The Secret Garden*, *The Wind in the Willows*, *The Adventures of Robin Hood*, and Rudyard Kipling's *Just So Stories*—all wonderful diversions for long car rides. Sprinkled through the other 30 pages are a number of stories that will be equally appealing to young listeners: *The Adventures of Tom Sawyer*, *The Call of the Wild*, *Treasure Island*, and *Kim*. Those a little older might go for *Rebecca*, *To Kill a Mockingbird*, or *Anne Frank: The Diary of a Young Girl*. These are all complete and unabridged narrations that run for several hours apiece.

Salad Days
P.O. Box 996, Harpers Ferry, WV 25425
Telephone: 800-248-3274 or 304-782-1106
Catalog price: $1, refunded with order
A finely honed collection of books for children and parents. Favorite authors and illustrators like Arnold Lobel, James Marshall, William Steig, Dr. Seuss, Tomie dePaola, A. A. Milne, and Maurice Sendak are well represented and easy to find. Long-popular titles like *The Story of Ferdinand*, *Goodnight Moon*, and *Island of the Blue Dolphins* hold cherished spots. And an effort has been made to seek out foreign-language editions of some well-known volumes. What's special about this catalog, however, is its excellent collection of newer titles and books that have not yet become standard fare on children's bookshelves. Bedtime books like *Sleepers*, alphabet books like *What's Inside?*, and storybooks like *Minerva Louise* add some fresh air to the library. Some nice photographic board books are offered for the youngest readers, along with works by Helen Oxenbury, Arnold Lobel, and Eric Hill. Older book worms will find such great reads as *The Wind in the Willows*, *The Borrowers*, and *The Phantom Tollbooth*. And parents can browse through a shelf of useful parenting guides.

Sensational Beginnings
Books for babies and young children. See entry under *Toys*.

Shibumi Trading, Ltd.
Origami guides and Japanese story- and coloring books. See entry under *Gifts*.

Spoken Arts
310 North Avenue, New Rochelle, NY 10802
Telephone: 914-636-5482

Catalog price: free

Spoken Arts has been in the business of recording stories, poems, and speeches for more than 30 years, and the dense catalog is rich with the fruits of that labor. Young children can hear Julie Harris read stories like *Rumpelstiltskin* and *Little Red Riding Hood* or Frances Sternhagen read the tales of Beatrix Potter. More advanced listeners can attend to Mark Twain, Rudyard Kipling, Hans Christian Andersen, Aesop, Shakespeare, and Edgar Allen Poe. The expansive poetry section includes Wordsworth, Gerard Manley Hopkins, Coleridge's "Rime of the Ancient Mariner," Percy Bysshe Shelley's "Ozymandias," Dorothy Parker, and an amazing collection of 100 modern American poets reading their own work. Speeches by Abraham Lincoln, John F. Kennedy, and Martin Luther King bring reality into the mix. Recordings in Spanish, French, Hebrew, and other languages broaden the horizons yet further.

Telltales
P.O. Box 614, Bath, ME 04530
Telephone: 800-922-READ or 207-443-3177

Catalog price: $2

A beautiful color catalog of children's books, mailed in seasonal editions throughout the year. The catalog's crisp reproductions show fine illustrated books to good advantage—works like Lisbeth Zwerger's *Thumbeline* and the new *Alice's Adventures in Wonderland* illustrated by Anthony Browne. Mixed in are books that deal with childhood problems (such as *Once When I Was Scared* and *Saying Good-bye to Grandma*), books on animals and nature, books that encourage young writers, adventure stories, read-aloud collections of poems, baby books, and works on architecture and science. In the spring of 1989 the catalog added some costumes to the library for book-related dress-up play. Among the outfits were a pirate suit (for *Treasure Island*), a hooded red cape (for *Little Red Riding Hood*), a Robin Hood costume, and a *Little House on the Prairie* dress.

Totline Books
P.O. Box 2255, Everett, WA 98203
Telephone: 206-485-3335 (800-334-4769 to order)

Catalog price: free

Activity books for preschoolers. The company's popular *Piggyback Songs* fits new lyrics to familiar tunes, a cookbook gives recipes for sugarless snacks, and a pattern book shows how to make inexpensive teaching toys. *Hug a Tree* and *Things to Do with Toddlers and Twos* present more ideas for activities and toys.

Superintendent of Documents, U.S. Government Printing Office
Washington, DC 20402
Telephone: 202-783-3238

Catalog price: free (request "U.S. Government Books")

The U.S. Government Printing Office puts about 700 of its most popular books into an easy-to-use mail-order catalog. Some of the titles are a little specialized for the average reader (*Biotechnology in Western Europe*, for example), but many others are useful, readable, and for the most part, inexpensive. *The Super Sitter* gives safety tips to prospective baby-sitters; *Your Child from One to Six* discusses patterns of development and offers advice on discipline, toilet training, school readiness, and other subjects. Books from NASA answer questions about space exploration. Several volumes of *The Back-Yard Mechanic* guide readers through basic car repairs.

A directory of free and inexpensive booklets put out by various government agencies can be had by writing to Consumer Information Center, P.O. Box 100, Pueblo, CO 81002. Several hundred titles present information on everything from the common cold to gardening. The four-page *Plain Talk About Raising Children* offers sensible advice for free. *A Look at the Planets* reproduces color pictures of the planets with basic information about each and details of the NASA flybys and landings that gathered the data. *Family Folklore* tells how to interview family members. *Where to Write for Vital Records* gives a state-by-state guide to tracking down birth, death, marriage, and divorce documentation.

Viking Penguin Inc.
40 W. 23rd Street, New York, NY 10010
Telephone: 212-337-5200

Catalog price: free (request children's backlist catalog)

Readers willing to take the time to scan Viking Penguin's backlist catalog will find such popular titles as *The Tale of Peter Rabbit*, *Make Way for Ducklings*, *Crow Boy*, *Madeline*, *The Amazing Bone*, *Corduroy*, and *Pippi Longstockings*. The company publishes all of the books of Beatrix Potter, Robert McCloskey, Don Freeman, and Astrid Lindgren, and offers most in both hardcover and paperback editions. The Puffin Classics are a good source of inexpensive paperback editions of such stories as *Black Beauty*, *The Adventures of Tom Sawyer*, *Little Women*, and *Treasure Island*. The Penguin backlist catalog is also worth a look for adult reading material, with its well-known orange-spined editions of many of the great works of literature.

Wilderness Press
Guidebooks for hikers and canoeists. See entry under *Outdoor Gear*.

Workman Publishing
708 Broadway, New York, NY 10003
Telephone: 800-722-7202 or 212-254-5900

Catalog price: free

Workman specializes in calendars, humor books, and works of entertainment and information, many of them for children. The list is not huge, but it's studded with such best-selling stars as *The Bug Book & The Bug Bottle*, *365 New Words for Kids Calendar*, *Steven Caney's Toy Book*, and *The Science Book* by Sara Stein. Cookbooks, activity books, and nature books offer more fun and learning for children. The parenting list includes *What to Expect When You're Expecting* and *Baby Let's Eat*.

Yellow Moon Press
P.O. Box 1316, Dept. 69, Cambridge, MA 02238
Telephone: 617-628-7894

Catalog price: free

Storybooks and cassettes for children and grown-ups. Rhyme and song recordings are tuned to the ears of toddlers; folktales, poems, and original stories are offered for older children. The collection takes in Indian myths and legends, the Bre'r Rabbit stories, Jewish fairy tales, and modern inventions like Bill Harley's *50 Ways to Fool Your Mother*. Adults can delve into Joseph Campbell's books on mythology, books on storytelling technique, and recorded stories aimed at mature listeners.

Book Clubs

Books of My Very Own
Div. of Book-of-the-Month Club, Inc., Camp Hill, PA 17011-9901
Telephone: 800-233-1066 or 717-697-1066

Free information

Four book clubs for children of different ages: Baby Books for babies and young children, Picture Books for preschoolers, Story Books for children five to eight, and Adventures, Mysteries, Puzzles, and Games for older readers. Each club offers a selection of the best books from many different publishers, in a good mix of old and new, fiction and nonfiction, and funny and serious stories. All are continuity clubs, which means that books are mailed out regularly—three or four books every five weeks—and subscribers can return them if dissatisfied. (Return too many, however, and your name disappears from the mailing list.) The books are sold at a substantial discount from publishers' list prices.

Children's Choice Book Club
P.O. Box 984, Hicksville, NY 11802
Telephone: 212-867-4300

Free information

A book club that specializes in inexpensive reprints of the great children's titles. The introductory offers are enticing (four books for $1, plus shipping and handling), and the prices of the follow-up books are low. Unfortunately the savings comes in part from manufacturing shortcuts—cheaper paper and lower-quality printing—and the club's books have in the past been noticeably inferior to the standard publishers' editions. If that trade-off doesn't bother you, the club is ready to send out titles like *Madeline*, *Corduroy*, *The Story About Ping*, and *Mike Mulligan and His Steam Shovel*.

Children's Reading Institute
Membership Center, Durham, CT 06422
Telephone: 800-243-3484 or 203-349-1014

Catalog price: free

Offers science and nature books for young readers through the Let's Read and Find Out book club. Two new books are mailed every six weeks. Members can either pay the bill and keep them or send them back. Titles include *How a Seed Grows*, *What I Like About Toads*, *My Five Senses*, and *How You Talk*. The same company at a different box number sells activity sets for teaching math and phonics at home. Write to Drawer 679 in Higganum to find out the details.

Early Advantage Programs for Children
The Children's Circle Video Reading Program, a video/book club that encourages children to read by pairing animation videos with popular books. See entry under *Videos*.

Golden Press
120 Brighton Road, Clifton, NJ 07012
Free information

Golden Press, known for inexpensive children's books like *The Poky Little Puppy*, also runs the Sesame Street Book Club, which mails a series of books featuring the characters from Sesame Street. Members receive two books each month, geared to ages three to seven. A new series in 1989, called Sesame Street Big Bird Beep Books, comes with a plastic device that beeps when the child points it to the "correct" spot on the page.

Grolier Enterprises Inc.
Sherman Turnpike, Danbury, CT 06816
Telephone: 203-797-3500
Free information

Grolier Enterprises offers four book clubs that operate on the two-books-a-month plan: two books arrive in the mail each month; you pay for them if you like them, return them if you don't. Disney's Wonderful World of Reading is geared to two- to six-year-olds and features books with Disney characters, published by Random House. Dr. Seuss & His Friends encompasses all of the Dr. Seuss titles plus the similar works of P. D. Eastman (*Are You My Mother?*), Stan and Jan Berenstain, Al Perkins, and Eric Gurney. First Time Books mails a series of Berenstain Bears books that deal with new experiences such as moving, getting a new baby-sitter, and going to the dentist. Help Me Be Good Books tackle subjects relating to behavior and manners. Specify which club you're interested in when you write.

Gryphon House
A book club for preschoolers that earns free bonus books for the child's school. See entry under *Books*.

L'Ecole des Loisirs Clubs
c/o Dorothy Guellec, P.O. Box 362, Croton, NY 10520
Telephone: 914-271-5644
Free information

French children's books in five series for children from 2 to 12. These are French editions, many of them translations of well-known American titles. Among the 1989 offerings were Beverly Cleary's *Ramona la peste* (*Ramona the Pest*), Chris van Allsburg's *L'épave du Zéphyr* (*The Wreck of the Zephyr*), and Judy Blume's *Super Sheila*. The books are mailed directly to members from the Paris publisher, and the price, including postage, compares well with foreign-language books in U.S. bookstores (when they can be found).

Parents Magazine Read Aloud Book Club
1 Parents Circle, P.O. Box 10264, Des Moines, IA 50336
Free information

Illustrated storybooks for children not yet reading but eager to have books read aloud. Titles include *Pigs in the House*, *Henry's Awful Mistake*, *Milk and Cookies*, and *Bicycle Bear*. Two books are sent each month, along with an issue of the newsletter *From Parents to Parents*. The books can be sampled for two weeks, then either paid for or returned. We haven't seen a copy of the newsletter, but according to the literature, it offers book reviews and parenting tips.

Weekly Reader Books
4343 Equity Drive, P.O. Box 16615, Columbus, OH 43216
Catalog price: free

Many parents remember *My Weekly Reader* from their own childhoods. The classroom newspaper for elementary schools is still going strong, and the publisher now offers a string of book clubs and learning programs for home membership. The book clubs sell inexpensive hardcover editions of familiar books in programs matched to a child's reading level. Members sacrifice a little in paper and printing quality in return for low prices—about $4 a book plus postage. The books are sturdy enough, and at that price most parents won't care about the brilliance of the illustrations.

A single catalog lays out all of the options, conveniently arranged by children's ages. The cornerstone of the operation is the Weekly Reader Children's Book Club, which satisfies the appetites of readers from age 4 to 12. Younger members receive titles like *Curious George Flies a Kite* and *The Day Jimmy's Boa Ate the Wash*. More advanced readers can settle into *Charlotte's Web* and *Ramona Forever*.

For prereaders there's a board-book club called Disney Babies Learn About Series. Several other clubs bridge the read-aloud and beginning-reading years, among them Muppet Babies Book Club, Fraggle Rock Book Club, the Just Ask Series (which answers basic questions about science and nature), I Can Read Book Club (which includes only books from Harper & Row's I Can Read series), and Ready-Set-Grow. Two sex-specific clubs, Just for Boys and Especially for Girls, offer teen novels, romances, adventure stories, and sports books that recognize the different reading tastes of boys and girls.

Two learning series offer games, activities, and learning cards that can be filed in a plastic storage box as they accumulate. The Sweet Pickles Preschool Program helps parents keep preschoolers busy in productive ways. The Illustrated Wildlife Treasury teaches older children (ages 4 to 12) facts about animals.

Childbirth

Childbirth Resources
Janet Isaacs Ashford, 327 Glenmont Drive, Solana Beach, CA 92075
Telephone: 619-481-7065
Free brochure
Books, slides, postcards, and prints relating to childbirth, particularly to home birth, midwifery, and traditional childbearing practices. Ms. Ashford also publishes the journal *Childbirth Alternatives Quarterly*, described here under *Parents' Magazines, Childbirth*.

Los Angeles Birthing Institute
Pregnancy, birth, and parenting books. See entry under *Books*.

Moonflower Birthing Supply
See entry under *Baby Needs*.

Naturepath
Pregnancy, birth, and parenting books in a birthing-supply catalog. See entry under *Baby Needs*.

Placenta Music, Inc.
Relaxation tape for childbirth. See entry under *Baby Needs*.

Children's Clothes

Aerie Design
141 Blackberry Inn Road, Weaverville, NC 28787
Telephone: 704-645-3285
Catalog price: free
Colorful wildlife designs silkscreened onto T-shirts. Children's shirts are available in sizes small (6–8), medium (10–12), and large (14–16). Adults can also order sweatshirts.

After the Stork
1501 12th Street N.W., Albuquerque, NM 87104
Telephone: 505-243-9100
Catalog price: $1
Sensible, attractive, and colorful cotton clothes for babies and bigger kids (to size 16). Soupçon outfits and baby sweat suits keep the littlest ones warm. At 12 months or so they can move into the specialties of the house: inexpensive long- and short-sleeved T-shirts in a rainbow of colors, turtlenecks, elastic-waist corduroy and denim pants, sweat suits, and brightly dyed long johns. Parents can bulk out the order with flannel shirts (from size 12 months), polo shirts, overalls, knee socks, tights, corduroy jumpers and skirts, cotton sweaters, raincoats, jackets, boots, shoes, and hats. The catalog tucks in a few toys, such as blocks, puzzles, and toddler trikes, as well as cotton diapers and Nikky diaper covers.

Elaine Aldrich
Rt. 2, Box 2675, Westford, VT 05494
Telephone: 802-879-4869
Catalog price: free
Cotton children's clothing with an old-fashioned country look. For girls Ms. Aldrich makes a wool coat with a matching bonnet, a corduroy jumper, a pleated blouse, a simple dress with lace collar, a pinafore, and bloomers. For boys she offers a coat, a shirt, and short dress overalls. A smock can be fashionably worn by either sex. The catalog comes with fabric swatches. All clothes are made to order, so requests for special modifications are welcomed.

Alice in Wholesale Land
140 Linden Street, Oakland, CA 94607
Telephone: 415-452-0507

Catalog price: $2

Children's clothes with pizazz at discounted prices. Alice outfits kids from birth to size 14 in fashionable brands like Anémone of Paris, Piccolo, Jordache, Base Kids, Cherokee, and Casual Time at great savings to parents. Shoppers will find play clothes, dress outfits, and winter coats.

Anatomical Chart Co.
T-shirts with anatomical designs. See entry under *Science and Nature*.

Hanna Andersson
1010 N.W. Flanders, Portland, OR 97209
Telephone: 800-222-0544, 800-346-6040, or 503-242-0920

Catalog price: $1

We admit to a pleasant addiction to Hanna Andersson clothes. The outfits are practically designed and well made of heavy, soft cotton in vivid colors. They cost a bit more than basic children's wear, but they last a lot longer. Not only does the fabric bear up under the abuses of childhood wear (we've found that use by a single child barely dulls their original brightness), but the outfits are designed with wide elastic cuffs and ample room from shoulder to crotch so that growing children get a few extra months of comfortable fit. The clothes are so durable that the company asks parents who don't hand them down to send used outfits back for a partial credit so that they can be donated to local charities.

Hanna Andersson's strength lies in clothes for babies and young children—simple, sturdy designs in bright solid colors and bold stripes. No appliqués or "cute" designs are found here. As its clientele has matured, the catalog has added clothes for older children as well, and now offers playwear, dress clothes, and outdoor gear for school-age kids (up to size 8-10). And each new catalog adds a few more outfits for grown-ups.

Annabetta
P.O. Box 1590, Cathedral Station, New York, NY 10025
Telephone: 212-666-7871

Free brochure

Icelandic sweaters in children's sizes from 2 to 12. Knit in Iceland using undyed Icelandic Lopi wool, the sweaters can be ordered as cardigans or pullovers with matching hats and mittens. A set (sweater, hat, and mittens) cost about $130 in 1989, but during spring and pre-Christmas sales better deals can be had.

Artisans Cooperative
Hand-knit children's sweaters. See entry under *Gifts*.

Laura Ashley by Post
1300 MacArthur Boulevard, Mahwah, NJ 07430-9990
Telephone: 800-367-2000 or 800-223-6917

Catalog price: free "Mother & Child" catalog, $5 for complete "Fashion and Furnishing" catalog

Adorable and expensive outfits for toddlers and children (to size 11/12), some of the girl's dresses available in matching sets for mother, daughter, and baby. Laura Ashley's specialty is simple, old-fashioned dresses made

Zippers and Stripes playsuits by Hanna Andersson

A multicolor lineup of cotton leggings, T-shirts, and socks from After the Stork

from beautiful calico prints, and these have a secure place in the children's line. Supplementing them are classic corduroy dungarees, velvet party dresses, blue blazers with plaid pants, and some wonderful wool sweaters with duck, rocking horse, and toy soldier patterns marching around the waist. Perhaps nicest of all are the wool coats, which look as if they've been plucked from family snapshots of the 1940s. A blue dress coat comes with velvet-trimmed collar and pockets; an everyday duffle coat is lined with cotton in bold stripes. The "Fashion and Furnishing" catalog offers some accessories for children's rooms, such as sheets, quilts, wallpaper, and animal-shaped pillows.

Babymax Angelwear
P.O. Box 777, Melville, NY 11747
Free flier
Cotton suits with soft satin wings make angels of even the most ornery babies. Choose from T-shirt and panty sets, one-piece romper suits, or long-sleeved and long-legged playsuits. Wings come in pink, yellow, or aqua with white suits; colored suits have white wings. Sizes are newborn (to 12 pounds), small (13–18 pounds), medium (19–26 pounds), and large (27–34 pounds).

Eddie Bauer
Fifth & Union, P.O. Box 3700, Seattle, WA 98130-0006
Telephone: 800-426-8020, 800-426-6253, or 206-885-5976
Catalog price: free
Selected items in the Eddie Bauer catalog are made in reduced scale for children, among them cotton sweaters, turtlenecks, jackets, fleece-lined slippers, and insulated boots. Children's clothes come in sizes from 4T to 14, the shoes and slippers from medium size 9 up. Some sidelines for children are listed as well: sleds, children's luggage, kites, and a few toys. The sideline collection changes completely with each new catalog.

Bear-in-Mind
T-shirts with bear designs. See entry under *Teddy Bears*.

Bimbini
2372 Boston Post Road, Larchmont, NY 10538
Telephone: 914-949-3306 or 212-994-4760
Free brochure
Children's clothes with a European look. A summer sundress in a country floral print (sizes 1/2 to 7/8) can be accented with a straw hat (sizes 2–4 years to 6–8

years). Elastic-waist corduroy pants and skirts can be dressed up with colored suspenders or a black cinch belt. Turtlenecks, jumpers, blouses, berets, and a corduroy jacket with big round buttons complete the picture. Prices are on the high side, but not at the top of the charts.

Biobottoms
P.O. Box 6009, Petaluma, CA 94953
Telephone: 707-778-7945 or 7152

Catalog price: $1

From its beginnings in breathable wool diaper covers, Biobottoms has expanded into all sorts of natural-fiber clothing for children. The diaper covers are still a mainstay of the business—they're available in a half-dozen styles, made of wool, cotton, and even waterproof nylon. Cotton terry prefold diapers are sold to fill the covers, as well as a pinless cotton diaper with Velcro closures. The clothing line includes cotton long johns in solids and prints, turtlenecks, polo shirts, sweats, jumpers, cotton knit dresses and skirts, flannel-lined corduroys, flannel shirts and blouses, jackets, rain gear, and a variety of coverall playsuits made of cotton knit, terry, and velour. To keep those little feet warm, Biobottoms sells lots of booties, moccasins, sneakers, and shoes (in sizes from newborn to big kids' 2). Prices are in the high to middling range. Child's room accessories include a play tent, flannel crib sheets, and a bed that fits a crib mattress.

Birth & Beginnings
Cotton clothing for babies and young children. See entry under *Baby Needs*.

Bliss Ridge
P.O. Box 590, Miranda, CA 95553
Telephone: 707-943-3632

Free brochure

Solid-colored cotton long johns at the best prices we've found. Available in sizes for infants up to 50-pound children.

Blue Balloon
8228 Starland Drive, El Cajon, CA 92021
Telephone: 619-561-0868

Catalog price: $1

Corduroy pants, Danskin leotards, Health-tex infant wear, fleece sweatshirts and suits printed with cartoon designs, and several styles of sleepwear. Sizes go up to 7.

Bohlings
159 Stark Street, Randolph, WI 53956
Telephone: 414-326-3533

Catalog price: free

OshKosh work clothes and children's wear, from footed bib overalls for infants to outfits for oversized adults (up to size 70). The 32-page catalog illustrates the wares with drawings and black-and-white photographs.

Boston Proper Mail Order
1 Boston Plaza, P.O. Box 7070, Mt. Vernon, NY 10551-7070
Telephone: 800-243-4300

Catalog price: free

Women's fashions dominate this glossy color catalog, with just a few children's outfits added to spice the mix. For the summer of 1989 the options included jungle-print overalls, a khaki fishing vest, a navy-striped sunsuit, and classic buckle-front raincoats.

Brights Creek
Bay Point Place, Hampton, VA 23653
Telephone: 800-622-9202 or 804-827-1850

Catalog price: free

Play and dress clothing at good prices, presented in a big color catalog. Simple knits, sweats, and corduroy outfits are the backbone of Brights Creek, in lots of colors and lots of sizes. There's plenty here for infants on up to teens (to size 14)—velveteen and taffeta party outfits for toddlers, novelty raincoats with duck and alligator hoods, acrylic sweaters, superhero sweatshirts, turtlenecks, polo shirts, blouses, jackets, sneakers, and dress shoes. Most of the clothes are made of acrylic or cotton-polyester blends, but some all-cotton items can be found.

Button Creations
3801 Stump Road, Doylestown, PA 18901
Telephone: 800-346-0233 or 215-249-3755

Catalog price: $2

A color catalog filled with 400 different buttons. A set with Beatrix Potter characters would make a wonderful finishing touch on a child's cardigan, as would a row of ceramic dinosaurs, enamel clowns, or plastic roller skates. Plastic letter buttons can be used to spell a child's name. There's *lots* more to choose from.

Buttons from Button Creations

Carrick Knitwear
2255 E. Commonwealth Drive, Charlottesville, VA 22901
Telephone: 804-974-6773

Free flier

Sweaters made to order by home knitters on the Isle of Man (a small island between England and Ireland). While the craftsmanship may be traditional, the patterns are decidedly up-to-date. Children can garb themselves in dinosaurs, city skylines, seahorses, fish, sailboats, snowflakes, and dancing Egyptians. All can be knit in either cotton or wool.

Channel Island Imports
P.O. Box 2995-DD, Evergreen, CO 80439-2995
Telephone: 800-622-2482

Catalog price: $1

Offers a traditional Guernsey sweater in children's sizes (from chest size 22). Cardigans, vests, and a Jersey sweater are also available, but in adult sizes only. All sweaters are knit in the English Channel Islands from heavyweight, lightly oiled wool. We like this firm's motto: "The sweaters, not the cows."

Chi Pants
120 Pearl Alley, Santa Cruz, CA 95060
Telephone: 800-331-2681

Catalog price: free

Several styles of casual pants for adults and one for children: the loose-fitting elastic-waist WeeChi. The child's pair is made of denim or muslin, with a wide knit cuff that keeps the legs up over the feet and allows parents to buy large for longer wear.

The Children's Collection
1717 Post Oak Boulevard, Houston, TX 77056-3882
Telephone: 713-622-4415 or 4350

Catalog price: free

Stylish clothes for special occasions and for children who demand the latest fashions. Dressy party outfits, funky prints for play and school, and fashionable rain gear all find a place in this color catalog. A herringbone double-breasted chesterfield coat with a matching cap could help some toddler stand out from the crowd (sizes 2T to 4T). Floral print dresses in cotton or wool challis might do the same for an older girl (sizes 4 to 14). The rotating collection changes completely with each catalog edition, so write for the catalog and see what's in style this season.

The Children's Shop
P.O. Box 625, Chatham, MA 02633
Telephone: 800-426-8716 or 508-945-4811

Catalog price: free

A broad assortment of quality clothes for children from infants to teens. Three 20-page color catalogs are mailed each year, with wear appropriate to each season. The spring catalog carries Izod Lacoste polo shirts (sizes 2T to 16), stone-washed denim jackets and shorts, sundresses, cotton playwear, terry bathrobes, and warm-weather dress outfits. The fall and winter catalogs offer coats, insulated vests, sweaters, corduroy jumpers, cotton turtlenecks, hand-painted sweat suits, sleepwear, and party clothes. The company offers American Widgeon rain gear year-round in sizes from infant 6M to child's 8. Made of yellow Supplex nylon, this looks like the ultimate in childhood rain protection. Infants can be fit with a one-piece rain suit; bigger kids can choose between a poncho and a jacket. A classic sou'wester hat tops off the outfit; Sabi rainboots keep feet and legs dry. Another four-season item is Kawamura cotton playwear, the softest infant wear we've found.

OshKosh overalls from Children's Wear Digest

Children's Wear Digest
3607 Maryland Court, Richmond, VA 23233

Catalog price: $2

Discounted OshKosh overalls (sizes 12M to 14) are one of the attractions in this color catalog, as are child-size Oxford shirts, zip-on neck ties, dress slacks and blazers, Belle France dresses (sizes 7 to 14), Izod Lacoste polo shirts (sizes 4 to 14), and Carter's sleepwear. American Widgeon rain slickers and baby rain suits are offered at slightly discounted prices. The spring catalog carries bathing suits, warm-weather playwear, and sale-priced winter gear. The fall and holiday catalogs offer sweaters, winter coats, and sweat suits. A fleece-line leather flight jacket lurks as a temptation to the wallet—at $160 we managed to resist.

Chock Catalog Corporation
74 Orchard Street, New York, NY 10002-4594
Telephone: 800-222-0020 or 212-473-1929

Catalog price: $1

Discounted underwear, sleepwear, socks, and stockings for the whole family, with special attention paid to babies and children. Carter's sleepwear and underwear is offered in every size from newborn up to a boy's 20 and a girl's 12. Duofold thermal underwear starts at boy's 6–8 and girl's 7–8, and Trimfit socks are sold in sizes to fit any child who hasn't left for college. Diapers, hooded bath towels, crib sheets, and receiving blankets are among the extras for baby.

Classics for Kids
P.O. Box 614, Silver Spring, MD 20901
Telephone: 301-587-5422

Catalog price: $2

Soupçon cotton infant and toddler wear, Benay's cotton playsuits, Gunze and Babysteps soft cotton long johns, Rubber Duckies diaper covers, American Widgeon rain and snow gear, and a selection of cotton turtlenecks, T-shirts, skirts, frocks, robes, and designer socks. Everything is illustrated by simple line drawings in a black-and-white catalog. The prices are good if you're looking for the comfort of cotton.

Clothkits
24 High Street, Lewes, E. Sussex BN7 2LB, England

Catalog price: write for current price

A big color catalog of children's and adults' clothing that can be ordered in kit form or ready-made. Sweaters, quilted jackets, pants, dresses, skirts, blouses, shirts, playsuits, cotton pajamas, and nightgowns make up the list. Shoes and rubber boots can be ordered (but not in kit form). Ask about the popper, a useful tool for attaching snaps. Ordering is easy if you use Visa or Mastercard.

Cloudburst Quiltworks
Rt. 1, Box 72B, Joseph, OR 97846
Telephone: 503-432-9431

Long self-addressed stamped envelope for brochure

Soft cotton suits for infants, cotton sweats for toddlers (some designs up to size 6), and appliquéd outfits of lighter cotton interlock for babies up to 12 months. The brochure shows black-and-white drawings of each style. Shoppers who want a more precise idea of what they're ordering can send a self-addressed stamped envelope with a request for swatches of specific fabrics and colors.

Colourwheel Designs
23158 Gonzales Drive, Woodland Hills, CA 91367
Telephone: 818-704-1865

Free brochure

Have your child's artwork reproduced on T-shirts. The artists at Colourwheel Designs hand-paint faithful copies onto cotton T-shirts, tank tops, or long-sleeved three-button baseball T-shirts. Only the standard crew-neck T-shirt comes in children's sizes.

Cot'ntot Fashions
P.O. Box 7495-S, Quincy, MA 02269
Telephone: 617-698-COTN

Catalog price: free

Cotton clothing for infants and bigger kids, up to size 8. The catalog carries almost the complete line of Soupçon baby clothes (made of soft cotton interlock in bright solid colors), plus other makers' playsuits, booties, dia-

A child's artwork reproduced on a T-shirt by Colourwheel Designs

The Brimmie sun hat by Crayon Caps

per covers, cotton socks, tights, sweat suits, Hawaiian shirts, Hawaiian-print overalls and shortalls, bathing suits, cotton long johns, sweaters, aprons, smocks, turtlenecks, T-shirts, skirts, and dresses. A good source of quality clothes for younger children.

Crayon Caps
16841 Franklin Road, Fort Bragg, CA 95437
Telephone: 707-964-7549
Long self-addressed stamped envelope for brochure
Makes two styles of children's hats: a brimmed cap of light cotton (in a choice of exquisite patterns), and a warmer cap of bright cotton interlock that pulls down to keep ears warm. Both styles are cute as can be, and the fabrics are exceptional.

Cumberland Crafters/Kid's Art
P.O. Box 128, Arthur, TN 37707
Catalog price: $1, refunded with order
Cumberland Crafters reproduces your children's artwork on T-shirts, sweatshirts, tote bags, calendars, and place mats. The T-shirts come only in children's sizes, the sweatshirts only in adult sizes.

Daisy Kingdom
Children's clothing kits. See entry under *Sheets, Blankets, and Sheepskins*.

Dallas Alice
8001 Cessna Avenue, Gaithersburg, MD 20879
Telephone: 800-777-0606 or 301-948-0400
Catalog price: free
Silly T-shirts are the order of the day at Dallas Alice. In children's sizes are glow-in-the-dark dinosaur shirts, a toddler shirt that says "Potty animal," and a shirt designed to be colored in by the owner. Scores more come only in adult sizes. Among the messages: "I can't be overdrawn, I still have checks left," "All the women moaning about finding husbands have obviously never had one," and "I don't do Mondays."

The Disney Catalog
Clothes emblazoned with Disney characters. See entry under *Gifts*.

Walt Disney World
Mickey Mouse T-shirts. See entry under *Gifts*.

Family Clubhouse
Cotton baby clothes from Kawamura and Fusen Usagi. See entry under *Baby Needs*.

Flap Happy
2322 Walnut Avenue, Venice, CA 90291
Telephone: 213-838-2757
Free flier
Makes a series of hats with wide brims and "camel flaps" around the back to shield ears and neck. A sun hat is made of cool cotton, a rain hat of vinyl, and a cold-weather hat of cotton flannel. All are made from the same pattern, which makes the wearer look like a little Lawrence of Arabia. The sun hats are made in sizes for newborns to large adults, the flannel and rain hats for infants and toddlers only. We bought one of the sun hats for our year-old son and can report that the whole family is happy with it. He loves to wear it, it protects his ears and neck wonderfully, and the hat's elastic should stretch to fit him for a couple more summers.

Friends
50305 S.R. 145, Woodsfield, OH 43793
Telephone: 614-472-0444

Catalog price: $1, refunded with order

Patterns for sewing traditional Amish clothes in sizes from 2 to large adult. Broadfall pants with button-on suspenders, an elastic-back broadfall vest, a round-collar shirt, and several dresses with aprons make up the basic wardrobe. A child's cap and an Amish bonnet add authenticity to the costume. The firm plans to expand its catalog to include cold-weather wear and some Amish-inspired clothes for the non-Amish.

Corduroy jumpers from Garnet Hill

Garnet Hill
Main Street, P.O. Box 262, Franconia, NH 03580
Telephone: 800-622-6216 or 603-823-5545

Catalog price: free

A catalog of natural-fiber clothing and bedding with some wonderful cotton playwear and some exquisite luxuries for babies and children. Soupçon baby wear, pima cotton dresses and polo shirts, Absorba cotton turtlenecks, cotton fleece jumpsuits, cotton tights, and sheepskin mittens are among the more affordable temptations. Long underwear made of silk and wool, pleated twill pants with plaid flannel cuffs, a reversible bomber jacket, and a hand-knit wool sweater coat provide more serious threats to the budget. The catalog offers an excellent selection of both summer and winter hats; some beautiful crib sheets of flannel, terry, and pima cotton; and a small array of cotton nursing wear.

GeorGee's
P.O. Box 19129, Minneapolis, MN 55419
Telephone: 612-861-5410

Free brochure

Disposable clothing made of Tyvek, a plastic-based paper that wipes clean. The company recommends the outfits as backup wear for vacations, nursery school, or anywhere a child may need a change of clothes. Included in the wardrobe are pants, a jumper, shorts, a tank top, a sunsuit, and a rain cape. A painting apron could be used as regular protection at home. Sizes run from 1/2 to 5/6.

Abbie Hasse Catalog
P.O. Box 1078, Cedar Hills, TX 75104
Telephone: 800-637-9806

Catalog price: free

Special clothes for special occasions. Abbie Hasse offers tuxedos for toddlers (sizes 6 months to 4T, complete with pants, jacket, pleated shirt, cummerbund, and bow tie), red velveteen and white lace Christmas dresses, and dress overalls made of raw silk. Shoppers will also find cotton long johns, cotton-lined vinyl raincoats, and such accessories as straw hats, lunch boxes, and personalized director's chairs.

Hazelwood
Airbrushed T-shirts. See entry under *Baby Needs*.

Holy Cow, Inc./Woody Jackson
P.O. Box 906, 52 Seymour Street, Middlebury, VT 05753
Telephone: 802-388-6737

Catalog price: $1, deductible from first order

T-shirts, sweatshirts, baseball caps, and other merchandise emblazoned with Woody Jackson's stylized cows. The shirts come in sizes from 6 months to adult large.

The Homespun Company
Rt. 2, Box 213, Greenwich, NY 12834
Telephone: 518-692-2441

Free brochure

Cotton long johns made of bold-striped cotton interlock. The brochure pictures the two design choices: a button-front union suit and a two-piece set. Attached fabric swatches let shoppers see and feel all the possibilities. The union suit comes in sizes 6 months to 4, the two-piece set in sizes 4 to 7.

Children's Clothes

Ident-ify Label Corp.
P.O. Box 204, Brooklyn, NY 11214-1204
Telephone: 718-436-3126

Free brochure

Sew-on name tapes to identify clothing headed off to school or summer camp, and personalized labels to sew into handmade clothes ("Hand knit by ...," "An original by ...," etc.).

Lands' End Roveralls
© Lands' End, Inc.

The Initial Place
634 E. Main Street, Barrington, IL 60010
Telephone: 312-381-0459

Free brochure

Sweatshirts, bibs, and baby blankets embroidered with messages like "I love Grandma" or with a child's name. The enterprise will custom-embroider any other design or message you may want for a small setup charge. The sweatshirts run from a child's size 2 to adult extra-large.

Initials+ Collection
Personalized sweaters, smocks, and raincoats. See entry under *Gifts*.

Justin Discount Boots and Cowboy Outfitters
P.O. Box 67, Justin, TX 76247
Telephone: 800-433-5704 (800-772-5933 in TX)

Catalog price: free

Cowboy boots in children's sizes (8½ to 13½ and youth 1 to 6), Wrangler cowboy-cut jeans (from size 1T up, including husky boys' sizes), and a child's duster. Adults have a much larger choice of jeans, boots, and hats.

KBS Designs
P.O. Box 844, Melrose, MA 02176
Telephone: 617-665-5270

Free brochure

Sweatshirts embroidered with pictures of animals, trains, and sports equipment, or monogrammed with a child's initials. Sizes run from a child's 2 to an adult large.

Lands' End
Lands' End Lane, Dodgeville, WI 53595
Telephone: 800-356-4444 or 608-935-9341

Catalog price: free

Lands' End has built its reputation on comfortable and durable leisure wear for adults, and it has used that background to launch an appealing and sensible line of children's clothes. Its overalls (sizes 2T to 7) are made with double knees and pockets front and back; choose from corduroy, twill, and denim in a dozen colors. Turtlenecks, sweats, polo shirts, rugby shirts, jumpers, knit dresses, cotton sweaters, and elastic-waist pants are all made with classic Lands' End styling in a broad palette of colors. Clothes are offered in sizes from 2T to 16, with the full line of adult clothes as the next step.

Laughing Bear
P.O. Box 732, Studio, Woodstock, NY 12498
Telephone: 914-246-3810

Free brochure

Cotton clothing hand-batiqued with hearts, flowers, and stars or with pictures of ducks, rabbits, whales, and other animals. The poster-format sales sheet shows drawings of the possibilities: long johns, T-shirts, knit dresses, baby buntings, and infant socks.

Les Petits
3200 S. 76th Street, P.O. Box 33901, Philadelphia, PA 19142-0961
Telephone: 215-492-6328

Catalog price: free

Fashionable children's clothing imported from France. Cotton polo shirts in pretty pastels and bold primary colors are a regular feature; they're available in both long and short sleeves. Sharp-looking sweaters, pleated twill pants, stone-washed denims with massive pockets, stylish sundresses, and luxurious terry bathrobes are all made a cut above the ordinary (and with prices a bit higher than average too). In its different seasonal editions the catalog presents bathing suits, polka-dot tennis shoes, canvas Mary Jane shoes, winter coats, snowsuits, fleece-lined high-top sneakers, and holiday party outfits. A good source for special back-to-school clothes and shoes that will stand out from the crowd.

Val Love
P.O. Box 1163, New York, NY 10028
Telephone: 212-534-4976

Catalog price: free

Cotton sweaters for infants, children, and adults, knit by hand in England with bold pictures on the front. The baby sweaters are decorated with penguins, seals, or zebras; the bigger sweaters with characters from nursery rhymes.

Metrobaby
Cotton flannel creepers and a terry-cloth robe (to size 3T). See entry under *Baby Needs*.

Milkduds
7277 Gumwood Lane, Raleigh, NC 27615
Telephone: 919-878-9680

Long self-addressed stamped envelope for brochure

Cotton playwear for babies and for girls to size 14. The brochure comes with fabric swatches so shoppers can see the actual colors and feel the weight of the cloth. Pleated overalls are sewn of cotton madras or lightweight stripes; drawstring pants are made in the same summer fabrics as well as in bright corduroys and pastel cotton interlock. Infant coveralls, toddler sweats, turtlenecks for all ages, skirts, and a corduroy jumper round out the list.

Modern Homesteader
1825 Big Horn Avenue, Cody, WY 82414
Telephone: 800-443-4934 or 307-587-5946

Catalog price: free

Housewares, tools, and supplies for rural families. The all-American visor cap is offered in children's sizes. Choose from denim, mesh, or insulated models with your child's name, 4-H Club, or sports team embroidered on the front. In 1989 the catalog sold miniature metal tractors, a mounted jackalope trophy, and a set of cookie molds for baking a gingerbread house.

Jim Morris Environmental T-Shirts
P.O. Box 831, Boulder, CO 80306
Telephone: 303-444-6430

Catalog price: free

Sweatshirts and T-shirts silkscreened with pictures of wildlife and messages about saving the environment. A lovely picture of a panda, for example, carries the message "Threatened by the destruction of its bamboo forest habitat." The text on the shirts is lettered in attractive calligraphy. Sizes start at a child's small (6–8) and span the range to a man's extra-large.

Mountain Shirts
P.O. Box 189, Dept. S, Redway, CA 95560

Catalog price: long self-addressed stamped envelope

Genuine tie-dyed clothing for children and adults. The children's line includes long- and short-sleeved T-shirts, leggings, and training pants. All are made of cotton, in sizes from 1 year to 18. A rainbow-dyed rayon scarf would make a wild accessory. A 10-foot silk parachute could make your child's room the envy of the neighborhood.

Tie-dyed T-shirts by Mountain Shirts

Mountaintop Industries
Wildlife T-shirts and sweatshirts. See entry under *Outdoor Gear*.

Munchkin Outfitters
913 Williamsburg Drive, P.O. Box 1684, Valdosta, GA 31603-1684
Telephone: 912-242-3610

Catalog price: free

A mix of cotton and cotton/poly children's clothes that reflects the split among the company's customers—some want the comfort of cotton, some the durability of cotton/poly. Infant coveralls from Gear Kids are made of soft cotton interlock. Sweatsuits for bigger children (2T to 16) come in a cotton/poly blend. Turtlenecks, flannel shirts, corduroy pants, overalls, and denim jumpers are all pure cotton. For cold weather protection the

The OshKosh Story

Say "overalls" and most people think OshKosh B'Gosh, so familiar has the manufacturer's label become. And in the past decade overalls have become the cornerstone of most children's wardrobes—nationally, they account for a quarter of all children's clothing sales. But where did OshKosh overalls come from? And why do so many of us buy them for our children?

Founded in Oshkosh, Wisconsin, in 1895 as the Grove Manufacturing Company, the enterprise at first made only adult-size overalls for farmers, railroad workers, and other laborers. A name change a year later, to the OshKosh Clothing and Manufacturing Company, put the familiar "OshKosh" on the label. Children's sizes were quietly added to the line in the first few years of this century. Like Levi Strauss & Co. and other makers, OshKosh sold both children's and adult overalls as rugged work garments. The 1902 edition of the Sears, Roebuck catalog, for example, features a pair of denim "Children's Brownie Overalls . . . for boys age 4 to 12 years" in among the men's work clothes, several pages removed from the rest of the children's offerings. By 1928 Sears had added overalls for toddlers, made of a lighter patterned fabric, and had moved the boys' denim overalls into the children's clothing section, with selling lines like "Inexpensive and durable" and "Let him try to wear 'em out." OshKosh chugged along, making its own rugged denim overalls in a range of sizes for youngsters and grown-ups, advertised as "Tough as a mule's hide."

The break came in 1968, when Miles Kimball, another business from Oshkosh, Wisconsin (see entry under *Gifts*), included a pair of OshKosh's hickory-striped children's overalls in its national catalog. The response was so positive that OshKosh launched a major sales effort to get its small-size garments into chic specialty stores. Bloomingdales placed a big order in 1972, and other retailers began to follow suit as the public's appetite for the little work pants became apparent. Parents recognized the OshKosh name as a mark of quality and bought the overalls as a practical, long-lasting outfit whose all-American country look matched the taste of the times.

© 1987 OshKosh B'Gosh, Inc., OshKosh, Wisconsin. Photo by Michael D. McCoy

As the baby boomers of the 1960s have turned into the parents of the 1980s, OshKosh's business has practically exploded. Where the children's line accounted for just 16 percent of the company's business in 1979, it now contributes 85 percent. The overalls come in a rainbow of colors and bold patterns, and they're complemented by OshKosh shirts, jackets, skirts, dresses, and infant wear.

Children's Wear Digest, Strauss' Country Wear, and Bohlings all sell the overalls in children's sizes. Olsen's Mill Direct offers the complete OshKosh line. Wobkins features OshKosh outfits decorated with appliquéd pictures.

T-shirt from Over the Moon Handpainted Clothing Co.

catalog sells Duofold long underwear, wool booties with sheepskin soles, mittens, gloves, and a "basset hat" with long earflaps that serve as a wrap-around scarf. Children's day packs are offered for toting everyday stuff, and a frame backpack for youngsters who carry their own gear on camping trips. A small collection of puppets, stuffed animals, and children's books closes out the catalog.

New World Trading Co.
P.O. Box 300, N. San Juan, CA 95960

Long self-addressed stamped envelope for brochure and fabric samples

Bloomer-like cotton pants for warm-weather wear. Just three sizes are offered: 6 months, 2T, and 4T. Fabrics range from pale solids to bold patterns and rainbow stripes.

Olsen's Mill Direct
1200 Highway 21, P.O. Box 2266, Oshkosh, WI 54903
Telephone: 414-685-6688

Catalog price: free

OshKosh clothing for the whole family. The famous overalls are displayed at every turn—in footed models for babies, hickory stripes for teens, white drill for grown-ups, and in the classic denim for everybody. But OshKosh also makes jeans, shirts, sweats, skirts, jackets, dresses, and shoes, all of which get their share of space in the catalog. Flannel-lined jeans and overalls are sold with matching plaid flannel shirts. Engineer's caps in stripes and solids top off an OshKosh outfit perfectly. At $5 they're one of the best bargains in the catalog.

Over the Moon Handpainted Clothing Co.
10224 Regent Street, #301, Los Angeles, CA 90034
Telephone: 213-838-2757

Catalog price: $1, refunded with first order

Cotton knit clothing for babies and children, camel-flap hats, diaper covers, bibs, and other accessories, available in plain solid colors or decorated with hand-painted designs. A splatter-paint background can be accented with stars, moons, sailboats, hearts, or other little pictures. Or the pictures can be placed by themselves on the fabric. All clothing, decorated or not, can be personalized with hand-painted lettering. Over the Moon starts with soft cotton knits, so whether you like the painted designs or not, this is a good source of basic, comfortable baby wear. One item that can't be painted is a Swedish moccasin for infants and toddlers. We've found its combination of leather sole and high knit upper to be the perfect thing for crawlers and toddlers. They keep feet warm, don't slip on polished floors, and fit snugly enough that babies can't take them off.

Patagonia Mail Order, Inc.
1609 W. Babcock Street, P.O. Box 8900, Dept. 485K7, Bozeman, MT 59715
Telephone: 800-638-6464

Catalog price: free (request "Kids" catalog)

The ultimate in tough and practical outdoor gear for children, some of Patagonia's clothes can also be rough on the family budget. A polo shirt, for example, comes in five exquisite colors, is designed for comfortable wear and complete freedom of movement, and will likely live through three of four children handily. But it costs about $20. Patagonia's Synchilla jackets offer more for the money. The fabric is washable, light, and incredibly warm (we wear an adult version comfortably when the mercury drops to 20°). Though the jackets are pricier than those at discount stores—they cost from $30 to $60—we think they're worth the extra money in warmth and rugged wear. Synchilla baby buntings offer the same cozy warmth to little ones at a price comparable to the widely sold Baby Bag. Shorts, jeans, sweats, and rain gear are all made of quality fabrics in designs that are both attractive and comfortable. And Patagonia's catalog is one of the best in the business. Its candid photographs of customers in action around the world are worth the price of admission, even if you never order a thing.

J. C. Penney Company, Inc.
See entry under *General Catalogs*.

Petit Pizzazz
2134 Espey Court, #10, Crofton, MD 21114
Telephone: 301-858-1221

Catalog price: free

Clothing and furnishings for babies and young children. A selection of American Widgeon outerwear includes a

Children's Clothes

poncho, a rain hat, a two-piece rain suit, waterproof mittens, a warm suit made of Polarfleece, and a jacket of Polarplus, in sizes from 6 months to 5T. Cotton knit playwear is offered from a number of makers in solids, stripes, and an occasional pastel print. Some are made in the U.S., some imported from Sweden. Baby Björn provides a few extras for babies and toddlers: an exceptional one-piece potty, a bouncing seat with an optional toy bar, and two wooden baby toys.

The R. Duck Company
650 Ward Drive, #11, Santa Barbara, CA 93111
Telephone: 805-964-4343

Free brochure

Rubber Duckies diaper covers made of brightly colored waterproof nylon. The company also makes nylon rain suits and ponchos for children to size 6, aprons and bibs of the same material, and elastic-waist shorts with bold stripes up the sides.

Rail Scene
P.O. Box 742, Bayonne, NJ 07002
Telephone: 201-823-4983

Catalog price: $1

T-shirts, sweatshirts, and hats printed with the trademarks of scores of railroad lines, both historic and current. The clothes can also be decorated with pictures of trains and with some nostalgic marks of coal, gas, automobile, and soft-drink companies. The children's shirts come in four sizes: 2–4, 6–8, 10–12, and 14–16.

Richman Cotton Company
529 Fifth Street, Santa Rosa, CA 95401
Telephone: 800-992-8924 (800-851-2556 in CA) or 707-575-8924

Catalog price: free

Cotton clothing for children and adults. On the children's side are rib-knit tank tops, gym shorts, long johns, sunsuits, turtlenecks, sweat suits, sweat overalls, Chinese shoes, tights (33 percent nylon), and straw garden hats. For babies the catalog offers lap-shouldered T-shirts, hooded baby sacks, rib-knit overalls with snaps down the legs, and Nikky and Rubber Duckies diaper covers. Extras include the Sara's Ride child carrier, the Taymore Outbound frame pack, a home ear scope, Papa Don's wooden baby toys, bug boxes with magnifying covers, face paints, capes, tutus, princess hats, percussion instruments, and stories and music on cassette. No color pictures dazzle the shopper, and the graphics are strictly homemade, but the selection here is terrific and the prices are excellent.

Saks Fifth Avenue
Folio Collections, Inc., 557 Tuckahoe Road, Yonkers, NY 10710
Telephone: 800-345-3454

Catalog price: free

Upscale children's wear. The collection spans up-to-date play clothes (like bomber jackets and one-piece sweat suits), fashionable school wear, and formal wear with a European look. An impeccable herringbone suit fits boys from size 8, playsuits with a high-fashion Tweedledum cut are offered for toddlers and preschoolers, snappy sweaters and beautiful lace slips are sold for girls.

Sears, Roebuck & Co.
Clothes for the whole family. See entry under *General Catalogs*.

Simply Divine
1606 S. Congress, Austin, TX 78704
Telephone: 512-444-5546

Catalog price: $1

Clothes for infants, children, and grown-ups, made of comfortable cotton interlock. The smallest members of the household can be clothed in snap-crotch T-shirts, footed leggings, and hooded sacks. Toddlers and children can move into classic button-front union suits, sweat suits, turtlenecks, leggings, skirts, dresses, and printed T-shirts. Parents can stock their own closets with tunics, skirts, dresses, union suits, and button-top T-shirts. The black-and-white catalog lists rather than shows the colors. Prices stack up well against other all-cotton merchants.

Simply Divine's cotton union suits

Soft as a Cloud
1355 Meadowbrook Avenue, Los Angeles, CA 90019
Telephone: 213-933-4417

Catalog price: $1, refunded with order

Cotton clothing for infants and toddlers, along with other natural-fiber baby needs. Soft as a Cloud makes about half of the clothes it sells, and supplements its own merchandise with outfits from Kawamura, Grasshopper, Creation Stummer, Alexis, Carters, Buster Brown, and other manufacturers. Small color pictures pack a deceptively big selection into what feels like a thin catalog. For infants the company sells a nice variety of gowns, playsuits, bonnets, booties, and undershirts. Older babies and toddlers can try on cardigans, turtlenecks, tights, thermal long johns, flannel shirts, tank tops, shorts, slips, and dresses. An exceptional collection of footwear includes socks (anklets, knee socks, and no-slip crew socks), Chinese silk shoes with floral embroidery, flannel-lined cord booties, and several styles of soft leather slippers. Bibs, hooded towels, receiving blankets, afghans, crib sheets, mattress pads, and a crib mattress are all made of pure cotton.

Southern Emblem Co.
P.O. Box 8, Toast, NC 27049
Telephone: 919-789-3348

Catalog price: free

Emblems to sew or iron onto caps, jackets, and shirts. Some proclaim special interests, such as ranching, camping, or Harley-Davidson motorcycles. Flags, state emblems, military and fraternal patches, and emergency rescue insignia offer more options. If you don't like what the company has to offer, create your own design—if you order a dozen or more, the price isn't unreasonable. Those who need something to attach the emblems to can order caps in six different styles, and jackets made of nylon or satin. The jackets come in sizes down to a child's small.

Special Ideas
2801-B Rodeo Drive, #218, Santa Fe, NM 87505
Telephone: 505-982-4800

Catalog price: free

T-shirts, posters, and buttons with messages promoting peace. A "Teddy Bears for World Peace" T-shirt comes in sizes down to a child's 2-4; "Baby Bears for World Peace" comes in infant sizes. The Earth as seen from space features in several of the designs. Some all-type messages sound off against the abuse of drugs and alcohol.

Strauss' Country Ware
100 George Street, Alton, IL 62002

Catalog price: free

A purveyor of work clothing for adults, Strauss sells some matching outfits for younger ones. OshKosh overalls in denim and hickory stripe are offered down to size 2. OshKosh jeans, complete with hammer loop and rule pocket, come in slim and regular sizes from 4T to adult. Hanes briefs, T-shirts, and thermal long johns provide protection underneath.

Toad'ly Kids
2428 Patterson Avenue, Roanoke, VA 24016
Telephone: 703-981-0233 (800-621-5809 to order)

Catalog price: $2

Quality children's clothes and a few sideline items. Dress up infants and toddlers in cotton fleece pants, knit coveralls, flannel shirts, taffeta rompers and dresses, and acrylic sweaters. An extraordinary christening gown might become a family heirloom. Older children can be clad in corduroy overalls, cotton turtlenecks, denim jackets and dresses, bathrobes, and an array of sweaters. Lace-trimmed velvet dresses, a red velveteen coat, cotton petticoats, and floral-print pants suits can make the girls shine at special occasions. Sailor suits are an option for the boys. Extras include lambskins, an electronic keyboard, a toy chest, and an easel.

Tortellini
23 E. 17th Street, New York, NY 10003
Telephone: 212-645-0286 (800-527-8725 to order)

Catalog price: $1

Children's clothes with a cosmopolitan flair. For play, Tortellini sells striped Breton sailor shirts from France, pleated shorts in seersucker and polka dots, cotton leggings, long-sleeved T-shirts, elastic-cuffed denim pants in blue and engineer stripes, denim skirts and jumpers, polka-dot suspenders, and a reversible baseball jacket ornamented with Chinese characters. Dressier duds include a polka-dot necktie and a lace-collar dress made from Liberty of London floral cotton.

Tortellini's Breton sailor shirt

U. B. Cool
350 Fifth Avenue, Suite 5120, New York, NY 10118
Telephone: 800-URCOOL2

Catalog price: $2

Jams, lightweight long pants, T-shirts, short-sleeved shirts, and jackets. Many of the outfits are decorated with distinctive U. B. Cool graphics—messages like "Island boy," "I gotta wear shades," and "How cool ru?" with cartoon drawings of sunglassed faces, waves, and "cool o meters." It took four months and a couple of phone calls to get our catalog, so b cool while u wait.

The Ultimate Outlet
A Spiegel Company, P.O. Box 88251, Chicago, IL 60680-1251
Telephone: 800-332-6000 or 312-954-2772

Catalog price: $2

Spiegel's discount and overstock catalog. Children's clothes generally merit a few pages, but some editions turn out to be for adults only. For the patient, deals can be found on jackets, OshKosh overalls, and other clothes for school and play. One word of caution, however: high shipping costs can sometimes bring what seems a bargain back up to a real-world price.

L. T. Walker Co., Inc.
762 E. 21st Street, Brooklyn, NY 11210
Telephone: 718-434-6948

$1 for brochure

T-shirts and sweatshirts silkscreened with designs from U.S. postage stamps. Only T-shirts decorated with the "Love" stamp are sold in children's sizes.

Wild Child
1813 Monroe Street, Madison, WI 53711
Telephone: 608-251-6445

Catalog price: free

Cotton play clothes in bright solid colors. Leggings, thermal long johns, turtlenecks, and knit cotton caps keep things warm during the cooler months. Tank tops, T-shirts (long- and short-sleeved), terry shorts, socks, and knit dresses provide year-round wear. Most of the outfits come in a choice of 24 colors; hickory-striped overalls and matching engineer hats are dyed turquoise, purple, pink, or scarlet. Sizes run from 3 months to 12 years.

Wobkins
1606-I Plaza Drive, Post Falls, ID 83854
Telephone: 208-773-2346

Catalog price: $1

OshKosh clothing decorated with appliquéd pictures. The hickory-stripe overall gets a train treatment, blue denim is ornamented with helicopters or hearts, and a pink jumper receives a pair of bunnies. Babies can be fitted with union suits and bibs, toddlers and children to size 8 in overalls, jumpers, sweat suits, and shortalls.

The Wooden Soldier
North Hampshire Common, North Conway, NH 03860-0800
Telephone: 603-356-7041 or 6343

Catalog price: free

Dresses with puffed sleeves, ruffles, and lace and boys' outfits with turn-of-the-century tailoring form the heart of this catalog. Mother-daughter dresses of white linen, coordinating sister outfits of blue stripes and lace, sailor suits for toddlers, and seersucker suits for young boys are a few of the warm-weather features. Winter brings out velvet dresses and suspender pants, wool vests, and flannel nightgowns. Seasonal extras include Halloween costumes, tropical-print beach wear, a reversible denim-fur jacket, and red cotton union suits with button drop seats.

World Wear
123 Alder Lane, Boulder, CO 80304
Telephone: 303-938-8404

Catalog price: free

The folks at World Wear launched their children's clothing business after noticing that their six-month-old daughter sucked her toes while she was being changed but couldn't get her feet to her mouth when dressed. They decided to design a line of clothes for infants and toddlers that would keep them warm while allowing their natural flexibility and curiosity free reign. Soft cotton interlock is one of the key ingredients; extra room at the shoulders and hips is another. One playsuit is adapted from a Japanese kimono—snaps along the legs keep it from flying open at the bottom. A jumpsuit was inspired by a traditional Russian costume; a tunic-and-bloomer set derives from Turkish wear. Sweatshirts, sweatpants, and turtlenecks are improvements on classic American garb.

Children's Magazines

For Young Children

Chickadee
56 The Esplanade, Suite 304, Toronto, ON M5E 1A7, Canada
Telephone: 416-364-3333
Annual subscription: $14.95, Can$17 in Canada
Sample issue: $2.95

A "see & do" magazine for children under nine, published monthly with a gap in July and August. Picture stories and activities relating to nature and animals take up the bulk of each issue, including connect-the-dots puzzles, mazes, and word puzzles. Bits of information about animals are offered in basic fashion—a few words of explanation on a page with a big color photograph. Non-nature activities have included instructions for writing in invisible ink and for making a simple printing stamp. *Chickadee* prints artwork sent in by readers and runs occasional contests—six-year-old Ainsley Greig entered one of these and her painting of a rabbit ended up as the cover of the February 1989 issue. Older children can graduate to *Owl*, put out by the same publisher.

Chuckles
Troll Associates, 100 Corporate Drive, Mahwah, NJ 07430
Telephone: 201-529-4000
Annual subscription: $9.95
Sample issue: not available

Troll Associates isn't interested in curious inquirers who want to find out more about its magazines. It will send out no sample issues and will talk by phone only to customers ready to give a credit card number and subscribe. *Chuckles* comes out in eight colorful issues a year, each filled with games that teach such basic concepts as colors, shape recognition, numbers, and the alphabet. The issue we saw offered instructions for making masks from paper bags and included some pretty photographs of baby birds, spring flowers, and kites. Troll also publishes *Prehistoric Times*.

Coulicou
Les Edition Héritage Inc., 300, avenue Arran, St. Lambert, PQ J4R 1K5, Canada
Telephone: 514-672-6710
Annual subscription: Can$22.50 in U.S., $17.95 in Canada
Sample issue: free

The French edition of *Chickadee* (see above). Ten colorful issues a year offer games, activities, and pretty pictures of animals.

Humpty Dumpty
Children's Better Health Institute, 1100 Waterway Boulevard, Box 567B, Indianapolis, IN 46202
Telephone: 317-636-8881
Annual subscription: $11.95, US$18.95 in Canada
Sample issue: 50¢

The third of the seven magazines put out by the Children's Better Health Institute, *Humpty Dumpty* is written for children aged three to seven. (Younger children can subscribe to *Stork* and *Turtle*, older readers to *Children's Playmate*, *Jack and Jill*, *Child Life*, and *Children's Digest*.) Simply written stories and poems, activities like connect-the-dots puzzles and word games, drawings by readers, and comics are blended with information about health, nutrition, and safety. Eight issues are mailed each year.

Mickey Mouse Magazine
Welsh Publishing Group, Inc., 300 Madison Avenue, New York, NY 10017
Telephone: 212-687-0680
Annual subscription: $7.80, US$10.80 in Canada
Sample issue: $3

A quarterly entertainment magazine for young children (ages two to six), starring Mickey Mouse and other Disney characters. Picture games and puzzles provide plenty of action, a story gives parents a chance to read

aloud, and jokes submitted by readers provide an interesting touch. The second half of the magazine is given over to a guide for parents, with advice on such topics as reading to children, holiday celebrations, and news of forthcoming films and videos.

Scienceland
501 Fifth Avenue, Suite 2102, New York, NY 10017
Telephone: 212-490-2180

Annual subscription: $13.95, $15.95 outside U.S.
Sample issue: $3

Scienceland guides young children (ages four to eight) in studies of basic science concepts. Huge color pictures, often spreading across two pages, add drama to the presentation. A recent theme issue on "The Secrets of Spring" laid out massive closeups of dandelions and ladybugs, while the text explained the flower's progression from bud to seed, the ability of the root to grow new plants from a small broken section, the predatory nature of ladybugs, and the insect's reproductive cycle. Another issue dealt with the sleep habits of wild animals, with similarly dramatic photographs. Each issue includes some hands-on collecting and observing activities. The subscription prices above are for eight issues of the "Student Edition," which is printed on ordinary magazine paper. For twice the price ($28, $32 outside U.S.), subscribers can receive the "Deluxe Edition," printed on heavier, stiffer, and glossier paper.

Sesame Street Magazine
Children's Television Workshop, One Lincoln Plaza, New York, NY 10023
Telephone: 212-595-3456

Annual subscription: $10.95
Sample issue: $1.50

Sesame Street Magazine is actually two publications bound separately, one for young children (ages two to six) and the other for their parents. The children's portion features read-aloud stories and rhymes, projects to cut out and assemble (such as paper dolls and measuring sticks), puzzles, games, riddles (sent in by readers), and simple reading and writing exercises. The bulging eyes of the characters from the television show peer out from almost every page. The 30-page parents' guide is as long as the children's magazine and is loaded with information and advice on raising young children. A couple of pages offer suggestions on getting more out of the children's issue. The rest of the guide is devoted to articles on parenting, child development, and children's health issues. Published monthly with breaks in February and August.

Snoopy Magazine
Welsh Publishing Group, 300 Madison Avenue, New York, NY 10017
Telephone: 212-687-0680

Annual subscription: $7.80, US$10.80 in Canada
Sample issue: $3

A quarterly publication for preschoolers (ages 2 to 5) and their parents. The front half of the magazine is given over to activities, stories, games, and riddles, all revolving around Snoopy and other characters from the *Peanuts* cartoon strip. After the centerfold—a pinup of Snoopy in a seasonal pose—the type shrinks and the "Parents' Pages" begin. Here the magazine serves up advice about such subjects as the importance of children's fantasy play, encouraging healthy eating habits, and indoor projects for bad weather.

Stork
Children's Better Health Institute, 1100 Waterway Boulevard, Box 567B, Indianapolis, IN 46202
Telephone: 317-636-8881

Annual subscription: $11.95, US$18.95 in Canada
Sample issue: 50¢

A good first magazine for the very youngest subscribers, ages one to three. Rhymes, crafts for children to make with parents, and simple puzzles are all appropriately geared to the toddler set. A recent issue included a "Peek-a-Boo Page," which when held to a light showed grandma coming through a door to join a family gathering. (The doorway was blank on the front side of the sheet; grandma was printed on the back.) For somewhat older readers it added counting activities, finger games, and a very simple "Which one is different?" game (eight turkeys and a pumpkin). Readers can move on from *Stork* to six other magazines from the Children's Better Health Institute. *Turtle*, for two- to six-year olds, is next; then *Humpty Dumpty*, *Children's Playmate*, *Jack and Jill*, *Child Life*, and *Children's Digest*.

Turtle
Children's Better Health Institute, 1100 Waterway Boulevard, Box 567B, Indianapolis, IN 46202
Telephone: 317-636-8881

Annual subscription: $11.95, US$18.95 in Canada
Sample issue: 50¢

The next step after *Stork* in the Children's Better Health Institute's magazine lineup, *Turtle* is geared for children

aged two to six. (*Humpty Dumpty* comes next, then *Children's Playmate*, *Jack and Jill*, *Child Life*, and *Children's Digest*.) Stories, rhymes, and such activities as hidden-picture puzzles, connect-the-dots drawings, and matching games hold readers' interest while they absorb the magazine's real message—information on health and nutrition for children. A medical question-and-answer column is included for parents.

Your Big Backyard
National Wildlife Federation, 8925 Leesburg Pike, Vienna, VA 22184-0001
Annual subscription: $10, $18 outside U.S.
Sample issue: $2

For children too young for the National Wildlife Federation's *Ranger Rick*, *Your Big Backyard* provides a gentle starting ground. Preschoolers aged three to five will delight in the spectacular color photographs of wild animals. Short lines of text in large type encourage curiosity about reading; questions, games, and project ideas provide something to do; and a longer story gives parents a chance to read aloud. The brief text in the magazine gives only limited information, but parents are armed for the inevitable barrage of questions with a long letter that supplies additional facts. The letter also offers an excellent guide to exploring the magazine fully with a child. Slimmer than *Ranger Rick*, *Your Big Backyard*'s monthly issues are about 20 pages long.

For School-Age Children

Adventures in Learning
Different Worlds Publications, 2814 19th Street, San Francisco, CA 94110
Annual subscription: $10, US$12 in Canada
Sample issue: $2.50, US$3 in Canada

A newsletter full of math puzzles and games, including play-by-mail games, logic puzzles, calculator problems, and more. Published primarily for classroom use, there's no reason why it can't be ordered at home. Five issues come out during the school year.

Alf Magazine
Welsh Publishing Group, Inc., 300 Madison Avenue, New York, NY 10017
Telephone: 212-687-0680
Annual subscription: $7.80, US$10.80 in Canada
Sample issue: $3 (see below)

A quarterly magazine starring Alf of television fame. The premier issue (spring 1989) featured mazes, jokes, recipes, a code puzzle, an article on weird pets, an "Ask Alf" column, and an Alf poster in the center spread. In color throughout, the 32-page magazine includes advertisements. Send requests for sample issues to the magazine at P.O. Box 10559, Des Moines, IA 50340.

Barbie Magazine
Welsh Publishing Group, Inc., 300 Madison Avenue, New York, NY 10017
Telephone: 212-687-0680
Annual subscription: $7.80, US$10.80 in Canada
Sample issue: $3 (see below)

A flashy magazine of fashion and entertainment for girls. Barbie, the well-known doll, graces the cover of each issue and shows up in a regular story inside, but she doesn't dominate every page. Fashion features, word puzzles, and party ideas are regular features. A section called "What's Happening" prints brief reviews of movies, television shows, records, and books. The 32-page magazine is printed in color throughout, comes out four times a year, and includes some advertising. Address requests for sample issues to Barbie Magazine, P.O. Box 10798, Des Moines, IA 50340.

The BASIC Teacher
Different Worlds Publications, 2814 19th Street, San Francisco, CA 94110
Annual subscription: $36 in U.S. and Canada, $42 in other countries
Sample issue: $3.50

A monthly newsletter for beginning computer programmers. BASIC and QuickBASIC vocabulary and syntax are covered in regular columns; more advanced programming problems are treated in special articles.

Boys' Life
Boy Scouts of America, 1325 Walnut Hill Lane, Irving, TX 75015
Telephone: 214-580-2000
Annual subscription: $13.20, US$18.20 in Canada
Sample issue: free

The official magazine of the Boy Scouts, *Boys' Life* prints articles on sports, history, scouting programs, and adventure. One fictional story in each issue balances the factual reporting. Comics offer both entertainment and instruction in scouting skills. Monthly columns give advice on stamp and coin collecting, pet care, hobbies, magic, health and safety, fishing, and electronics. Parents who subscribed as children will be amazed to find the same "Think & Grin" joke column at the back, and many of the same classified ads: readers can still order live sea horses and sea monkeys by mail, along with X-ray glasses, animal traps, and incubators complete with quail eggs, and they can still sell greeting cards for "famous name prizes." Subscribers get 12 issues a year, each approximately 70 pages long.

Child Life

Children's Better Health Institute, 1100 Waterway Boulevard, Box 567B, Indianapolis, IN 46202
Telephone: 317-636-8881

Annual subscription: $11.95, US$18.95 in Canada
Sample issue: 50¢

The sixth of the Children's Better Health Institute's publications, *Child Life* is geared to readers aged 7 to 11. (It follows after *Stork, Turtle, Humpty Dumpty, Children's Playmate,* and *Jack and Jill.* Readers move from it to *Children's Digest.*) A typical issue combines four or five illustrated stories with several pages of games and puzzles; a few pages of jokes, poems, and pictures sent in by readers; a recipe for a healthy snack; and a question-and-answer column about food and health. The publication's goal is to instruct readers on health-related issues, and it succeeds by blending the information with entertaining reading. Subscribers receive eight issues a year, each about 50 pages long.

Children's Album

EGW Publishing Co., 1300 Galaxy Way, Concord, CA 94520
Telephone: 415-671-9852

Annual subscription: $15, US$20 in Canada
Sample issue: $2.95

Stories and poems written by children, and craft projects for children to make. A few book reviews and word puzzles add variety to the mix. Six issues are published a year; each prints roughly 15 stories submitted by readers, and as many poems. For children 8 to 14.

Children's Playmate

Children's Better Health Institute, 1100 Waterway Boulevard, Box 567B, Indianapolis, IN 46202
Telephone: 317-636-8881

Annual subscription: $11.95, US$18.95 in Canada
Sample issue: 50¢

The fourth of the seven magazines put out by the Children's Better Health Institute, *Children's Playmate* is written for children aged five to nine. (Younger children can subscribe to *Stork, Turtle,* and *Humpty Dumpty,* older readers to *Jack and Jill, Child Life,* and *Children's Digest.*) Stories, jokes and riddles, activities like connect-the-dots puzzles and word games, pictures and poems from readers, and comics combine with information about health, nutrition, and safety. Eight issues are mailed each year.

Classical Calliope

Cobblestone Publishing Inc., 20 Grove Street, Peterborough, NH 03458
Telephone: 603-924-7209

Annual subscription: $14.95, $19.95 outside U.S.
Sample issue: $3.75

Subtitled *The Muses' Magazine for Youth, Classical Calliope* delves into archaeology and ancient history. That may sound like unlikely territory for a children's magazine, but Cobblestone Publishing has met the challenge brilliantly. Informative and well-written articles draw readers into the subject, bringing out the similarities and differences between the ancient world and our own. A recent issue on "Children in Ancient Times" (each issue revolves around a new theme) held articles on ancient toys and games, the origins of words, the Phi Beta Kappa society, traditional rites of passage, and famous rulers who were adopted. An excerpt from a 3rd century B.C. play on Greek school life must have made readers glad to be born in the late 20th century. Crossword puzzles and other games provide some light diversion at the back of each issue. Four black-and-white issues are mailed, each running to about 40 pages (the folios in roman numerals, of course). The articles are aimed at readers from age 8 to 14.

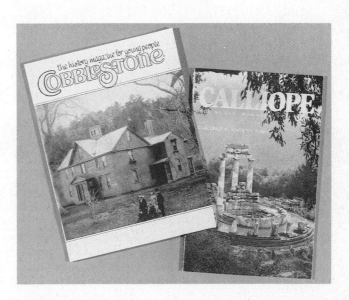

Cobblestone: The History Magazine for Young People

20 Grove Street, Peterborough, NH 03458
Telephone: 603-924-7209

Annual subscription: $19.95, $24.95 outside U.S.
Sample issue: $4.95

One of our favorites of the current crop of children's magazines, and one that should be on every parent's list for consideration. Each monthly issue deals with a different theme. One on "American Clothing: Then and Now" included articles on the clothing industry, clothing for the disabled, and clothing as a statement of rebellion (from Amelia Bloomer to hippies). An issue devoted to Louisa May Alcott discussed her parents and her upbringing, her little-known mystery stories, her life as an army nurse, and how she came to write *Little Women*. A play allowed readers to play the parts of the Alcott sisters and some of their Concord neighbors; a "how-to" article gave advice on keeping a journal. Suggestions for further reading, places to visit, a recipe for a white taffy such as the Alcotts might have made, and an entertaining quiz all added further dimension.

Creative Kids
GCT Inc., P.O. Box 6448, Mobile, AL 36660-0448
Telephone: 205-478-4700

Annual subscription: $23.97, $29.97 outside U.S.
Sample issue: $3

A magazine written and illustrated by its young readers (ages 5 to 16). Travel reports, poems, stories, artwork, and photographs form the core of the publication. The drawings and photographs are reproduced in color. Games and puzzles, all created by readers, add some extra fun. Perhaps the most extraordinary thing about the magazine is the international range of the submissions. Each issue holds writing and pictures by children from such countries as Australia, Saudi Arabia, Colombia, Thailand, and West Germany. Eight monthly issues are published during the school year (October to May).

Cricket
P.O. Box 52961, Boulder, CO 80322-2961
Telephone: 800-435-6850 (800-892-6831 in IL)

Annual subscription: $24.97, $30.97 outside U.S.
Sample issue: $2 (see below)

An exceptional magazine filled with stories, informative articles, activities, and great illustrations. A recent issue concentrated on trains and books, with articles about the origins of words, word collecting (or vocabulary building), the first steam engine, and a baboon trained to operate a railroad switching tower. In between were instructions for sewing a cloth book jacket and for making a fall-proof Humpty Dumpty (by softening an egg's shell in vinegar), articles on the behavior of wild gorillas and on George Washington's inauguration, a Philippine folk tale about a firefly and an ape, and a story about a boy who won a rabbit in a pet shop raffle. The winning entries in monthly writing and poetry contests are published as a regular feature, as are letters from readers. In a wonderful extra touch, the margins are dotted with tiny-type definitions of the uncommon words used in the text. *Cricket* comes out 12 times a year and is geared to readers aged 6 to 13. For a sample issue, send $2 to *Cricket*, Box 300, Peru, IL 61354.

Disney's DuckTales Magazine
Welsh Publishing Group, Inc., 300 Madison Avenue, New York, NY 10017
Telephone: 212-687-0680

Annual subscription: $7.80
Sample issue: $3

A quarterly entertainment magazine for children from 6 to 12, based on Donald Duck and his family. A four-page cartoon strip is a regular feature, as is a brief adventure story, a cartoon drawing lesson, and a "Really Weird Invention" from Gyro Gearloose. A column called "What's Quackin'" brings news of new movies and television shows, with an emphasis on Disney productions.

Dolphin Log
The Cousteau Society Membership Center, 930 W. 21st Street, Norfolk, VA 23517
Telephone: 804-627-1144

Annual subscription: $28 for family membership
Sample issue: not available

Family members of the Cousteau Society receive both the newsletter *Calypso Log* and the bimonthly children's magazine *Dolphin Log*. Running to about 16 pages, *Dolphin Log* prints expedition news, illustrated articles on unusual sea creatures, and reports on such ocean topics as signal flags and fish migration. A regular column lets readers know how they can become involved in specific conservation efforts. For ages 7 to 15.

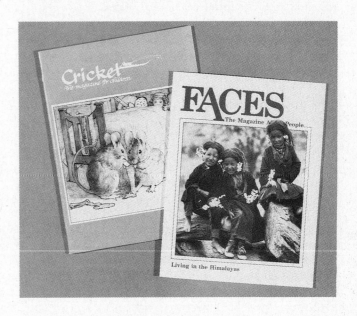

Faces: The Magazine About People
20 Grove Street, Peterborough, NH 03458
Telephone: 603-924-7209

Annual subscription: $18.95, $23.95 outside U.S.
Sample issue: $3.95

Another fascinating magazine on a serious topic from Cobblestone Publishing, creators of *Classical Calliope* and *Cobblestone*. *Faces* deals with anthropology in a "theme issue" format—each new issue concentrates on a single subject, such as movable houses, hats, or the use of shells in different cultures. An issue on life in the Himalayas roped in a Nepalese folk tale and articles about the Tibetan Dalai Lama, mandalas, the Sherpa people of highland Nepal, and the practice of polyandry (a marriage of one wife and several husbands). Activities included a recipe for dal bhat, a traditional Nepalese dish, and instructions for playing a game called pan-das. *Faces* comes out monthly, with a break for July and August, in black-and-white issues that average 40 pages. The writing is aimed at children aged 8 to 14.

Hibou

Les Edition Héritage Inc., 300, avenue Arran,
St. Lambert, PQ J4R 1K5, Canada
Telephone: 514-672-6710

Annual subscription: Can$22.50 in U.S., $17.95 in Canada
Sample issue: free

The French-language edition of *Owl* (see below), a magazine of entertainment and information with a loose concentration on animals and nature. A recent issue included articles on baby elephants, the microscopic creatures that live in an ordinary house, and winter sports.

Highlights for Children

P.O. Box 269, Columbus, OH 43272-0002
Telephone: 717-253-1080

Annual subscription: $19.95, US$24.95 in Canada
Sample issue: $2.25 and an extra-large self-addressed stamped envelope

We remember *Highlights* from our own childhood, and were pleased to find that the magazine hasn't strayed from its basic formula of stories, games, and puzzles, though it's been livened up with color and more attractive graphics. A normal issue holds eight to ten stories and factual articles for beginning and intermediate readers, and lots of fun activities like craft projects, rhebuses, hidden-picture puzzles, mazes, and crosswords. Subscribers get 11 issues a year, each about 40 pages long. For children to age 12.

Jack and Jill

Children's Better Health Institute, 1100 Waterway Boulevard, Box 567B, Indianapolis, IN 46202
Telephone: 317-636-8881

Annual subscription: $11.95, US$18.95 in Canada
Sample issue: 50¢

The fifth in the Children's Better Health Institute's series of magazines, preceded by *Stork, Turtle, Humpty Dumpty,* and *Children's Playmate,* and followed by *Child Life* and *Children's Digest. Jack and Jill* is geared to readers between the ages of six and ten. It combines stories, jokes, poems, puzzles, and games with information about health and nutrition. The magazine prints recipes, poems from readers, and a question-and-answer column that lets readers pose health questions to a doctor. Eight issues are mailed each year.

Kid City

Children's Television Workshop, One Lincoln Plaza, New York, NY 10023
Telephone: 212-595-3456

Annual subscription: $13.97, $19.97 outside U.S.
Sample issue: $1.50

Children who've outgrown *Sesame Street Magazine* can graduate to *Kid City*. Written for an audience from 6 to 12 years old, it contains short articles and stories, projects, games, comics, puzzles, recipes, jokes, book reviews, and contests. Each issue focuses on a different theme. A recent one on "Hidden Treasure" included an article about a real sunken treasure ship, a cut-and-paste pirate costume (complete with decoder), strange facts about money, and instructions for making a secret jewelry box that looks like a book. *Kid City* comes out in ten monthly issues (with gaps in February and August).

KidsArt News

P.O. Box 274, Mt. Shasta, CA 96067
Telephone: 916-926-5076

Annual subscription: $8, US$10 in Canada
Sample issue: $2.50, US$3 in Canada

A quarterly newsletter filled with ideas for art projects that can be made at home, and with articles about famous artists. Recent issues have explained how to make clay masks, duck decoys, puppets with clay heads, and aluminum foil relief art, how to draw a perching bird, and how to make a bird mobile. Mixed in with the project instructions were stories about wildlife artist John James Audubon, the lion on Babylon's Ishtar Gate, and Lorenzo Ghiberti's Baptistery doors in Florence.

Koala Club News

Zoological Society of San Diego, P.O. Box 551, San Diego, CA 92112
Telephone: 619-231-1515

Annual subscription: $9
Sample issue: free

A quarterly newsletter, illustrated with wonderful black-and-white photographs, for children under 15. The issue we saw featured the hatching of the first California condor egg ever laid in captivity, and included a great picture story on the zoo's Avian Propagation Center (where baby birds are hatched and raised). A coloring page and a craft project are regular features; a "Critter of the Quarter" rates a full-page portrait suitable for pinning on a bulletin board. The $9 fee brings the newsletter and membership in the Koala Club, which includes free entrance to the San Diego Zoo for a year and one pass to the Children's Zoo.

The McGuffey Writer

400A McGuffey Hall, Miami University, Oxford, OH 45056
Telephone: 513-529-6462

Annual subscription: $5
Sample issue: $2

A magazine of stories and poems written by its young readers, aged 6 to 14. Submissions by first and second graders are generally succinct; stories by older children can stretch to two tightly spaced pages. Three issues come out each year, each about 16 pages.

Muppet Magazine
Welsh Publishing Group, 300 Madison Avenue, New York, NY 10017
Telephone: 212-687-0680

Annual subscription: $7.80, US$10.80 in Canada
Sample issue: $3

From the publisher of *Barbie* and *Mickey Mouse Magazine*, another character-based magazine of pictures and entertainment. A "Miss Piggy's Advice" column answers questions in a lighthearted way; a recipe page offers ideas for snack and party food that children can make; and a "Muppet Round-Ups" column keeps readers current on new books, records, movies, and videos. The center spread is given over to an oversize color photograph suitable for hanging on the refrigerator (in a recent issue it featured Kermit and Miss Piggy in a spoof poster for "Nerdy Dancing"). Four issues a year, for readers 7 to 14 years old.

National Geographic World
17th and M Streets N.W., Washington, DC 20036

Annual subscription: $10.95, Can$17.75 in Canada
Sample issue: $1.40, Can$2.20 in Canada

A beautiful monthly magazine from the National Geographic Society, filled with gorgeous color photographs of animals, wild places, and phenomena like soap bubbles. A huge fold-out poster comes in the center of each issue, along with ideas and instructions for craft projects, book reviews submitted by readers, and a column called "Why in the World?" that answers such questions as "Why do trees have dark rings?" and "Why is ice slippery?" While the photographs are top-notch, the text does not go into much detail when compared to other magazines for this age group. For children 8 to 13.

Noah's Ark
8323 Southwest Freeway, Suite 250, Houston, TX 77074
Telephone: 713-771-7143

Annual subscription: $8, US$10 in Canada
Sample issue: 80¢

A newspaper for Jewish children. Ten four-page issues are published, each with an assortment of jokes ("What could Noah put in the ark to make it lighter? A candle."), puzzles, recipes, contests, and articles. A pen-pal column prints the names, addresses, and special interests of readers looking for pen pals. For 6- to 12-year olds.

Odyssey
Kalmbach Publishing Co., 1027 N. 7th Street, Milwaukee, WI 53233
Telephone: 414-272-2060

Annual subscription: $16, US$20 in Canada
Sample issue: $3.50

A monthly magazine of space exploration and astronomy for children between 8 and 14. Beautiful color photographs illustrate feature articles on the planets, stars, and missions of space exploration. A news section keeps readers posted on the latest happenings in the sky, from meteor showers to plans for rocket launches, and a monthly star chart helps locate the constellations. Occasional puzzles and quizzes add some interaction to the blend, and articles sometimes wander as far afield as a report on how to fake UFO photographs.

Owl
56 The Esplanade, Suite 304, Toronto, ON M5E 1A7, Canada
Telephone: 416-364-3333

Annual subscription: $14.95, Can$17 in Canada
Sample issue: $2.95

A magazine for older children (ages 8 to 14) from the publisher of *Chickadee*. A recent issue printed an article about an 11-year-old Toronto girl's sailing trip around the world, a collection of body tricks to play on friends ("How can you stop someone from getting out of a chair, using the strength in one finger?"), a diagram of classic body proportions used by artists, a quiz on the effects of weightlessness in space, and a question-and-answer column about pets. Subscribers receive ten issues a year (monthly, with a summer break). A French-language edition is also available; see *Hibou*, above.

Pack-O-Fun
14 Main Street, Park Ridge, IL 60068
Telephone: 312-825-2161

Annual subscription: $6, $8 outside U.S.
Sample issue: $2.50

A quarterly magazine of craft projects, games, recipes, and puzzles. Each issue contains more than two dozen things to make, all from materials readily available in most homes. The Christmas 1988 issue, for example, held instructions for building a log cabin from pretzel sticks, for making a wreath from sandwich bags, and for creating a crèche from plastic drinking cups. Crossword puzzles, connect-the-dots, and word games are regular features, as is an extensive pen-pal directory with the names and addresses of willing correspondents from as far away as Ghana.

Penny Power
P.O. Box 54861, Boulder, CO 80322-4861
Telephone: 914-667-9400

Annual subscription: $11.95, $15.95 outside U.S.
Sample issue: free

Consumer Reports for kids. *Penny Power* comes out every other month with analyses of misleading advertisements, advice on money management, readers' reviews of video movies, and evaluations of children's products. The product reviews cover everything from jeans to toys—two recent issues rated boom boxes, fast-food chicken, cereal premiums, electronic games, and a children's video camcorder. And *Penny Power* encourages readers to play an active role. When a subscriber

wrote to complain about a deceptive ad for a toy called Army Ants, the magazine forwarded the complaint to the Children's Advertising Review Unit of the Council of Better Business Bureaus, which used it as evidence in an appeal to the manufacturer. The result? The ad was dropped, and Jennie Randall got her picture printed in a *Penny Power* article. To keep matters from centering on the material and commercial, the magazine also includes "life-style" advice, ranging from tips for having fun at parties to guidance on handling wardrobe conflicts with parents.

Plays
120 Boylston Street, Boston, MA 02116
Telephone: 617-423-3157

Annual subscription: $23, $31 outside U.S.
Sample issue: $3

Royalty-free plays for school-age children. Seven issues come out in a year, each with 8 to 12 one-act plays and skits. Holiday plays are joined by comedies, dramas, mysteries, monologues, science fiction, puppet plays, and dramatized classics and fairy tales.

Pockets
1908 Grand Avenue, P.O. Box 189, Nashville, TN 37202-0189
Telephone: 615-340-7333

Annual subscription: $12.95
Sample issue: $1.70

A children's magazine written from a Christian perspective. Stories deal with religious and moral themes without being heavy-handed—one in a recent issue told of a boy who broke his father's fishing pole; after a talk with his grandfather he's ready to face up to his father, knowing that punishment will be followed by forgiveness. Some of the stories end with questions asking readers how they would respond to a dilemma, such as a sibling's drug problem or a conflict with a friend, and responses are printed in the following issue. The word games and puzzles yield scriptural answers and a recipe column is called "Loaves and Fishes," but not everything connects directly to the Bible. An activity page in a recent issue lets readers design their own coats of arms.

Prehistoric Times
Troll Associates, 100 Corporate Drive, Mahwah, NJ 07430
Telephone: 201-529-4000

Annual subscription: $9.95
Sample issue: not available

A slim magazine (about eight pages per issue) that explores the subject of dinosaurs eight times a year. Puzzles and games share space with brief articles on important dinosaur finds and facts about dinosaurs. The writing never goes into great detail, and passionate dinosaur fans may be disappointed. The company has a "subscribe or don't bother us" approach, so don't look for sample issues or information.

Prism
1040 Bayview Drive, Suite 223, Ft. Lauderdale, FL 33304
Telephone: 305-563-8805

Annual subscription: $19.95, $25 outside mainland U.S.
Sample issue: $4

A bimonthly magazine in a newspaper format, written "by and for the gifted and talented." Stories, poems, and letters from young readers are mixed with articles intended for educators (on subjects such as "Developing a Rigorous Curriculum for the Gifted: A Pressing Need") in such a way that we can't help but wonder if even gifted young subscribers might be confused. For readers six years old and up.

Ranger Rick
National Wildlife Federation, 8925 Leesburg Pike, Vienna, VA 22184-0001

Annual subscription: $14, $22 outside U.S.
Sample issue: $2

The National Wildlife Federation's monthly nature magazine for children aged 6 to 12. Beautiful color photographs of wild animals illustrate the articles, which in one recent issue covered grizzly bears, winter birds, sea anemones, creatures of the Antarctic, and carousel animals. The factual reports are supplemented by puzzles, mazes, and other entertaining activities, and by a monthly story about the cartoon raccoon, Ranger Rick. One of the better children's magazines, *Ranger Rick* deserves a close look. Children too young to read the articles may want to try *Your Big Backyard*, also published by the National Wildlife Federation.

Reflections
P.O. Box 368, Duncan Falls, OH 43734
Telephone: 614-674-4121

Annual subscription: $5, US$8 in Canada
Sample issue: $2, US$4 in Canada

A national poetry and literary magazine published twice a year by seventh- and eighth-grade journalism students at Duncan Falls Junior High. Though the editorial work is done in Ohio, the poems and stories come from all over the country. In the issues we saw, the youngest writer was 7, the oldest 18. Most of the pieces are accompanied by photographs of the authors.

Rubberstampmadness
P.O. Box 6585, Ithaca, NY 14851

Annual subscription: $15 in U.S. and Canada, $30 in other countries
Sample issue: $3

Rubber-stamp enthusiasts can catch up on the latest techniques, see the work of fellow stampers, and shop for dozens of stamp catalogs in the bimonthly pages of *Rubberstampmadness*. The tabloid format allows plenty of room for detailed articles and for displaying remarkable impressions. The September/October 1988 issue was devoted to stamping with children. Filled with project ideas, suggestions for teachers and parents, and sources for stamps and materials, it would be worth writing for a back issue along with a subscription.

Seedling Short Story International
P.O. Box 405, Great Neck, NY 11022
Telephone: 516-466-4166

Annual subscription: $14, US$17 in Canada
Sample issue: $3.95

Subscribers receive four short-story collections a year, bound in a format closer to a paperback book than a conventional magazine. The editors make an effort to include writers from diverse backgrounds in each issue. The copies we saw held tales from Malta, China, Australia, Sri Lanka, Korea, France, Greece, and several other countries. For children aged 9 to 13.

Shoe Tree
215 Valle del Sol Drive, Santa Fe, NM 87501

Annual subscription: $15, $18 outside U.S.
Sample issue: $5

Fiction, poetry, book reviews, and art by children aged 6 to 14. Three well-designed issues come out each year, cleanly typeset and printed on quality paper. The stories range from horror tales to family sagas, and all of the writing is first-rate. Subscribers are encouraged to submit their own work for publication.

Shofar
Senior Publications Ltd., 43 Northcote Drive, Melville, NY 11747
Telephone: 914-638-0333

Annual subscription: $14.95, $18.95 outside U.S.
Sample issue: $3

A magazine "for Jewish kids on the move," published monthly with breaks in January, May, and the summer months (six issues are mailed in all). Interviews with Jewish stars and sports figures are a regular feature, as are fictional stories, plays, book and record reviews, games, puzzles, cartoons, and contests. A "Dear Ira" column answers readers' questions about problems at home or with friends, a poetry page prints work submitted by readers, and a page titled "This Month in History" lays out a chart of important Jewish dates.

Sports Illustrated for Kids
P.O. Box 830606, Birmingham, AL 35282-9487
Telephone: 800-632-1300

Annual subscription: $15.95, US$18.95 in Canada
Sample issue: $1.75

A weekly sports magazine for children from the publishers of *Sports Illustrated*. Articles are aimed at readers ages 8 to 13, with features on both adult and child athletes. The definition of "sports" is expanded here to include such activities as roller skating, surfing, and dog shows. Launched in the summer of 1989, the magazine's first issues featured Michael Jordan, Janet Evans, and a stellar young skateboarder, along with an array of other junior athletes.

Stone Soup: The Magazine by Children
Children's Art Foundation, P.O. Box 83, Santa Cruz, CA 95063
Telephone: 408-426-5557

Annual subscription: $20, $24 outside U.S.
Sample issue: $4

The most beautiful of the children's literary magazines, *Stone Soup* comes out in five exquisite issues during the school year. Stories, poems, book reviews, songs, and artwork are all crisply printed on fine paper, some of the paintings reproduced in color. Writers as young as 4 have had their poems published, but most of the writing comes from children aged 6 to 13. An insert suggests activities based on the contents of the issue; these are generally encouragements to write and draw, using the stories and artwork in the magazine as a jumping-off point.

Surprises: Activities for Today's Kids and Parents
P.O. Box 236, Chanhassen, MN 55317
Telephone: 612-937-8345

Annual subscription: $12.95, US$22.95 in Canada
Sample issue: $2

A 40-page magazine crammed with coloring pages, codes to break, mazes, writing projects, crosswords, travel games, board games, connect-the-dots, trivia quizzes, calculator problems, and hidden-picture puzzles. A new issue is published every other month, and we're sure subscribers wish it came out more often. For ages 4 to 12.

3-2-1 Contact
Children's Television Workshop, One Lincoln Plaza, New York, NY 10023
Telephone: 212-595-3456

Annual subscription: $15.97, $21.97 outside U.S.
Sample issue: $1.50

A glossy color magazine of science and technology for readers aged 8 to 14. Two recent issues carried articles on hunting for lost treasure, the *MacGyver* television series, efforts to save endangered sea turtles, skateboarding, the ozone layer in the Earth's atmosphere, and animal species that have survived for more than 100 million years. A "Square One" section (based on the television show) features interactive math puzzles and games. Short news takes fill readers in on scientific discoveries, technological innovations, and facts about animals and nature (Did you know that people's teeth are shrinking 1 percent every 1,000 years?). A column called "Any Questions" fields such queries from readers as "Why do we have different time zones?" and "Why do people laugh when they get tickled?" Activities such as mysteries, mazes, craft projects, and word puzzles take up three or four pages of each issue. Published monthly except February and August.

Wee Wisdom
Unity School of Christianity, Unity Village, MO 64065
Telephone: 816-524-3550

Annual subscription: $8, $12 outside U.S.
Sample issue: free

Founder Myrtle Filmore wrote in the 1890s that "the mission of *Wee Wisdom* is not to entertain children but to call them out." Close to a century old now, the magazine still blends stories, activities, and puzzles in an entertaining format that strives to "call out" the best in its readers. Most of the stories deliver a moral or biblical message, though generally a soft-edged one—often just the simple advice that it is good to treat others kindly. Published ten times a year for children to age 12.

Zoobooks
P.O. Box 85271, Suite 6, San Diego, CA 92138
Telephone: 619-745-0685

Annual subscription: $14.95, US$17.95 in Canada
Sample issue: $3.45

Each of the ten monthly issues of *Zoobooks* could stand on its own as a complete book about a particular wild animal. A different creature gets the spotlight in each new edition—it might be giant pandas one month, elephants the next. Text and pictures delve thoroughly into the animal's behavior, its anatomy, its prehistoric ancestors, the history of its relationship with humans, its growth from birth to adulthood, and its current range in the wild. Excellent color photographs are supplemented by paintings that graphically demonstrate points made in the text. A discussion of the eating habits of elephants, for example, was backed up by a two-page picture of an animal standing among the mountains of hay, alfalfa, grain, potatoes, and fresh vegetables that it consumes in a year. For ages 6 to 12.

For Older Children and Teens

American Spacemodeling
A monthly rocketry magazine. See entry for National Association of Rocketry under *Associations, Hobby*.

Astronomy
AstroMedia/Kalmbach Publishing Co., 1027 N. 7th Street, Milwaukee, WI 53233
Telephone: 414-272-2060

Annual subscription: $24, US$24.75 in Canada
Sample issue: $3.50, US$4 in Canada

A popular magazine of astronomy, illustrated with beautiful photographs and paintings. Teenagers should have no trouble with the articles; younger readers may struggle. A monthly almanac gives advance notice of stellar and planetary happenings, with sky maps to guide stargazers.

Children's Digest
Children's Better Health Institute, 1100 Waterway Boulevard, Box 567B, Indianapolis, IN 46202
Telephone: 317-636-8881

Annual subscription: $11.95, US$18.95 in Canada
Sample issue: 50¢

The final stop on the progression of Children's Better Health Institute magazines, this one aimed at readers aged 8 to 12. (It trails *Stork*, *Turtle*, *Humpty Dumpty*, *Children's Playmate*, *Jack and Jill*, and *Child Life*, all for younger children.) Following the formula of the other publications, *Children's Digest* serves up stories, informative articles, word games, jokes, and a couple of comic strips, and slips in a question-and-answer column and an article or two dealing with health and nutrition issues. A year's subscription brings eight issues, each about 50 pages long.

Classic Toy Trains
Kalmbach Publishing Co., 1027 N. 7th Street, Milwaukee, WI 53233-1471
Telephone: 414-271-2060

Annual subscription: $14.95, US$16.95 in Canada
Sample issue: $4.95

If toy train play turns serious, a collector's magazine might be in order. Published quarterly, *Classic Toy Trains* prints nicely illustrated articles on exceptional layouts and historic rolling stock, and gives practical tips for construction and repairs.

Current Consumer & Lifestudies
General Learning Corp., 60 Revere Drive, Northbrook, IL 60062-1563
Telephone: 312-564-4070

Annual subscription: $14.95, US$16.95 in Canada
Sample issue: free

"The Practical Guide to Real Life Issues," reads the subtitle to this magazine, and the articles in our sample issue matched the claim. The articles dealt with managing money, obtaining and handling credit, finding a service-related job, dealing with child abuse, and choosing athletic shoes. Also included were an interview with a woman who'd suffered and recovered from anorexia nervosa, a history of the Republican and Democratic parties, and an introduction to stir-fry cooking. Subscribers receive nine monthly issues during the school year.

Dance Magazine
33 W. 60th Street, New York, NY 10023
Telephone: 800-227-7585

Annual subscription: $24.95, $36.95 outside U.S.
Sample issue: free

A monthly magazine for adults that may be of interest to teenagers serious about dance. The magazine reviews performances, interviews dancers, and publishes a national schedule of upcoming events. A directory of schools, colleges, and teachers is a regular feature. The magazine also publishes a college guide for young people considering dance as a career.

Doll Reader
Hobby House Press, 900 Frederick Street, Cumberland, MD 21502
Telephone: 301-759-3770

Annual subscription: $24.95, $35.95 outside U.S.
Sample issue: $4.95

A fat magazine for doll collectors, loaded with illustrated articles on rarities and antiques. Not for little girls stuck on Barbie or My Little Pony, this is a publication for those who've turned doll play into a more serious endeavor.

Dragon Magazine
P.O. Box 111, Lake Geneva, WI 53147
Telephone: 414-248-3625

Annual subscription: $30, US$30 in Canada
Sample issue: free

A thick and beautiful magazine for fans of science fiction and fantasy role-playing games. Stories and game reviews fill the bulk of each issue; dramatic works of fantasy art make the pages fun to browse through, even without reading a word.

Fiberarts
50 College Street, Asheville, NC 28801
Telephone: 704-253-0467

Annual subscription: $18, US$20 in Canada
Sample issue: $4

A bimonthly magazine of textiles, with an emphasis on the work of innovative modern artists. Past issues have dealt with blueprinting on fabric, contemporary basketry, rock video costuming, and woven and quilted portraiture. With a break for the summer, subscribers receive five color issues.

FineScale Modeler
Kalmbach Publishing Co., 1027 N. 7th Street, Milwaukee, WI 53233
Telephone: 414-272-2060

Annual subscription: $17.95, US$20.95 in Canada
Sample issue: $3.75

For readers deeply into plastic models. A recent issue held a story on how plastic kits are made and step-by-step guides to adding detail to a ship model, transforming a plastic Firebird into a Batmobile, and building a miniature Lockheed F-104 Starfighter. Eight color issues are mailed each year.

First Strike
A hobby magazine for young coin collectors. See entry for American Numismatic Association under *Associations, Hobby*.

Flying Models
P.O. Box 700, Newton, NJ 07860
Telephone: 201-383-3355

Annual subscription: $21, US$25 in Canada
Sample issue: $2.50

A monthly magazine for fans of radio-controlled models. Though the name suggests planes, the pages delve into boats and cars as well, with "how-to" articles, reports on gatherings of modelers, new product reviews, and scores of ads for kits and equipment.

Foxfire
P.O. Box B, Rabun Gap, GA 30568

Annual subscription: $9, US$12 in Canada
Sample issue: free

An extraordinary magazine of folklore, crafts, and traditions of southern Appalachia, published by the students of Rabun County High School in Clayton, Georgia. Since 1966 the students have steadily put out four issues each year; they will reach number 90 in early 1989. We can remember spending long hours dreaming over the instructions for building a log cabin in an early issue. Recent editions have dealt with hand-carved wooden bowls, recollections of the Depression years, and the history of a mountain railroad line (showing how to hew ties with an axe). Always fascinating, *Foxfire* also contributes to scholarships and student programs in Clayton.

Karter News
International Kart Federation, 4650 Arrow Highway, B-4, Montclair, CA 91763
Telephone: 714-625-5497

Annual subscription: $18, US$20 in Canada
Sample issue: $2

A publication for racing kart enthusiasts. (For those not in the know, karts are low-to-the-ground four-wheeled vehicles powered by two- or four-cycle gasoline engines. Most of the racers are adults, but children compete in special races with restrictions on engine power.)

Merlyn's Pen
P.O. Box 1058, E. Greenwich, RI 02818
Telephone: 800-247-2027 or 401-885-5175

Annual subscription: $14.95
Sample issue: free

"The National Magazine of Student Writing," written by and for students in grades seven to ten. The quality of the writing is excellent and makes fun reading even for adults. While the magazine is published primarily for use in school writing classes, it can also be ordered at home. Four issues come out a year.

Miniatures Showcase
1027 N. 7th Street, Milwaukee, WI 53233
Telephone: 414-272-2060

Annual subscription: $12.95, $16.95 outside U.S.
Sample issue: $4.25

A quarterly for dollhouse and miniature enthusiasts, each issue devoted to a period in history or a popular theme. The articles show exceptional creations, interview artisans and collectors, and explain where the pieces were purchased or how they were made. Lots of color photographs show the heights to which dollhouse decorators can climb.

Model Aviation
1810 Samuel Morse Drive, Reston, VA 22090
Telephone: 703-435-0760

Annual subscription: $18, US$29 in Canada
Sample issue: $1.75

A monthly magazine for fans of radio-controlled and rubber-band-powered model airplanes. Hundreds of models are pictured in each issue, some with building instructions, some in stories on gatherings of fliers. With almost 200 pages to scan in each issue, hobbyists will find much to read and lots of mail-order shopping in the advertising section.

Model Railroader
1027 N. 7th Street, Milwaukee, WI 53233
Telephone: 414-272-2060

Annual subscription: $27.95, US$33.95 in Canada
Sample issue: $3.95

A hefty monthly magazine (more than 200 pages an issue) for model train buffs. Photographs of amazing layouts and real-life railroads form the heart of the magazine, with plans, "how-to" articles, and product reviews given plenty of attention as well.

New Expression
Youth Communication/Chicago Center, 207 S. Wabash Avenue, Chicago, IL 60604
Telephone: 312-663-0543

Annual subscription: $10, US$15 in Canada
Sample issue: $2

A newspaper written by high-school students for other teenagers. The paper confronts tough issues like youth violence and uninsured teenage drivers, and gives youthful insight in record reviews. The issue we saw contained lots of useful information on choosing and applying to colleges, including a schedule of tests and application deadlines. Subscribers get eight issues a year.

Nutshell News
1027 N. 7th Street, Milwaukee, WI 53233
Telephone: 414-273-6332

Annual subscription: $29, US$36 in Canada
Sample issue: $4.25

A smaller-format monthly magazine from the publisher of *Miniatures Showcase* (see above). *Nutshell News* covers the world of dollhouses and miniatures with reports on trade shows, picture stories on collections and museums, profiles of craftspeople, and techniques and plans for the home builder.

The Pen and Quill
A bimonthly journal for autograph collectors. See entry for Universal Autograph Collectors Club under *Associations, Hobby*.

Plays
Royalty-free plays. See entry in *Children's Magazines, For School-Age Children*.

Railroad Model Craftsman
P.O. Box 700, Newton, NJ 07860
Telephone: 201-383-3355

Annual subscription: $23, $27 outside U.S.
Sample issue: $2.50

A monthly for model railroaders, packed with pictures of realistic layouts and full-size trains, articles on modeling techniques, plans for miniature rolling stock and buildings, and advertisements for everything a modeler could want. Each issue spans some 150 pages, roughly a third of which is illustrated in color.

Rubberstampmadness
For rubber stampers. See entry under *Children's Magazines, For School-Age Children*.

S Gaugian
Heimburger House Publishing Co., 310 Lathrop Avenue, River Forest, IL 60305
Telephone: 312-366-1973

Annual subscription: $22, $27 outside U.S.
Sample issue: $4.75

Heimburger House publishes books on model railroading, as well as two specialized magazines: *S Gaugian* for hobbyists who work in S scale, and *Sn3 Modeler* for collectors of S-scale narrow-gauge trains.

TG Magazine, Voices of Today's Generation
202 Cleveland Street, Toronto, ON M4S 2W6, Canada
Telephone: 416-487-3204

Annual subscription: Can$14 in U.S., Can$10 in Canada
Sample issue: $2

Career information, fashion, music reviews, opinions, articles on social and health issues, fiction, photography, and poetry, almost half of the contents produced by *TG*'s readers (aged between 12 and 18). Self-esteem and decision-making tools and tips are included in every issue. The magazine comes out six times a year.

Teddy Bear & Friends
Hobby House Press, Inc., 900 Frederick Street, Cumberland, MD 21502
Telephone: 301-759-3770

Annual subscription: $14.95, US$18.95 in Canada
Sample issue: $3.50

A bimonthly magazine for teddy bear collectors. Photographic portfolios share space with price guides to older bears, new bear reviews, clothing patterns, and articles about collectors and makers. Each colorful issue runs to more than 100 pages.

Threads
63 S. Main Street, P.O. Box 355, Newtown, CT 06470
Telephone: 203-426-8171

Annual subscription: $20, US$24 in Canada
Sample issue: $3.95

A stellar magazine for people who sew, knit, weave, quilt, braid rugs, or work in any other way with textiles and threads. The oversize format, excellent color photography, good writing, and interesting choice of articles make this one of the best hobby publications around. Subject matter hops around considerably, given the broad range of the magazine. The January 1989 issue held patterns for making soft-sculpture elephants, tips on "free-motion" embroidery with a sewing machine, instructions for knitting wheel-pattern Fair Isle tammies, streamlined sewing methods for making clothes, rug-braiding techniques, improved cast-ons and bind-offs for machine-knit sweaters, a guide to fabric dyes, an article on garment underlinings, and a report on fur repair.

Trains
Kalmbach Publishing Co., 1027 N. 7th Street, Milwaukee, WI 53233-1471
Telephone: 414-272-2060

Annual subscription: $27.95, US$33.95 in Canada
Sample issue: $3.75

A monthly magazine for railroad enthusiasts, with dramatic picture stories on historic rail lines, engines, and cars, and news of preservation efforts and impending changes. Articles on operating passenger lines, with maps and schedules, make the publication useful as a travel guide as well.

Treasure
Jess Publishing Company, Inc., 6745 Adobe Road, Twentynine Palms, CA 92277
Telephone: 800-321-3333 or 619-367-3531

Annual subscription: $23.40 in U.S. and Canada, $31.40 in other countries
Sample issue: $1.95

A monthly magazine for treasure hunters and for those who dream of finding hidden fortunes. Each issue publishes several articles with information on lost caches that have yet to be uncovered—stolen loot, sunken ships, and Spanish treasure—and a few reports of actual

finds. Two other magazines from the same publisher, *Treasure Search* and *Treasure Found!*, offer even more reading on the subject.

Tropical Fish Hobbyist
One TFH Plaza, Neptune City, NJ 07753
Telephone: 201-988-8400

Annual subscription: $25, $36 outside U.S.
Sample issue: $3

A fat (150-page) monthly magazine with lots of beautiful color photographs of aquarium fish. Regular sections offer articles on goldfish, live-bearers, characoids, saltwater fish, garden pools, and new foods and feeding methods. Book reviews, ads, and an extensive question-and-answer column give readers expert advice and access to a wealth of equipment and accessories.

Venture
Christian Service Brigade, P.O. Box 150, Wheaton, IL 60189
Telephone: 312-665-0630

Annual subscription: $8, $10 outside U.S.
Sample issue: $1.50 plus a 9 x 11 self-addressed envelope with 85¢ postage affixed

A bimonthly magazine with a Christian message, aimed at 10- to 15-year-old boys. Profiles of celebrities and sports figures who are graduates of the Christian Service Brigade are a regular feature, as are articles about wildlife, fictional stories, cartoons, and jokes.

Computers and Software

Chaselle, Inc.
Educational software. See entry under *Educational Supplies*.

Cheatsheet Products, Inc.
P.O. Box 8369, Pittsburgh, PA 15218
Telephone: 412-731-2460
Catalog price: free

Software and accessories for Commodore 64 and 128, Amiga, Apple, IBM, and compatible computers. Games, both educational and not, are offered in abundance, as are drawing and painting, desktop publishing, and financial planning programs.

Educational Activities, Inc.
Educational software, mostly for IBM and Apple computers, but also for PET, Commodore 64, Atari, and TRS-80 systems. See entry under *Educational Supplies*.

Electronic Arts Direct Sales
P.O. Box 7530, San Mateo, CA 94403
Telephone: 415-571-7171
Catalog price: free

Dozens of computer games, from Scrabble to The Bard's Tale, and a handful of early education programs. Software is offered for owners of IBM PC, Tandy, Macintosh, Amiga, Atari, Apple II, and Commodore computers.

The Family Software Catalog
Evanston Educators, 915 Elmwood Avenue, Evanston, IL 60202
Telephone: 312-475-2556 (800-972-5855 in IL)
Catalog price: $2

Games, educational software, graphics packages, and writing aids like Webster's On-Line Thesaurus. Preschoolers can learn their way around the computer with programs like Kid's Stuff, Mixed-Up Mother Goose, and Facemaker: Golden Edition. Older children can play Scrabble against Monty, try action games like Gauntlet, or adventure games like Ultima V: Warriors of Destiny. All software is sold at a discount.

Golem Computers
P.O. Box 6698, Westlake Village, CA 91360
Telephone: 805-499-7785
Free price list

Software and accessories for Apple computers. Shoppers will have to know their way around to order from this list. There are no descriptions or explanations, but there's lots to choose from. The hardware includes printers, memory, monitors, modems, mice, and game controls. The software spans more than 50 games, as many educational programs, and two dozen drawing, clip art, and home newsletter disks.

Good Apple
Educational software. See entry under *Educational Supplies*.

The Learning Company
6493 Kaiser Drive, Fremont, CA 94555
Telephone: 800-852-2255 or 415-792-2101
Free brochure

Ten educational software programs that help children learn reading, writing, math, and problem-solving skills. The math and reading programs are geared to younger children, ages four to seven. A writing and publishing program is aimed at children nine and up. All are available for Apple II and IBM-compatible machines; some can be run on Commodore 64 or 128 computers.

Learning Gifts
P.O. Box 56-1006, Miami, FL 33256-1006
Telephone: 800-874-7588 or 305-252-0455
Catalog price: free

Computer software for education and entertainment in a big 200-page catalog. Educational programs are offered in the standard fields—reading, language arts, math, science, social studies, and foreign languages—along with dozens of programs for younger keyboarders. The entertainment section spans several hundred computer games.

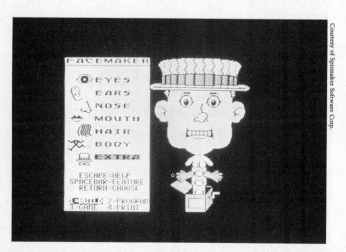

Spinnaker Software's Facemaker program, available from The Family Software Catalog

Manna Computing Concepts
P.O. Box 527, Woodstock, GA 30188
Telephone: 404-479-7178

Catalog price: $1

"Software for Christian computing." The catalog is stocked with educational programs and computer games relating to the Bible, Christian computer graphics, and Bible texts on disk for quick word and phrase searches. Several dozen secular software disks are added in smaller type at the back of the catalog, including some of the best educational and practical applications programs.

Micrograms
1404 N. Main Street, Rockford, IL 61103
Telephone: 800-338-4726 or 815-965-2464

Catalog price: free

Educational software for Apple computers. Shapes and Patterns is geared to young children who don't yet read. A dozen other programs teach basic reading skills and more advanced reading comprehension, spelling, grammar, science, and math.

Opportunities for Learning, Inc.
20417 Nordoff Street, Chatsworth, CA 91311
Telephone: 818-341-2535

Catalog price: free (request "Microcomputer Software" catalog)

A specialized catalog from a general school-supply company (see separate entry under *Educational Supplies*). The 90-page software booklet offers hundreds of programs that aid in the learning of reading, writing, grammar, math, and science. Graphics programs help children create intricate computer drawings and design newsletters, calendars, banners, and other printed works. SAT- and ACT-preparation programs may be a help to college-bound high-school students.

The Public (Software) Library
P.O. Box 35705, Houston, TX 77235-5705
Telephone: 800-242-4PSL or 713-665-7017

Catalog price: free

Public-domain software for use on IBM PC and compatible computers. The authors of these programs allow them to be copied and used without restriction or payment, so the software can be purchased for about the cost of blank disks. Games, graphics, and educational programs are available, along with home-business software, genealogy programs, and much more. The $2 fee buys a sample of the monthly newsletter and list, which generally holds more than 1,000 offerings. A year's subscription costs $18.

Random House Media
Dept. 460, 400 Hahn Road, Westminster, MD 21157
Telephone: 800-638-6460, ext. 5000, or 301-848-1900

Catalog price: free

Early learning software for Apple II series computers.

Some programs feature the characters from the Peanuts cartoon strip. Subjects include the alphabet, shape recognition, telling time, opposites, reading, counting, and typing.

Software-of-the-Month Club
The Learning Advantage, 101 Poor Farm Road, Princeton, NJ 08540
Telephone: 609-921-6100

Free brochure

Software for Apple and IBM computers at good prices. A few preschool programs teach the alphabet, numbers, and colors. Elementary software delves into typing, reading comprehension, writing, and diverting entertainments. Twenty junior-high programs mix games and adventures with physics, writing, and other educational pursuits. The minimum order is two programs, with no obligation for future purchases.

Strawberry Kite Collection
15466 Los Gatos Boulevard, Suite 109-306, Los Gatos, CA 95032
Telephone: 408-867-1329

Free brochure

A storybook program for IBM PC computers that combines animation, music and sound effects, and text in a retelling of *Jack and the Beanstalk*. The pictures can be printed in outline form for coloring. Special learning modes teach counting and reading skills.

Telegames USA
P.O. Box 901, Lancaster, TX 75146
Telephone: 214-227-7694

Catalog price: $1

A vendor of many of the biggest names in video games. More than 100 games are presented in the catalog, for ColecoVision, Nintendo, Sega, Apple, Atari, Commodore, and IBM PC systems. The company sells Coleco, Nintendo, and Sega hardware, and joy sticks and controllers for other machines. A page of radar detectors moves beyond the realm of games.

Weekly Reader Software
Optimum Resource, Inc., 10 Station Place, Norfolk, CT 06058
Telephone: 800-327-1473 or 203-542-5553

Catalog price: free

Educational software programs that teach everything from numbers and letters to computer programming in Basic. Most programs feature the Stickybear character, attractive color graphics, and an open-ended approach that lets parents enter their own stories, questions, and exercises to meet the specific needs of the learner. A picture library, a couple of games, a car-design program, and a program called Codes and Cyphers add some fun to the mix.

Cooking

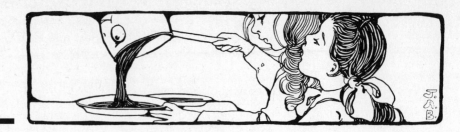

Chaselle, Inc.
9645 Gerwig Lane, Columbia, MD 21046
Telephone: 301-381-9611

Catalog price: free (specify "Pre-School & Elementary School Materials" or "General School & Office Products" catalog)

The "Pre-School and Elementary School Materials" catalog carries two classroom cooking sets that include pans, mixing bowls, beaters, measuring cups and spoons, spatulas, cookie cutters, colanders, and rolling pins. A child-powered ice-cream maker might also be a kitchen hit. See main entry under *Educational Supplies* for a description of Chaselle's other offerings.

The Chef's Catalog
3215 Commercial Avenue, Northbrook, IL 60062-1920
Telephone: 800-338-3232

Catalog price: $3

A catalog for grown-up cooks with some cookie molds that might make for fun projects with children. One mold makes a herd of dinosaurs, another a family of teddy bears, and another all the pieces for a gingerbread house.

Clothcrafters, Inc.
P.O. Box 176, Elkhart Lake, WI 53020
Telephone: 414-876-2112

Catalog price: free

Sells a child's apron made of striped denim with a wide front pocket. The firm also sells tote bags, laundry bags, flannel sheets (crib, twin, full, queen, and king), gardening and kitchen aprons for grown-ups, and a chef's hat. The prices are remarkably fair throughout.

Cumberland General Store
Cookie cutters. See entry under *Toys*.

Just for Kids!
Cookie cutters, cake molds, and ice-cream makers. See entry under *Toys*.

The Left-Handed Complement
Left-handed kitchen tools. See entry under *Art Supplies*.

The Little Fox Factory
931 Marion Road, Bucyrus, OH 44820
Telephone: 419-562-5420

Catalog price: 25¢

Handcrafted cookie cutters in lots of novel shapes, from trains to dinosaurs to the maps of several states. The list spans more than 300 designs, all priced under a dollar.

The Lyon & Gryphon
2000 Pereira Road, Martinez, CA 94553
Telephone: 415-372-0151

Catalog price: $4, refunded with order

Hundreds of cookie cutters in a catalog crammed with amazing little outline pictures. Why settle for dime-store angels and Santas when you can have all this? The Lyon & Gryphon sells cutters for making cookie insects, dinosaurs, castles, trains, whales, mythological beasts, and zoo animals. The company sells sets for carving

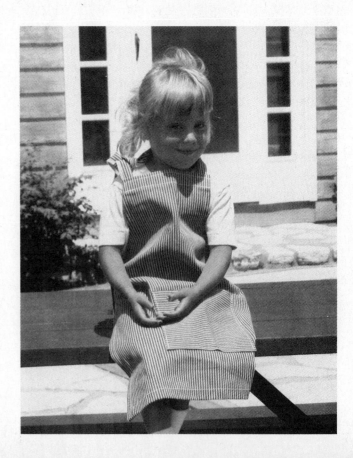

A denim apron from Clothcrafters, Inc.

cookie characters from Oz, *Winnie the Pooh*, *Alice in Wonderland*, *Star Wars*, and *The Wind in the Willows*. It sells cutters in the shapes of musical instruments, carpenter's tools, chess pieces, fairies, and of course the required bears and holiday standards. Order the catalog and have some fun.

Maid of Scandinavia
Cake-decorating and candy-making supplies. See entry under *Party Supplies*.

Maine Baby
A child's apron. See entry under *Baby Needs*.

Williams-Sonoma
Mail Order Department, P.O. Box 7456, San Francisco, CA 94120-7456
Telephone: 415-421-4242 or 4555

Catalog price: $1

A catalog for people who take cooking and entertaining seriously. Elegance and utility are the catchwords here, with such kitchen luxuries as Calphalon cookware, blue-and-white porcelain, fine balsamic vinegar, and an electric pizelle iron. Family fare has included zoo- and barnyard-animal cookie mold sets, an alphabet cookie cutter, a no-stick waffle iron, and an old-fashioned ash picnic basket.

Costumes

ABC School Supply
See entry under *Educational Supplies*.

Allstar Costume
125 Lincoln Boulevard, Dept. C-300, Middlesex, NJ 08846
Telephone: 201-805-0200

Catalog price: $1

Costumes and dress-up accessories for children and adults. Young masqueraders can disguise themselves as angels, brides, princesses, dinosaurs, skeletons, tomatoes, Ninja warriors, or California Raisins. A rabbit-in-a-hat costume is particularly cute: it's a huge top hat from the stomach down and a rabbit on the top. Masks, makeup, and such accessories as magic wands and fake mustaches may add just the touch you're after.

Childcraft
See entry under *Toys*.

Cracker Barrel Old Country Store
Cowboy and Indian costumes. See entry under *General Catalogs*.

Grey Owl Indian Craft Co.
American Indian costumes. See entry under *Hobby Supplies*.

Just for Kids!
See entry under *Toys*.

Kaplan School Supply Corp.
See entry under *Educational Supplies*.

Richman Cotton Company
Face paints, capes, tutus, princess hats, and other garb for make-believe. See entry under *Children's Clothes*.

F. A. O. Schwarz
See entry under *Toys*.

Smithsonian Institution
Astronaut and baseball costumes. See entry under *Gifts*.

Special Effects Merchandise
A *Star-Trek* costume. See entry under *Videos*.

Photo: George Leisey

Suzo's dragon suit

Suzo
P.O. Box 186, Pleasant Street, Grafton, VT 05146
Telephone: 802-843-2555

Catalog price: $2

Quality costumes for year-round dress-up play. The line includes bee suits, princess garb, bat capes, a long-tailed spiny dragon outfit, feather boas, pirate duds, a regimental soldier's uniform, and suits for turning children into cats, pigs, turtles, rabbits, and mice. Suzo's outfits are made of tough velour with cotton rib cuffs, not the thin polyester of the one-night Halloween costumes. Special materials are used on accessories—bee wings are sewn of foam-stuffed silver lamé, for example, and the regimental coat is made of pure wool with a cotton lining.

Taffy's-by-Mail
701 Beta Drive, Cleveland, OH 44143
Telephone: 216-461-3360

Catalog price: see below

Taffy's sells costumes for dancers, gymnasts, cheerleaders, and majorettes in several specialized catalogs. "The Basics," available for $3, offers shoes for ballet and tap, and basic dance and exercise wear for practice sessions and workouts. "On Parade," also $3, presents dozens of dressy parade outfits, including sequined leotards, fringed skirts, parade boots, pom-poms, batons, and hats. More amazing yet is the "Showstoppers" catalog, priced at $5. It holds costumes for dance performances, from princess dresses and Rockette simulations to rabbit suits and butterfly wings.

Telltales
Book-related costumes. See entry under *Books*.

Toys to Grow On
Costume sets. See entry under *Toys*.

Troll
Dinosaur costumes and other make-believe outfits. See entry under *Toys*.

Tryon Toymakers
Magic wands. See entry under *Toys*.

Dollhouses and Miniatures

Armor Products
Dollhouse kits. See entry under *Woodworking*.

Back to Basics Toys
A dollhouse open on all sides. See entry under *Toys*.

Boston & Winthrop
A hand-painted doll cradle. See entry under *Furniture*.

Cherry Tree Toys
Wooden dollhouse and scale-furniture kits. See entry under *Woodworking*.

Country Cupboard
Primitive country miniatures. See entry under *Dolls*.

Doll Domiciles
P.O. Box 91026, Atlanta, GA 30364
Telephone: 404-766-5572

Catalog price: $2

Dollhouse plans for the skilled home craftsperson. Some are reproductions of actual buildings—a 19th-century Louisiana plantation home, for example—while others are composites with lots of period detail. Architectural styles span four centuries, from a 17th-century English Tudor townhouse to a North Carolina mountain farmhouse built in 1920 and a contemporary home from the 1960s.

Favorites from the Past
P.O. Box 561, Stone Mountain, GA 30086-0561

Catalog price: $3; $12 for full catalog

The shorter edition of this catalog holds all that most children or parents will ever need. In 80 pages it lays out kits and plans for dozens of dollhouses, from colonials to contemporaries; construction tools for miniature carpenters; fixtures like doors, stained-glass windows, fireplaces, curving staircases, and working chandeliers; wallpaper in a choice of 50 patterns; rugs; parquet flooring; food; and furniture enough to outfit a miniature town. The full catalog runs to almost 400 pages and has much more of the same—more than 300 wallpaper patterns (some pictured in color), hunting trophies, garden ornaments, framed paintings, and 40 pages of lights and wiring. This is a veritable mail-order department store for the miniature enthusiast. If you can't find what you're looking for here, you'll probably end up making it yourself.

HearthSong
Dollhouses open for play on all sides and rustic furniture made from sticks. See entry under *Toys*.

Highway 70 Chair Shop
Rt. 2, Box 294, Woodbury, TN 37190
Telephone: 615-563-5155

Free brochure

Handcrafted doll chairs with woven seats, and doll beds and cradles. This furniture is too big for dollhouses (the beds are 18 inches long), but just right for larger dolls.

Hill's Dollhouse Workshop
9 Mayhew Drive, Fairfield, NJ 07006
Telephone: 201-226-3550

Free brochure

Dollhouses sold finished or in kit form. The choices include several Victorian homes, a three-story mansard-roof townhouse, an English Tudor house, and a plantation house with columns in front. These dollhouses are practical play structures: some of the realistic detail has been pared away to make them sturdy and affordable. All but a couple of the kits are priced under $100.

The Country Victorian dollhouse, made by Hill's Dollhouse Workshop

Hobby Surplus Sales
Dollhouse furniture kits. See entry under *Model Trains, Planes, Cars, and Boats*.

Jan's Small World
3146 Myrtle, Billings, MT 59102
Telephone: 406-652-2689

Catalog price: $3

Like a trip down a mouse hole, this color catalog leads us through room after room of miniature dollhouse furniture and accessories. Modern, old-fashioned, and country kitchens can be stocked with tiny Fiesta Ware dishes, seed-sized fruits and vegetables, and shrunken canned goods. Nurseries, bathrooms, bedrooms, and living rooms can be similarly outfitted in a wide array of styles and with a broad choice of realistic knickknacks. A doll-sized log cabin can be furnished with rustic benches and beds, and with accessories like a bearskin rug, a gun rack, and bales of straw. Dollhouse builders will find scaled-down siding, shingles, door frames, windows, gutters, staircases, porch posts, railings, wiring supplies, and much more.

Marvelous Toy Works
Dollhouses and doll cradles. See entry under *Toys*.

Mill Pond Farms
A simple dollhouse and several larger-scale pieces of doll furniture. See entry under *Toys*.

The Miniature Shop
1115 Fourth Avenue, Huntington, WV 25701
Telephone: 304-523-2418

Catalog price: $10.00

The Miniature Shop sends out the same encyclopedic catalog as Favorites from the Past (see entry above), and like that company is a source of anything and everything for the dollhouse. Here the huge catalog is a bit cheaper, but there's no option of sending for a reduced version. From wallpaper to wiring to spiral staircases, The Miniature Shop has the small things in life. The dollhouse inventory will satisfy the most ambitious doll developer.

Mountain Craft Shop
Handmade doll furniture and a doll log cabin. See entry under *Toys*.

Mueller-Wood Kraft, Inc.
A crib, a cradle, and a high chair for dolls. See entry under *Toys*.

My Sister's Shoppe
1671 Penfield Road, Rochester, NY 14625
Telephone: 716-381-4037

Catalog price: $2

A color catalog of miniatures that holds some exceptional little pieces. Landscape the grounds around your house with diminutive silk rosebushes and artificial grass. Set a picnic table with brand-name mayonnaise, ketchup, potato chips, and soft drinks. If it rains, move inside to ogle a fully dressed Victorian Christmas tree, a plump upholstered sofa, or a country kitchen with a brass-trimmed wood stove. Two dollhouse kits are sold: an affordable six-room Victorian and a larger turn-of-the-century home.

My Uncle
133 Main Street, Fryeburg, ME 04037
Telephone: 207-935-2109

Catalog price: free

Makes a New England saltbox dollhouse kit. The roof and walls are hinged both front and back for easy access to the nine rooms. The windows are made of Plexiglas and crowned with curtain rods. The exterior is finished with realistic-looking clapboard siding, tiny wood shingles, and detailed window and door trim.

Northeastern Scale Models Inc.
P.O. Box 727, Methuen, MA 01844
Telephone: 508-688-6019

Catalog price: $1

A mail-order lumberyard for modelers and scale builders. Basswood and mahogany strips are sold in stepped

A dollhouse from Rose's Dollhouse Store

sizes down to 1/80-inch thick and 1/32-inch wide. Shrunken moldings are offered for building dollhouse doors and windows, along with miniature gutters, tiny turned porch balusters, and scaled-down Victorian trim. Model railroaders will find wooden car parts in the standard scales (HO, S, O, and N). Ship modelers will find ship decking and lumber for cabins and hulls. If you want to inspect the quality of the products, send $2 for a packet of samples (the fee is refunded with your first order). Specify whether you're interested in samples for dollhouses, ships, architectural models, or railroads. The model train samples come in four scales, so let them know your preference.

Rose's Dollhouse Store
5826 Bluemound Road, Milwaukee, WI 53213
Telephone: 414-259-9965

Catalog price: $1

Miniatures, dollhouse kits, and dolls like Ginny, Kewpie, and Amanda Jane. The miniatures take up the biggest part of the catalog, ranging from chic decor for a city apartment to a 1950s den with a jukebox and wet bar. Several outdoor scenes are presented for landscaping around a dollhouse. Lawn furniture, wading pools, birdhouses, edging bricks, trash cans, even lawn flamingos, can be ordered to liven up the grounds. The dollhouse kits include Victorian, contemporary, and simple country models, all priced reasonably and designed for easy play access.

Scientific Models, Inc.
340 Snyder Avenue, Berkeley Heights, NJ 07922
Telephone: 201-464-7070

Catalog price: $1

Wooden ship models, dollhouse furniture kits, and hardware for dollhouse builders. Most of the 25 ship models are 19th-century sailing vessels. A Mississippi steamboat, a Spanish galleon, and a few other warships of the 16th, 17th, and 18th centuries round out the selection. Dollhouse carpenters will find such turned lumber as porch posts, balusters, newel posts, spindles, and finials, and a hardware collection that spans hinges, door knockers, doorknobs, weather vanes, towel rods, and miniature nails.

Trend-Lines, Inc., Woodworking Supplies
Victorian dollhouse kits. See entry under *Woodworking*.

U-Bild
Patterns and kits for building dollhouses. See entry under *Furniture*.

Dolls

ABC School Supply
See entry under *Educational Supplies*.

Artisans Cooperative
Sock monkeys and Raggedy Ann and Andy dolls. See entry under *Gifts*.

Back to Basics Toys
Raggedy Ann and Andy dolls. See entry under *Toys*.

Bellerophon Books
Paper-doll books. See entry under *Books*.

Cabin Creek Quilts
Country calico dolls. See entry under *Sheets, Blankets, and Sheepskins*.

Chaselle, Inc.
9645 Gerwig Lane, Columbia, MD 21046
Telephone: 800-CHASELLE (800-492-7840 in MD) or 301-381-9611

Catalog price: free (specify "Pre-School & Elementary School Materials" catalog)

Offers an assortment of dolls with ethnically accurate features and coloring. The set includes white, Native American, black, Hispanic, and Asian dolls, both boy and girl. The dolls can be ordered with or without anatomically correct features. Another set of life-size newborn dolls looks extremely realistic. These come in a choice of white, black, or Hispanic. See main entry under *Educational Supplies* for a description of Chaselle's other offerings.

Cotton Patch Crafts
Rt. 3, Box 790, Mansfield, LA 71052
Telephone: 318-697-5745

Catalog price: $2 (specify retail brochure)

Country-style rag dolls sold as kits or in finished form. Both black and white dolls are offered in several styles and sizes, along with rabbits, ducks, and cats.

Country Cupboard
143 E. Main Street, Zeeland, MI 49464
Telephone: 616-772-1523

Catalog price: $2

Country craft rag dolls, miniatures, and decorations presented in a 30-page color catalog. Two-inch mini dolls can play house with the primitive miniatures; larger dolls (in heights up to 14 inches) can rest on the combination shelves and peg racks.

Creatively Speaking
P.O. Box 27683, Salt Lake City, UT 84127-0683
Telephone: 801-531-0531

Catalog price: free

A color catalog of country decorations that takes a detour through the children's room to offer old-fashioned dolls and the furniture to make them comfortable, wood name and number puzzles, personalized children's T-shirts, and wooden trucks.

The Doll Cottage
427 Meeting Street, West Columbia, SC 29169
Telephone: 803-794-2119

Catalog price: $3

The $3 catalog fee buys a year of occasional newsletters that will keep you current on the latest in the world of fine dolls. Only a few pictures are printed with the descriptive listings, but in the spring edition a number of manufacturer's catalogs are offered for sale, and these illustrate everything in color. Among the lines represented are Robin Woods, Madame Alexander, Furga, Steiff (both dolls and bears), and Yolando Bello. Barbie and Ginny make the list, though only in unusual or discontinued editions.

The Doll Factory
1953 S. Military Trail, West Palm Beach, FL 33415
Telephone: 305-967-9772

Catalog price: $2

Infant dolls that are so realistic they're almost scary. A newborn doll (available in three skin colors) comes with a hospital wristband, its eyes open just a slit, and the wrinkles on its neck and feet just like the live article. Some other newborn and baby dolls are made without quite the attention to detail, but with prices down where you won't mind your children engaging them in rugged play. (The priciest vinyl dolls here cost about $60, the cheapest under $20.) The company sells several toddler dolls, a couple of anatomically correct baby dolls, an infant doll that giggles and coos when you rub a pacifier on her cheek, and another that coughs and cries when you put a stethoscope to her chest. At the top of the line are two infant dolls with carved wooden faces and hands.

Doll Repair Parts, Inc.
9918 Lorain Avenue, Cleveland, OH 44102-4694
Telephone: 216-961-3545

Catalog price: $1

Parts, tools, and supplies for the repair of ailing dolls, and patterns for making new ones. The company sells

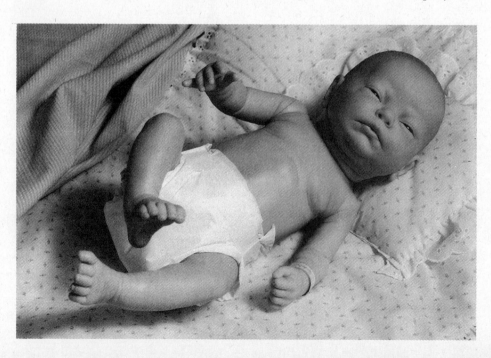

One of The Doll Factory's newborn dolls

wigs in several styles and sizes (choose human hair or Dynel), round and oval eyes, lashes by the strip, loose teeth by the dozen, shoes, buttons, paint, glue, and various fillers for chips and dents. Books on doll collecting and repair offer guidance and inspiration. Those who want to make new dolls can shop from the list of china parts, clothing patterns, and Byron doll molds. The firm buys and sells antique dolls, though these aren't listed in the repair catalog.

Dollsville Dolls & Bearsville Bears
461 N. Palm Canyon Drive, Palm Springs, CA 92262
Telephone: 619-325-2241

Catalog price: $5

A collectors' newsletter of dolls and bears that arrives in the mail several times a year with the latest temptations. With porcelain Barbie dolls and other limited edition bears and dolls, the offerings here will break the budgets of most children. Prices start at about $40 and rocket up over $1,000. Makers include Steiff, Karin Heller, Nisbet, and American Beauty.

The Donnelley Corp.
1213 Harrison Boulevard, Boise, ID 83702
Telephone: 208-342-1461

Free brochure

Makes several dolls called Baby Sweet Beats that pulse with "subtle, soothing, continuous heartbeats," a sound that the firm claims will help children relax and fall asleep. The faces and bodies are of soft cloth construction, similar to Cabbage Patch dolls.

Dover Publications, Inc.
Paper-doll books. See entry under *Books*.

The Enchanted Doll House
Route 7, Manchester Center, VT 05255-0697
Telephone: 802-362-1327

Catalog price: $2

A color catalog of dolls, accessories, and toys that's really meant for children, not just collectors. A number of the dolls bring children's stories to life—dolls like Paddington Bear, Curious George, Robin Hood, Snow White and the Seven Dwarfs, and a set created from Rumer Godden's book *Four Dolls*. Baby boomers will linger over a section with Troll dolls, Gumbi, Ginny, and a Howdy Doody hand puppet. Several Madame Alexander dolls are offered, as well as dolls by many other makers. Prices start at about $20 (even less for paper dolls) and climb up over $500. Teddy bears, Brio trains, Lauri construction sets, a miniature castle, a table soccer game, and other toys are offered for children whose hearts aren't set on dolls.

Hawk Meadow of New England
Porcelain dolls. See entry under *Gifts*.

Matreshka dolls from HearthSong

HearthSong
Matreshka dolls and dollhouse-scale dolls with wooden heads and limbs. See entry under *Toys*.

Hoover Brothers, Inc.
See entry under *Educational Supplies*.

Judi's Dolls
P.O. Box 607, Port Orchard, WA 98366
Telephone: 206-895-2779

Catalog price: long self-addressed stamped envelope with two first-class stamps

Patterns and kits for making dozens of different soft dolls. The smallest measures 9 inches tall, the largest almost 3 feet. All are clearly pictured in the color catalog.

Love-in-Idleness Dolls
P.O. Box 272, Hermosa Beach, CA 90254

Catalog price: free

In 1986 the Effanbee Doll Corporation closed its factory, ending 75 years of doll production. Love-in-Idleness is selling the remaining dolls from the warehouse. Most are made of vinyl, though one baby doll has a cotton body. Another baby doll is made with drink-and-wet action.

Milk & Honey
P.O. Box 1315, Bloomington, IN 47402

Free flier

A mother-and-baby breastfeeding doll set, designed by a practicing midwife. The mother's hands are padded with Velcro, allowing her to cuddle her baby and hold it to her breast. The set comes with a baby carrier and a blanket. Dolls can be ordered with white, brown, tan, or cream complexions.

Mountain Craft Shop

Traditional country dolls like sock monkeys, Raggedy Ann and Andy, and clothespin dolls. See entry under *Toys*.

The Mouse Hole Workshop
Mrs. Lois Polaski, 524 Kinderkamack Road, Westwood, NJ 07675
Telephone: 201-666-1263

Catalog price: 25¢

Felt mice made in England by Mrs. Polaski's daughter, Kathleen Maseychik. The felt bodies are adorned with scores of different outfits and accessories to represent fairy tale characters, famous British personalities, ballerinas, clowns, cowboys, doctors, and farmers. A mouse football team is decked out in uniforms with the team name "MICE" across the back.

Patterncrafts
P.O. Box 25370, Colorado Springs, CO 80936-5370
Telephone: 719-574-2007

Catalog price: $2

Patterns for sewing soft dolls, Christmas ornaments, and other household decorations. The color catalog shows some 200 possible projects.

Pieces of Olde
P.O. Box 65130, Baltimore, MD 21209
Telephone: 301-366-4949

Catalog price: $2

Soft dolls and calico bears, available finished or in kit form. Pieces of Olde draws its inspiration from historic designs, but the dolls are not copies of antiques. One topsy-turvy doll is actually two dolls in one: turn it over, flip the skirt down, and the second doll appears.

Pleasant Company
P.O. Box 497, Dept. 2008, Middletown, WI 53562
Telephone: 800-845-0005

Catalog price: $2

The American Girls Collection of dolls and books, created with the idea of bringing history alive for today's children. Three dolls form the foundation of the collection: Kirsten, an immigrant girl from Sweden who settles on a frontier farm in the 1850s; Samantha, raised by her wealthy grandmother in the first years of this century; and Molly, who grew up during World War II. Each of the dolls comes with a set of books that delve into her life and times, and a small warehouse of historically accurate clothes, toys, furniture, and accessories. The complete Samantha set ran about $800 in 1989,

Pleasant Company's Samantha doll, with a set of school accessories

so be forewarned that this can be a costly road to start down. The dolls really are exceptional, however, and the clothes and furnishings nicely detailed. All are well photographed in the oversize catalog, so buyers can get a good look before they take that first step.

A Real Doll

P.O. Box 1044, Sebastopol, CA 95472

$1 for brochure and samples

Natural-fiber doll-making supplies. Cotton interlock in peach or brown is offered for the main body, mohair and wool yarns for hair, washed and carded wool for stuffing, and tubular gauze to fill the heads. Kits are sold for those who want some guidance.

Rose's Dollhouse Store

Ginny, Kewpie, and Amanda Jane dolls. See entry under *Dollhouses and Miniatures*.

Sears, Roebuck & Co.

Barbie and Ken dolls, sold in the "Toys" catalog. See entry under *General Catalogs*.

Shoppe Full of Dolls

39 N. Main Street, New Hope, PA 18938

Telephone: 215-862-5524

Catalog price: $1

A color catalog of dolls plain and fancy. A few of the dolls can be had for less than $20, among them Kewpie and Katie dolls, nesting matrushka dolls, and some smaller figures. Most, however, fall in the $40 to $100 price range, with some going for considerably more. Brand names include Madame Alexander, Royal House of Dolls, Suzanne Gibson, Ginny, Effanbee, and Robin Woods.

Standard Doll Company

23-83 31st Street, Long Island City, NY 11105

Telephone: 718-721-7787

Catalog price: $3

A one-stop source for those who make their own dolls and sew their own doll clothes. Dozens of kits for making china and porcelain bisque dolls are offered, along with inexpensive unclad vinyl dolls, heads and limbs made of plastic and porcelain, wigs, hats, shoes, acrylic eyes, and patterns for making doll clothes.

The Sycamore Tree

Biblical action figures. See entry under *Educational Supplies*.

A Humpty Dumpty doll from The Toy Works, Inc.

The Toy Works, Inc.

Fiddler's Elbow Road, Middle Falls, NY 12848

Telephone: 518-692-9666

Catalog price: $2 (request retail catalog)

The Toy Works started life in 1973 by reproducing turn-of-the-century rag toys—dolls that were originally made of silkscreened muslin stuffed with sawdust. An old-fashioned tabby cat, a rabbit, a chicken, and other heirloom creatures are still on the roster, joined now by characters from Mother Goose, *Goldilocks and the Three Bears*, *The Wind in the Willows*, and *The Velveteen Rabbit*. The Cheshire Cat makes a curiously happy pillow, Humpty Dumpty a silly beanbag. All of the dolls can be ordered ready-made or as sew-them-yourself kits.

Unknown Products, Inc.

Gumby and Troll dolls. See entry under *Toys*.

Educational Supplies

ABC School Supply
6500 Peachtree Industrial Boulevard, P.O. Box 4750, Norcross, GA 30091
Telephone: 404-447-5000

Catalog price: free (request full catalog, "The Preschool Source," or "Early Learning Materials")

This company's full catalog is a 400-page encyclopedia of materials for early childhood education. From art supplies to wooden blocks, few gaps are left. Shoppers will find climbing structures both simple and massive, records, musical instruments, construction sets, games, wooden puzzles knobbed and plain, costumes, toy trucks, dolls, baby and toddler toys, cribs, nap mats, children's furniture, balls and other sports equipment, children's books, maps and globes, magnets and other science supplies, dinosaur toys, math toys and games, and wire-and-bead roller-coaster mazes. If you're not up to sorting through the 15,000 possibilities in the main catalog, a 64-page edition, "The Preschool Source," culls the most popular toys and furnishings and leaves out the duplicating masters. Or choose a middle road: the 200-page "Early Learning Materials" catalog should have more than enough for the average parent.

American Montessori Consulting
P.O. Box 5062, Rossmoor, CA 90721-5062
Telephone: 213-598-2321

Catalog price: long self-addressed stamped envelope

Montessori techniques for the home. The company sells a home learning kit that includes the book *Montessori at Home* and samples of teaching aids.

Audio-Forum
96 Broad Street, Guilford, CT 06437
Telephone: 800-243-1234 or 203-453-9794

Free brochure

Audio and video self-instruction tapes for learning foreign languages. Special programs for children use games, stories, and songs to draw listeners in and make the lessons fun. Foreign-language editions of Monopoly and Scrabble give children a chance to practice their language skills. Audio-Forum also offers foreign-language tapes for adults and audio self-instruction courses in typing, reading music, vocabulary building, and preparing for the SAT.

Bluestocking Press/Educational Spectrums
P.O. Box 1014, Dept. SP, Placerville, CA 95667-1014
Telephone: 916-621-1123

Catalog price: long self-addressed stamped envelope with two first-class stamps

Reference books for parents. Twenty titles offer advice on choosing books for children, finding an alternative school, supplementing classroom education, and working at home. *How to Stock a Home Library Inexpensively* looks like a sound investment at $4.50 postpaid.

Calvert School
Tuscany Road, Dept. MOS, Baltimore, MD 21210
Telephone: 301-243-6030

Catalog price: free

Home-study courses for children from kindergarten through eighth grade. The curriculum is approved for home schooling by the Department of Education of the State of Maryland. Parents in other states should check local laws. Any parent can use the material to supplement a classroom education. For children who can't attend school for stretches of time due to travel or illness, the program may provide a useful bridge to span the missed class work.

The Center for Early Learning
P.O. Box 250, Amherst, NH 03031
Telephone: 603-882-8688

Free brochure

A home education program for preschoolers, kindergartners, and first graders, designed to prepare children for and supplement their schooling. When parents sign up for the Special Time-Together series they receive a wipe-clean activity book, a set of "Helpcards," and a monthly idea booklet with suggestions for games and projects that make use of the materials.

Educational Supplies

Chad's Newsalog
50 Business Parkway, Richardson, TX 75081
Telephone: 800-262-CHAD or 214-680-9787

Free sample issue; $10 for 12 monthly issues

An ever-changing array of books and educational toys and games, sent with a four-page monthly newsletter. The catalog part of the monthly mailing takes the form of a few color pages from one or two manufacturers or publishers. Knobbed wooden puzzles from The Puzzle People made up one month's offering in 1988, wooden kits and working science models from Woodcrafter another. The items for sale tend to be educational, made primarily for use in schools, and not the kind of things found in ordinary toy stores. The goal at Chad's Newsalog is to give parents an opportunity to buy toys and games that are ordinarily sold to schools but are just as helpful and fun at home. The newsletter that accompanies these monthly mailings gives a brief explanation of the learning value of the month's offerings, presents a calendar with holidays and birthdays of famous people, and offers short articles on such subjects as reading readiness, the role of grandparents in child rearing, and how children learn. These articles are generally insubstantial when compared to those in other newsletters, and parents who subscribe should do so mainly for the merchandise.

Chaselle, Inc.
9645 Gerwig Lane, Columbia, MD 21046
Telephone: 800-CHASELLE (800-492-7840 in MD) or 301-381-9611

Catalog price: free (specify "Pre-School & Elementary School Materials," "Art & Craft Materials," "General School & Office Products," or "Educational Software" catalog)

Chaselle sells virtually anything a day-care center, preschool, elementary school, middle school, special education class, or art class could ask for, and offers any and all of it to parents as well. The only limitation is a $25 minimum order.

In the 200-page "Pre-School & Elementary School Materials" catalog, shoppers will find hardwood blocks, Lego, Duplo, and Waffle block sets, dolls, books, trikes, wagons, children's furniture, educational games, magnets, stethoscopes, wooden puzzles, child-size tools for carpentry and cooking, art and craft supplies, simple musical instruments, toy trucks and trains, costumes, and all sorts of balls. Much of this merchandise is not readily available through normal retail stores, and since the material is designed for classroom and group use by many children, much of it is built to standards higher than parents may be used to. This is truly a valuable shopping source.

The "Art & Craft Materials" catalog runs to almost 400 pages and is filled with supplies for projects as simple as coloring or as advanced as kiln-fired pottery. Markers, paints, brushes, papers, glues, and easels are here in abundance. Deeper in the catalog, readers will find 50 pages of pottery supplies, from clays and glazes to potter's wheels and kilns. Delve further and you'll discover sculpting and woodworking tools, screen-printing and etching supplies, and all the necessities for metal enameling, mosaics, stained glass, weaving, and leather craft.

Children's Reading Institute
Activity sets for teaching math and phonics at home. See entry under *Book Clubs*.

Community Playthings
Route 213, Rifton, NY 12471
Telephone: 914-658-3141

Catalog price: $1

Toys, furniture, and climbing sets for day-care centers and elementary schools. The merchandise is built to bear up under active use by lots of children, and most of it will be too expensive for a family budget. Wooden blocks are offered in family-size sets at fair prices, however, and hardwood chairs in a number of sizes can be had for under $25. A plastic potty costs less than $5, and a low high chair about $60. Those who can afford to invest in durable goods may want to look at the hardwood riding trucks, the workbench, the wagons and tricycles, the foam blocks, and the hardwood indoor climbing sets.

Constructive Playthings
1227 E. 119th Street, Grandview, MO 64030-1117
Telephone: 816-761-5900

Catalog price: free "Home Edition;" $3 for complete "School Edition"

A school supplier that culls some of its most popular items into a smaller catalog for families. The "Home Edition" features wooden puzzles, costumes, a child-size upholstered sofa-and-chair set, maple blocks, hollow cardboard construction blocks, Brio train sets, bath toys, sand toys, art supplies, percussion instruments, and science toys. The "School Edition" offers all that in even

Cardboard brick blocks from Constructive Playthings

greater abundance, plus playground equipment (mostly large and expensive sets), athletic balls, heavy-duty trikes, ethnic dolls, and educational games, books, and records.

Creative Publications
788 Palomar Avenue, Sunnyvale, CA 94086
Telephone: 408-720-1400 (800-624-0822 to order)

Catalog price: free (specify "Early Childhood" or "K–12" catalog)

Educational games, puzzles, software, and books in two color catalogs. The "Early Childhood" catalog contains such teaching tools as pattern blocks, Colorforms, hand puppets, play money, and plastic dinosaurs. The "K–12" catalog adds balance scales, flash cards, artificial pizzas for learning fractions, child-safe compasses, metric measuring rules and beakers, Cuisenaire rods, logic and strategy games, and polyhedral dice.

Cuisenaire Company of America, Inc.
12 Church Street, New Rochelle, NY 10802
Telephone: 800-237-3142 or 914-235-0900

Catalog price: free

Cuisenaire rods, pattern blocks, counting and sorting materials, dominoes, geometric models, play money, scales, measuring sticks and cups, and other games and tools for teaching math concepts. The big color catalog goes on to offer a wealth of science kits and instruments in its last 40 pages: hot-air balloon kits, seed-starting supplies, magnifying lenses, microscopes (including a couple under $10), test tubes, stethoscopes, electric switch and light kits, magnets, compasses, dinosaur models, and chemistry sets.

Didax Educational Resources
1 Centennial Drive, Peabody, MA 01960
Telephone: 800-458-0024 or 508-532-9060

Catalog price: free

Supplies and handbooks for teaching math, science, and reading to young children (grades K–6). Much of the material is designed for classroom use, but some kits are specifically aimed at the home. Unifix cubes are offered in a home set for teaching basic mathematical concepts. Pattern blocks, plastic counter pieces, fraction shapes, play money, balances, wooden puzzles, and geometric construction sets might make good home supplements to a child's schooling.

EBSCO Curriculum Materials
P.O. Box 486, Birmingham, AL 35201
Telephone: 800-633-8623 or 205-991-1208

Catalog price: free

Workbooks, software, and learning activity kits, most of which are sold in classroom sets too expensive to consider for the home. A brightly colored stacking toy for learning fractions looks like a possibility, however, as do a couple of Audubon software programs that follow the activities of a grizzly bear and a whale to teach environmental concepts.

Early Learning Centre
Educational toys for young children. See entry under *Toys*.

Educational Activities, Inc.
P.O. Box 87, Baldwin, NY 11510
Telephone: 800-645-3739 or 516-223-4666

Free brochure

Records, cassettes, videos, and software. The records and cassettes are for younger children, and most are collections of songs and musical games. Hap Palmer, Ella Jenkins, Joe Scruggs, and Rick Charette are among the performers. The videos and software programs teach computer skills, reading, writing, geometry, math, science, social studies, and health at levels from early childhood to high school.

Educational Materials Library
P.O. Box 415, Wilmington, CA 90748

Catalog price: $1

A teacher's catalog that can be mined by parents willing to flip through the many pages of workbooks and duplicating masters. Pattern blocks, wood-and-peg construction kits, dinosaur coloring books, maze books, color-your-own posters, and a domino set with bugs instead of dots are among the rewards for those who look carefully.

Cuisenaire rods from Environments, Inc.

Educational Teaching Aids
199 Carpenter Avenue, Wheeling, IL 60090

Telephone: 312-520-2500

Catalog price: $2

A thick school-supply catalog stocked with children's furniture, art supplies, ethnic dolls, toy cars, trikes, playground balls, infant toys, stacking toys, pegboards, construction sets (Lego, Duplo, Octons, and others), pattern blocks, math manipulatives, and aids for learning to tell time. The company sells a big selection of knobbed wooden puzzles featuring maps, alphabets, numbers, and pictures of animals and trucks. It also offers a good lineup of music for young children (records only), featuring recordings of Raffi, Ella Jenkins, Hap Palmer, and others. Simple percussion instruments, a ukelele, an Autoharp, and an African slit-log drum are offered for children who want to play along.

Environments, Inc.
P.O. Box 1348, Beaufort Industrial Park, Beaufort, SC 29901-1348

Telephone: 800-EI-CHILD

Catalog price: free

Educational supplies for preschool and early grades. Nicely organized and designed, this is one of the easiest of the big educational-supply catalogs for parents to work with. Its 200 pages offer an excellent selection of toys, supplies, and furniture for productive play and early learning. The sections on "Blocks & Accessories" and "Fine Motor Skills," for example, hold some of our favorite toys: maple blocks sold by the piece or in sets (good sets at good prices), Lincoln Logs, Bill Ding stacking figures, the Arcobaleno arch blocks, lots of Lego and Duplo building sets, big brick cardboard blocks, and Waffle blocks in several sizes. A browse through the rest of the catalog turns up an abundance of stacking toys, shape sorters, and simple puzzles; early math and science toys like number sorters, Cuisenaire rods, plastic dinosaurs, and giant magnets; toy cars, trucks, and farm sets; an ample collection of art supplies; sand and water toys; nap mats; tables and chairs; even a child-size workbench with a set of starter tools and a carpenter's toolbox. Everything is nicely photographed in black and white and clearly described. The wooden blocks, for example, are explained with a chart that shows exactly how many of each piece come with each set.

GCT Inc.
350 Weinacker Avenue, P.O. Box 6448, Mobile, AL 36660-0448

Telephone: 205-478-4700

Catalog price: $2

Books, activities, games, and software for "gifted, creative, and talented children and youth." Most parents would agree that their children qualify, and indeed there probably is something in this catalog that will interest most children. Riddle and puzzle books, games from Avalon Hill, and how-to books like *The New Joy of Photography* seem appealing. But *My Very First Preschool Workbook*, in a section called "Parents' Pages," raises a red flag. In the hands of parents who push normal children to be exceptional, many of the items in this catalog could do more harm than good.

Hardwood blocks from Environments, Inc.

Good Apple
P.O. Box 299, Carthage, IL 62321
Telephone: 800-435-7234 or 217-357-3981

Catalog price: free

Activity books that explore subjects like reading, writing, creative thought, math, science, and social studies. Several dozen software programs help with reading, math, grammar, and early learning.

J. L. Hammett Co.
30 Hammett Place, Braintree, MA 02184
Telephone: 800-672-1932 or 617-848-1000

Catalog price: free

A massive catalog of school supplies that may fill some needs at home. Art supplies are a specialty at Hammett; they take up a third of the 500-page catalog. From scissors and crayons to specialty papers, enameling kilns, and potter's wheels, the catalog has more than any child is likely to want. The company sells toys for infants and toddlers, wooden blocks and construction sets for preschoolers, and a small warehouse of puzzles, miniatures for imaginative play, easy-to-play musical instruments, indoor and outdoor climbing sets, tricycles, wagons, children's furniture, wall maps, science apparatus, and sports equipment.

High Q Products
Educational toys and games. See entry under *Toys*.

John Holt's Book & Music Store
See entry under *Books*.

Home Education Press
P.O. Box 1083, Tonasket, WA 98855
Telephone: 509-486-1351

Catalog price: free

Books for home schoolers and for parents who are considering alternatives to public schools. The *Home School Primer* delves into the legal aspects of home schooling, discusses the advantages and the potential problems of teaching at home, and lists support groups and sources of information and supplies. *Alternatives in Education* presents articles on various alternative schools, both public and private, and discusses the ideas of some important alternative educators. Home Education Press also publishes *Home Education Magazine* (see entry under *Parents' Magazines, Education and Child Care*).

Hoover Brothers, Inc.
2050 Postal Way, P.O. Box 660420, Dallas, TX 75266-0420
Telephone: 214-634-8474

Catalog price: free

An 800-page grand tour of the educational-supply world. Shoppers will find everything from flagpoles to auditorium seats. Much of the merchandise has no place in the home, but parents will find plenty that is of interest. Construction toys, children's furniture, musical instruments, science kits, wooden blocks, art supplies, climbing sets, sports equipment, books, records and tapes, puzzles, maps, globes, flash cards, math toys, scales, dolls, tricycles, and games are all offered in profusion. The company's primary sales region covers the states of Texas, New Mexico, Arizona, Oklahoma, Missouri, Kansas, Nebraska, and Colorado, though orders can be shipped to any address.

Kaplan School Supply Corp.
1310 Lewisville-Clemmons Road, Lewisville, NC 27023
Telephone: 800-334-2014 (800-642-0610 in NC) or 919-766-7374

Catalog price: free

A one-stop shopping center for day-care centers and preschools that could also be a resource for parents. For babies Kaplan sells cribs, mobiles, early toys, and safety aids like gates and outlet covers. For toddlers the company offers riding and push toys, sorting boxes, simple puzzles, nesting cups, and small climbing sets. Older children will find dolls, kitchen sets, costumes, furniture, playhouses, wooden blocks and other construction sets (including Lego, Duplo, Waffle blocks, and others), toy cars and trucks, water and sand toys, tricycles, sports equipment, larger swings and climbing sets, games, puzzles, math and science toys, and a library of several hundred books and records.

KolbeConcepts, Inc.
P.O. Box 15050, Phoenix, AZ 85060
Telephone: 602-840-9770

Catalog price: free

Games, tapes, puzzles, and books aimed at sharpening logical, critical, and creative thought. Collections of brain teasers and problem-solving activities fill much of the list. Science projects, math kits, political games, writing encouragements, and inventor's kits are among the other mind honers.

Learning At Home
P.O. Box 270-MO, Honaunau, HI 96726
Telephone: 808-328-9669

Catalog price: free

Books, teaching guides, workbooks, art supplies, and such educational supplies as colored wooden blocks, science kits, and math manipulatives. The book list includes songbooks, the *Rand McNally Classroom Atlas*, and *Festivals, Family and Food*, a book of family celebrations. The catalog is aimed at home schoolers, but other parents may want to take a look.

The Left-Handed Complement
Booklets and tools for teaching left-handed children. See entry under *Art Supplies*.

Leonard Bear Learning, Ltd.
319 S. 7th Street, St. Charles, IL 60174
Telephone: 312-377-9322

Free brochure

This company makes two teddy bears that are active participants in a learning program to teach children basic concepts and social skills. Leonard Bear's owner receives a letter from Leonard's mother each month asking the child to teach him about colors, numbers, shapes, rhyming, opposites, and other basic concepts. In teaching the bear, the child reinforces her own learning. Maynard Bear's owner is mailed a similar set of monthly letters, asking help in teaching the bear manners and basic social skills.

Math Games with Manipulatives
Math games. See entry under *Games and Puzzles*.

Nasco
901 Janesville Avenue, Ft. Atkinson, WI 53538
Telephone: 414-563-2446

Catalog price: free (specify "Learning Fun" or "Math" catalog)

Nasco, one of the country's largest educational supply houses, offers several thick catalogs. (See additional entries under *Art Supplies* and *Science and Nature*.) "Learning Fun" is stocked with preschool and elementary classroom supplies, from furniture to software to wooden blocks. Among the building sets are foam shapes, Lincoln Logs, Wee Waffles, Tinkertoys, Octons, Bill Ding stacking clowns, Duplo, Lego, and cardboard bricks. Baby dolls, play kitchens, playhouses, and miniature animals are sold for imaginative play. For infants and toddlers the catalog sells mobiles, stacking toys, push and pull toys, potties, car seats, baby gates, and other supplies. Balls, wagons, tricycles, water tables, and metal climbing sets are offered for active play. Games, records, musical instruments, art supplies, science equipment, and math manipulatives are offered in similar profusion. More math teaching aids can be found in the "Math" catalog, more art supplies in the "Arts & Crafts" catalog, and more science equipment in the "Science" catalog.

Oak Meadow
P.O. Box 712, Blacksburg, VA 24060
Telephone: 703-552-3263

Free brochure

Home study courses for grades K through 12, and a course for mom and dad called "Parent Sensitivity Training." Home schoolers can enroll their children in the Oak Meadow correspondence program (after getting the approval of local school officials); children who attend school can supplement their learning with the Oak Meadow curriculum. The brochure gives brief descriptions of the various courses. A more detailed description of the program can be had in *The Oak Meadow Curriculum Overview*, available for $2.95.

Opportunities for Learning, Inc.
20417 Nordoff Street, Chatsworth, CA 91311
Telephone: 818-341-2535

Catalog price: free (request "Materials for Early Education" catalog)

Opportunities for Learning splits its school-supply mailings into several different catalogs, so be sure to ask for the one you want. "Materials for Early Education" is stocked with 40 pages of puzzles, activity books, records, plastic sorting and building sets, and learning software. More software and science offerings can be found in the company's specialized catalogs (see separate entries under *Science and Nature* and *Computers and Software*). Three of the company's other catalogs might be of interest to parents: "Much Ado About Math," "The Right Selections for the Gifted (grades 4–12)," and "Materials for the Gifted (grades K–8)."

Palmer Method Handwriting
Macmillan Publishing Company, Front and Brown Streets, Riverside, NJ 08075-1197
Telephone: 800-323-9563 or 312-894-4300

Catalog price: free

Handwriting aids using the century-old Palmer Method. Workbooks, writing pads, and special pencils and pens are laid out in an orderly color catalog.

R & G Products
1952 Angling Road, North Fairfield, OH 44855

Free flier

The multiplication tables set to music. An audiocassette comes with an activity book to reinforce the listener's learning with both audio and visual images.

Research Concepts
1368 E. Airport Road, Muskegon, MI 49444

Free price list

Sells a checklist and a handbook to help parents determine if their child is ready for kindergarten or first grade.

The Sycamore Tree
2179 Meyer Place, Costa Mesa, CA 92627
Telephone: 714-650-4466

Catalog price: $3, refunded with purchase

A comfortable, homey catalog of "educational materials for Christian home and school teaching." A fish symbol in the margin draws browsers' attention to Bible-based materials, which include games, books, dolls, and music. The McGuffey Readers rate the fish sign, as do a series of Christian sex-education texts and several books on dinosaurs and fossils that fit the information into the biblical model of creation and flood. Dozens of science activity books, experiment kits, and optical devices allow for the exploration of science without reference to religion. A big selection of art supplies, coloring books, birdhouse kits, geography study aids, writing guides, and history books are similarly free of religious ties.

The Timberdoodle
E. 1610 Spencer Lake Road, Shelton, WA 98584
Telephone: 206-426-0672

Catalog price: free

Books, supplies, and learning kits for home educators. The math section takes in pattern blocks, balances, and Cuisenaire rods. Fischertechnik construction kits teach about motors and gears, electronics, pneumatics, robotics, and electromechanics. Parent-teachers can shop through the rest of the catalog for foreign-language tapes, maps, Lauri puzzles, coloring books, Cray-pas, magnet kits, bug viewers, learning games, flash cards, workbooks, and Bible felts.

Trend Enterprises
P.O. Box 64073, St. Paul, MN 55164
Telephone: 612-631-2850

Catalog price: free

Flash cards, wipe-clean workbooks, stickers, tracing stencils, and a massive collection of educational bulletin-board sets. Special Bingo and Lotto games teach colors, numbers, letters, and other early concepts.

Visual Education Association
581 W. Leffel Lane, P.O. Box 1666, Springfield, OH 45501
Telephone: 800-543-5947 or 513-864-2891

Catalog price: free

Study card sets for learning foreign-language vocabulary, chemistry nomenclature, history, legal terminology, and algebra.

Weekly Reader Books
Two monthly learning activity clubs: the Sweet Pickles Preschool Program and the Illustrated Wildlife Treasury. See entry under *Book Clubs*.

Zaner-Bloser
1459 King Avenue, P.O. Box 16764-6764, Columbus, OH 43216-6764
Telephone: 614-221-5851 or 486-0221

Catalog price: free

Workbooks, software, and guides for teaching spelling, handwriting, and reading skills. Zaner-Bloser gears programs to the learning strengths of the child—whether visual, auditory, or kinesthetic. A book explains the theory, and a kit can be used to identify a child's strongest learning mode. Tactile and kinesthetic learners, for example, may benefit from handwriting cards designed to be touched—oversize letters are recessed into their surfaces.

The Fischertechnik Robotic Arm from The Timberdoodle

Furniture

ABC School Supply
Nap mats and children's furniture. See entry under *Educational Supplies*.

All But Grown-Ups
P.O. Box 555, Berwick, ME 03901
Telephone: 800-448-1550

Catalog price: free

Maple furniture, blocks, and roller-coaster bead mazes from Kinderworks Corporation. Several table-and-chair sets are offered, one with Windsor-style chairs and others with a more contemporary look. The tabletops are laminated with a hard Formica-like surface to cope with the wear of active children. Bunk beds, beds with storage drawers, trundle beds, toy chests, and clothes chests complete the furniture section. An easel is made with wooden legs, a markerboard surface on one side and a chalkboard on the other. Kinderworks' bead maze is made with a solid wood base, not the dowel bottom found on competing toys.

Amish Country Collection
Rt. 5, Sunset Valley Road, P.O. Box 5085, New Castle, PA 16105
Telephone: 412-458-4811

Catalog price: $5

Rustic twig furniture in the style of the 19th-century Adirondack camps. Two children's rockers are offered: a tiny one for a two- or three-year-old and a larger version for a school-age child. A rocking cradle, made from woven oak and bent hickory, will set a baby to sleep in down-to-earth style. And a spectacular high chair seems almost too nice for the abuse it's sure to receive. A selection of adult rockers, porch swings, tables, chairs, benches, and beds completes the inventory. Prices are on the high side, but not outrageous. The smaller child's rocker sold for $75 in 1989 (plus $55 for shipping one to four pieces).

Robert Barrow, Furnituremaker
412 Thames Street, Bristol, RI 02809
Telephone: 401-253-4434

Catalog price: $3

Handmade Windsor chairs, including a few designed for children. Mr. Barrow offers a low children's chair in three styles, two high chairs, and a child's settee. The chairs are on the pricey side, from $150 to $350, but might be the answer for someone looking to complement a colonial dining room set.

The Bartley Collection, Ltd.
3 Airpark Drive, Easton, MD 21601
Telephone: 800-BARTLEY or 301-820-7722

Catalog price: free

Antique furniture reproductions sold finished or in pre-cut kits. Among the designs are a child's Windsor chair, a Windsor-style high chair (without a tray), a child's table-and-stool set, and a rocking horse with a hand-carved head and cushioned seat.

Biobottoms
A crib-size bed. See entry under *Children's Clothes*.

A child's rocker from The Amish Country Collection

A sampling of the Kinderworks furniture sold by All But Grown-Ups

Boston & Winthrop
2 E. 93rd Street, New York, NY 10128 or 35 Banks Terrace, Swampscott, MA 01907
Telephone: 212-410-6388 or 617-593-8248
Catalog price: $3

Hand-painted children's furniture that manages to be charming without being cute. Anita Boston Dana and Hope Brock Winthrop start with simple country furniture and brush on floral patterns, ducks, rabbits, sailboats, teddy bears, trains, balloons, or anything else a customer requests. Patterns and colors are all custom-designed to the purchaser's desires. The artists will paint to specific request or will suggest styles and colors to complement walls, curtains, or a child's quilt. Customers can choose from beds, bureaus, chairs, rockers, tables, stools, and such accessories as a rocking horse, a toy chest, a mirror, a peg rack, a shelf unit, and cradles sized for babies and dolls. Prices range from about $30 for the peg rack and $140 for the chairs and stools to more than $1,000 for a four-poster canopy bed.

Bright Future Futon Co.
Futons. See entry under *Sheets, Blankets, and Sheepskins*.

Cabin North
P.O. Box 56, Eldridge, IA 52748
Catalog price: $1

Country pine furniture and wall decorations. Shaker peg racks might find a place in a child's room. A rocking horse with mane and tail of acrylic yarn looks like a sturdy mount. A cradle could be the start of a family tradition.

Cane & Basket Supply Co.
A child's rocker kit. See entry under *Hobby Supplies*.

Chaselle, Inc.
9645 Gerwig Lane, Columbia, MD 21046
Telephone: 800-CHASELLE (800-492-7840 in MD) or 301-381-9611
Catalog price: free (specify "Pre-School & Elementary School Materials" or "General School & Office Products" catalog)

Sturdy wood, plastic, and steel furniture built to take life in the classroom. The "Pre-School & Elementary School Materials" catalog offers furniture (and other school supplies) for younger children: chairs, tables, desks, storage cabinets, cots, mini couches, cribs, and mirrors. The 700-page "General School & Office Products" catalog holds furniture and supplies for middle-school and high-school children. See main entry under *Educational Supplies* for a description of Chaselle's other offerings.

Childcraft
Child-size tables, chairs, and fold-out sofas. See entry under *Toys*.

The Children's Room
318 E. 45th Street, New York, NY 10017
Telephone: 212-687-3868
No catalog

Sensible furniture that grows with your child. Most of the pieces are modular in design and can be stretched or added to as needs change. Chair heights are adjustable; beds can be stacked into bunks; desks can be expanded to new widths or combined with bureaus, shelves, and storage compartments. A novel crib with three removable side bars allows a mobile child to crawl out the side rather than climb over the top. In addition to simple wooden chairs, the firm sells child-size desk chairs that roll and swivel. The beds, desks, bureaus, and chairs are available in teak, oak, birch, or pine construction, painted or with a natural finish.

Cohasset Colonials by Hagerty
Cohasset, MA 02025
Telephone: 617-383-0110

Catalog price: $3

Reproduction colonial furniture, including a child-size chair with a woven rush seat and a Windsor high chair. All furniture is available finished or at considerable savings in kit form. The child's rush-seat chair can be ordered with a name painted across the back.

Community Playthings
Hardwood chairs and tables, benches, high chairs, and other nursery- and elementary-school furniture. See entry under *Educational Supplies*.

Conran's Mail Order
475 Oberlin Avenue S., Lakewood, NJ 08701-1053
Telephone: 201-905-8800

Catalog price: $2

Smartly designed and reasonably priced furniture for children and adults. For the children's room Conran's sells an inexpensive wood and laminate table-and-chair set, a brightly painted bunk bed made of tubular steel, and a white laquered captain's bed with a storage drawer that can be used as a trundle bed. At the top of the line is an elegant ash bed with arched slatted head- and footboards, available as a bunk or twin bed. Conran's also offers shelving, rugs, linens, bureaus, wardrobes, and lamps that will serve equally well for small or full-size people.

Constructive Playthings
See entry under *Educational Supplies*.

Country Workshop
95 Rome Street, Newark, NJ 07105
Telephone: 800-526-8001 or 201-589-3407

Catalog price: $1

Ready-to-finish hardwood furniture. A bunk bed features three deep storage drawers below the bottom bed. A four-drawer chest can be topped with frame and pad for use as an elegant changing table. Country Workshop's crib is among the most sensible and attractive on the market. The platform lowers as the baby grows, with its lowest setting near the floor. The rails are thus at a comfortable height for a parent's reach, and don't require the squeaky raising and lowering that often wakes a drowsy baby. When the child outgrows the crib, it converts easily to a youth bed, and can again be converted into loveseat. We own one and swear by it. Tables, stools, shelf units, and desks are also available. Everything can be ordered finished or unfinished. (We finished our crib ourselves, and recommend that you let County Workshop do it.)

Country Workshop's crib

Crate & Barrel
P.O. Box 3057, Northbrook, IL 60065-3057
Telephone: 312-272-3112

Catalog price: $2

Crate & Barrel's distinctive stock changes with each season, so don't expect to find everything the company carries in your first catalog. In the summer, look for child-size lawn furniture and beach chairs. During the cooler months, the firm offers plastic tables and chairs, a puppet theater, easels, toy boxes, rockers, desks, lamps, and an excellent oak and sycamore table-and-chair set. One recent catalog offered an amazing table-and-stool set: the tabletop was shaped and painted to look like an artist's palette, the legs sculpted to look like giant paintbrushes, and the stool legs dipped in paint to look like stirring rods. See additional entry under *Toys*.

Cumberland General Store
A high chair that folds down to a baby rocker. See entry under *Toys*.

Educational Teaching Aids
See entry under *Educational Supplies*.

The Enchanted Child
Personalized children's stools. See entry under *Baby Needs*.

Environments, Inc.
Basic furniture for younger children. See entry under *Educational Supplies*.

A table-and-stool set from First Class

First Class
3305 Macomb Street, N.W., Washington, DC 20008
Telephone: 202-363-3449

Catalog price: free

This color catalog sells several durable table-and-chair sets, including a Windsor set made of maple, an oak butcher-block set, one with giant colored pencils for legs, and another of molded plastic. A white wicker rocker, an upholstered rocker, and an intricately carved Chippendale chair add to the seating roster.

Fun Furniture
8451 Beverly Boulevard, Los Angeles, CA 90048
Telephone: 213-655-2711

Catalog price: $1

Furniture that can transform a child's room into a fantasy world. The color catalog shows storage shelves made to look like castles, skyscrapers, or palm trees; a trundle bed with a dinosaur headboard; a desk with baseball bats for legs; a clothes rack shaped like a cactus; and a toy box in the guise of a taxicab. Put it all together and you get a bedroom that will be the envy of the neighborhood. The furniture is practical as well as fanciful. It's made of easy-to-clean plastic laminates, and the pieces are designed to use space efficiently.

Furniture Designs
1827 Elmdale Avenue, Glenview, IL 60025
Telephone: 312-657-7526

Catalog price: $2

Furniture plans for do-it-yourselfers. The catalog shows small pictures of the possibilities: a high chair, a crib, a toy chest, children's chairs and tables, a bunk bed, a trundle bed, and dozens of adult-size pieces. The plans run from $7 for smaller and simpler projects to almost $20 for an armoire. Styles range from colonial to modern. Plans are full-size, so no enlargement is necessary.

Goods from the Woods
Sweet Home Farm, Onchiota, NY 12968-0114
Telephone: 518-891-0156

Free information

A beautiful handcrafted child's rocker. The frame is built of maple or cherry, the back slats of ash, and the seat is woven rush. Companion straight-back chairs are made without the arms and the rockers.

GYM*N*I Playgrounds, Inc.
A child's picnic table. See under *Swings and Climbing Sets*.

H.U.D.D.L.E.
11159 Santa Monica Boulevard, Los Angeles, CA 90025
Telephone: 213-836-8001

Catalog price: $2

Innovative children's furniture that looks as fun as it is practical. Most of the pieces are of modular construction so that they can be expanded, stacked, combined, and even completely transformed as their occupants grow. A crib can be converted into a larger bed as baby turns into toddler, then rebuilt into a bunk bed. An appealing-looking set called Big Toobs is built of gigantic brightly colored plastic tubes: the child climbs in a window on the side of the stacking bunk bed and into her own tiny room; shelf units are made from upended tubes with the fronts cut out. Desks, chairs, bureaus, toy boxes, and shelving complete each of the design series. One exceptional bed deserves special mention: the Night Racer is a child's bed made to look just like a miniature racing car. It has wheels, headlights, an aerodynamic fiberglass body, and dozens of decals.

J. L. Hammett Co.
A good source of school-tough furniture—chairs, desks, tables, lockers, and storage cabinets. See entry under *Educational Supplies*.

Hancock Toy Shop
Chairs, toy chests, tables, and storage shelves made of maple. See entry under *Toys*.

Hangouts
1328 Pearl Street Mall, Boulder, CO 80302
Telephone: 800-HANGOUT or 303-442-2533

Free brochure

Colorfully woven hammocks based on a Mayan design. They make great nests for napping, reading, or swinging. And according to the company's advertisements, "millions of babies have been born in Mayan hammocks."

Kids Quarters' table with storage benches

Abbie Hasse Catalog
Personalized director's chairs. See entry under *Children's Clothes*.

Heir Affair
Tike Hike furniture for toddlers and a chintz-upholstered chaise lounge. See entry under *Gifts*.

Hold Everything
Mail Order Dept., P.O. Box 7807, San Francisco, CA 94120-7807
Telephone: 415-421-4242

Catalog price: free

A catalog of storage solutions that always offers a few items for the children's room. Recent issues have carried child-size tables and chairs, desk sets, toy boxes, and doll trunks. The company, which specializes in making storage space more efficient, offers a "closet doubler" which does just that: nylon straps hang extra bars from the main closet pole so that twice or three times as many short children's clothes can be stored in the same space. An added advantage of the system is that children can reach the lower clothes racks themselves.

Hoover Brothers, Inc.
See entry under *Educational Supplies*.

Just for Kids!
A foam chair in the shape of a dinosaur that folds out to make a nap mattress. See entry under *Toys*.

Kaplan School Supply Corp.
See entry under *Educational Supplies*.

Kids Quarters Inc.
229 Cedar Trail, Winston-Salem, NC 27104
Telephone: 919-765-8631

Free brochure

Kids Quarters makes an attractive and practical table-and-bench set, with matching bed and bureau. All the pieces are made of either maple or red oak, the tabletop finished with colored plastic laminate that wipes clean after crayon and paint projects. The benches open to reveal a storage compartment for paper, art supplies, or whatever else children need to keep handy.

Laura D's Folk Art Furniture
106 Gleneida Avenue, Carmel, NY 10512
Telephone: 914-228-1440

Catalog price: $2

Wacky hand-painted furniture that will tickle any child's funny bone. Laura Dabrowski does not confine her painting to restrained borders and corner ornaments. Her "Ogden Dog" chair is a personable, bow-tied bull terrier; his lap is the chair's seat. Just as appealing are the cat, rabbit, frog, elephant, and bird chairs, and the tables painted with flowers and alphabets. A rocking frog makes an inviting alternative to the traditional horse. A rocking stegosaurus takes the joke a step further. Ms. Dabrowski, who paints everything in brilliant and finely detailed patterns, will tailor colors and designs to her customers' needs. As might be expected for such exceptional work, prices are on the high side—about $400 for a chair, $600 for a rocking animal.

Chair, stool, and storage bench by Laura D's Folk Art Furniture

Maine Baby
A crib-size futon and a child's first bed made to fit a crib mattress. See entry under *Baby Needs*.

Mill Pond Farms
A table-and-chair set. See entry under *Toys*.

Nasco
Furniture for preschool and elementary-age children. See entry under *Educational Supplies*.

Out of the Woodwork
7509 Fiesta Way, Raleigh, NC 27615
Telephone: 919-847-7182

Free brochure

Wooden name puzzles taken an extra step. In addition to the usual board puzzles, Out of the Woodwork builds the puzzles into footstools and clothes racks. Names are limited to nine letters, with one exception: "Christopher" is acceptable by popular demand.

Pine Specialties
Rt. 2, Box 276, Randleman, NC 27317

Catalog price: $2

Pine furniture with a rustic country look. A child's chair has its back and sides cut from wide pine boards, and a heart cut out at the top. A loveseat and rocker are made in the same design. A stool, toy chest, peg board, and bureau complete the children's offerings. Prices are reasonable, the chairs and rockers less than $50.

Premarq
P.O. Box 840, Astoria, OR 97103
Telephone: 800-826-3562

Free brochure

Sells the Teddy Bed corner hammock for storing toys and dolls. Mounted on a wall or in the corner of a room, the hammock provides a tidy spot for toys that keeps them out from underfoot but still in sight. One Teddy Bed holds up to 20 average-size dolls.

The Right Start Catalog
High chairs, portable cribs, children's table-and-chair sets, and booster seats. See entry under *Baby Needs*.

Sears, Roebuck & Co.
A wide selection of plastic, metal, and wooden furniture, from traditional pieces to modular systems. Cribs, playpens, high chairs, and changing tables can be found along with furniture for older kids. See entry under *General Catalogs*.

Shaker Workshops
P.O. Box 1028, Concord, MA 01742
Telephone: 617-646-8985

Catalog price: $1

Reproductions of Shaker furniture sold finished or in kit form for home assembly. A rocker might appeal to a nursing mother. For children the catalog offers a high chair, a rocker, and two low chairs, one with arms and the other without. Stools, beds, peg racks, and hanging

A child's chair and rocker from Shaker Workshops

shelves would also look at home in a child's room. Prices range from about $50 for a chair kit to more than $500 for a finished twin bed.

Smith & Hawken
A child-size Adirondack chair. See entry under *Gardening*.

Squiggles and Dots
P.O. Box 870, Seminole, OK 74868
Telephone: 800-937-KIDS

Catalog price: free

Hand-painted children's furniture in perky animal and geometric forms. A holstein set features table, chairs, and rocker in cow shapes with bold black-and-white markings; the chairs have pink wooden udders hanging discreetly from the seats. A geometric set draws on 1980s design to combine circles and triangles in bold contrasting colors. A pair of cat chairs pull up to a tabby-striped table, with lamp, bookends, and a picture frame in a matching cat motif. Prices are surprisingly low for furniture that requires so much hand work.

Storage Concepts, Inc.
3199-B Airport Loop Drive, Costa Mesa, CA 92626
Telephone: 714-432-1968

Catalog price: free

Makes Toy Organizers storage shelves. The basic unit measures 6 feet tall, 7 feet wide, and 1 foot deep. A child's desk is built into the middle of each unit, and storage drawers are added to some models.

Sturbridge Yankee Workshop
Dept. NI-119, Blueberry Road, Westbrook, ME 04092
Telephone: 207-774-9045

Catalog price: $2

Reproductions of colonial and 19th-century home furnishings. A child's rocker with a heart carved out of the back, a toy chest, and a Windsor youth's chair (between a high chair and an adult chair) were among the listings in a recent catalog. Stock changes, so other items are likely to be offered in the future.

SweetGrass
445 Bishop Street, N.W., Suite A, Atlanta, GA 30318
Telephone: 404-875-3754

Catalog price: free

A catalog with its heart in the American country style. Wicker chairs, corner tables, bent-willow beds, and reproductions of antique folk art pieces are offered for grown-ups. For children SweetGrass sells a bent willow rocker, an upholstered wing chair, and a rustic pine table-and-chair set. The firm also sells an assortment of dolls, stuffed animals, and larger doll furniture.

Tabor Industries
8220 W. 30th Court, Hialeah, FL 33016
Telephone: 305-557-1481

Free brochure

Tabor Industries sells cribs, changing tables, dressers, and rocking chairs through stores that handle baby-room furniture. Customers who can't locate the pieces in stores can order them by mail directly from the company. A small child's rocker shaped like a smiling teddy bear looks like a good value at about $25, and it's small enough for UPS shipment.

Tapestry
Hanover, PA 17333-0046
Telephone: 717-633-3333

Catalog price: free

A gift catalog that concentrates on home furnishings. A few offerings for the children's room fit on a couple of pages: a personalized toy chest, a child's sofa that folds out into a bed, a wooden rocker, and linens featuring dinosaurs and Disney characters.

Think Big!
390 W. Broadway, New York, NY 10013
Telephone: 800-221-7019 or 212-925-7300

Catalog price: free

Outrageously oversized versions of everyday objects, some of which could be used to furnish or decorate a child's room. Browsers here will find 5-foot Crayola crayons (in six different colors), 6-foot pencils, a wardrobe built and painted to look like a giant's crayon box, a desk with a top made of a huge school notebook, a floor lamp in the shape of a flashlight, an outscale Kodak slide mount masquerading as a picture frame, a gigantic light switch that actually works, and a toy chest made from one of Paul Bunyan's blocks. Any of these shockers would make a nifty addition in a household that doesn't take itself too seriously.

Toy boxes by Think Big!

Timbers Woodworking
Patterns and plans for making children's furniture. See entry under *Woodworking*.

Treasured Toys
Sells a Kinderworks table-and-chair set. See entry under *Toys*.

Tryon Toymakers
A two-step stairway. See entry under *Toys*.

U-Bild
P.O. Box 2383, 15233 Stagg Street, Van Nuys, CA 91409-2383
Telephone: 818-785-6368

Catalog price: $3.95

U-Bild's "Patterns for Better Living" catalog offers hundreds of plans for the home woodworker. Each comes with a full-size traceable pattern, a list of needed materials, and step-by-step instructions with photographs. Cradles, cribs, and changing tables can get new parents started. When children are ready for action, turn a few more pages to find rocking chairs, pull toys, dollhouses (kits or plans), playhouses, children's boats, and riding toys. Some of these projects can yield results more spectacular than anything in stores. A riding dump truck moves by pedal power and has a working dump lever, a 5-foot sailboat really sails, and a backyard playhouse is built to look like a giant shoe.

The Woodworkers' Store
Plans, parts, tools, and lumber for making children's furniture. See entry under *Woodworking*.

Yield House
Route 16, North Conway, NH 03860
Telephone: 603-356-3141 (800-258-4720 to order)

Catalog price: $3

Reproduction 18th- and 19th-century furniture, with a few modern adaptations thrown in (such as an antique-look microwave cart). Parents may be interested in the braided rugs, the storage chests, or a child's rocker.

Games and Puzzles

ABC School Supply

Games and wooden puzzles for young children. See entry under *Educational Supplies*.

Ampersand Press

691 26th Street, Oakland, CA 94612
Telephone: 415-832-6669

Free brochure

Science games to help children learn about such subjects as food chains, marine biology, pollination, electrical circuits, geometry, astronomy, and computer programming.

Animal Town Game Company

P.O. Box 2002, Santa Barbara, CA 93120
Telephone: 805-682-7343

Catalog price: free

Children's games chosen for their emphasis on cooperation and learning, along with toys, books, and tapes. Among the games are Pictionary, Save the Whales, Funny Face, the Royal Game of Goose (played since 1725), and Ampersand Press's science games. Many of the games can be won only when children realize that they have to work together rather than against each other. Colored chalk for sidewalk games is sold, as well as marbles, tiddlywinks, and a book of string games.

Among the non-game offerings are a Radio Flyer wagon and wheelbarrow, a set of children's gardening tools, a child-size tepee, and an umbrella with a duck handle. Tapes include old-time radio classics, nature sounds, and children's music by Raffi and Tickle Toon Typhoon. Books range from classics like *Goodnight Moon* and *The Story of Ferdinand* to *Games of the North American Indians* and *Four Arguments for the Elimination of Television*.

AristoPlay

P.O. Box 7028, Ann Arbor, MI 48107
Telephone: 800-634-7738

Catalog price: free

Beautifully designed board and card games that add learning to play. Music Maestro teaches the sound, shapes, and functions of 48 instruments with the help of cards, a board, and an audio tape; By Jove is an adventure board game set in ancient Greece; Made for Trade introduces players to everyday life and historical events in colonial America; Main Street and Good Old Houses help players identify different architectural styles and features. The games have won praise from *Parents' Choice*, *Games Magazine*, and *Parents* magazine.

Avalon Hill Game Co.

4517 Harford Road, Baltimore, MD 21214
Telephone: 301-254-9200

Catalog price: $1

Board games, many with sports and military themes. Dinosaurs of the Lost World puts players in a prehistoric world to search for lost treasure; Aquire lets them take charge of competing hotel chains; and Diplomacy offers the shrewdest player a chance to rule the world.

Back to Basics Toys

Tabletop soccer, baseball, and shuffleboard games. See entry under *Toys*.

Music Maestro II by AristoPlay

Bits & Pieces
1 Puzzle Place, B8016, Stevens Point, WI 54481
Telephone: 800-JIG SAWS or 715-341-3521

Catalog price: free

Jigsaw puzzles take up the most space in this 32-page color catalog, but a browse through also turns up sculpture puzzles, ring puzzles, kaleidoscopes, books of puzzles and games, a desktop basketball game, and a radio in the form of a 1930s-style microphone. Some of the jigsaw puzzles are simple enough for children; others, with several thousand pieces, look like matches for any adult. Blank jigsaw puzzles can be used as stationery to make unusual letters, birthday messages, or Christmas cards.

Brookstone Company
787 Vose Farm Road, Peterborough, NH 03458
Telephone: 603-924-9511 or 9541

Catalog price: free

Brookstone's main catalog concentrates almost exclusively on tools and useful gadgets for the home and shop. The slimmer fall and summer gift catalogs offer croquet sets, cribbage boards, a tabletop shuffleboard game, and a miniature billiard table.

Chaselle, Inc.
A big school-supply catalog with a good selection of games and puzzles. See entry under *Educational Supplies*.

Childcraft
Hand-held electronic games, alphabet and map puzzles, and tabletop sports games. See entry under *Toys*.

Family Pastimes' Harvest Time

Claire's Bears & Collectibles
Ravensburger games and puzzles. See entry under *Toys*.

Constructive Playthings
Wooden puzzles and educational games. See entry under *Educational Supplies*.

Creative Publications
See entry under *Educational Supplies*.

Early Learning Centre
Simple puzzles for young children. See entry under *Toys*.

Educational Materials Library
Maze books and a domino set with bugs instead of dots. See entry under *Educational Supplies*.

Educational Teaching Aids
Knobbed wooden puzzles. See entry under *Educational Supplies*.

Emily's Toy Box
Lauri and Ravensburger puzzles for starting puzzlers, and games from Family Pastimes, Ravensburger, AristoPlay, and other makers (including a Chinese checkers set made of oak). See entry under *Toys*.

The Enchanted Child
Wooden name, number, and alphabet puzzles. See entry under *Baby Needs*.

Every Buddies Garden of Puzzles
P.O. Box 778, Corvallis, OR 97339

Catalog price: free

A small color catalog of attractive wooden puzzles. Some of the simplest come with peg grips on the pieces to make them easier for little fingers to handle. More complicated designs for older children are sold without the knobs. Among the pictures are puffins, pandas, an Indian village, a group of whales, a vegetable garden, and a stegosaurus. The firm makes alphabet and number puzzles and will custom-make name puzzles—your child's name alone on a board or added to a larger picture puzzle.

Family Pastimes
R.R. 4, Perth, ON K7H 3C6, Canada

Catalog price: $1

Cooperative games that encourage children to play together rather than against each other. In the board game Harvest Time, players plant gardens, then try to harvest the crops before winter comes. The game can be won only if players help one another. In Max, players move little creatures and a tomcat (Max) around the board, consulting with each other before each play to prevent the cat from making a meal of the other pieces. A dozen other games provide fun while encouraging cooperation. We can attest to the fact that children really do love these games. Most young children hate to lose, and many will cheat to win in normal competitive games. In Family Pastimes games there's no need for dishonesty, and everyone goes away happy.

Geode Educational Options
Cooperative board games. See entry under *Books*.

HearthSong
Tiddlywinks, pick-up-sticks, Chinese checkers, and board games for young children. See entry under *Toys*.

Heir Affair
Tiddlywinks, pick-up-sticks, and Chinese checkers. See entry under *Gifts*.

Herron's Books for Children
Lauri puzzles. See entry under *Books*.

High Q Products
Educational games and puzzles. See entry under *Toys*.

Hoover Brothers, Inc.
Puzzles, flash cards, and games. See entry under *Educational Supplies*.

Judy/Instructo
4325 Hiawatha Avenue S., Minneapolis, MN 55406
Telephone: 800-523-1713 or 612-721-5761
Catalog price: free

An educational catalog with hundreds of wooden inlay puzzles. Most are simple designs for young children, with 15 pieces or fewer; many have knobs for easy manipulation. The company distributes Disney School House products, so a good number of the puzzles feature Disney characters. Alphabet and map puzzles offer the most complex challenge. Curriculum units and flannel-board aids won't have wide application in the home.

Julia & Brandon
Wood, crepe-rubber, and cardboard puzzles. See entry under *Toys*.

Kaplan School Supply Corp.
See entry under *Educational Supplies*.

KolbeConcepts, Inc.
Brain teasers and learning games. See entry under *Educational Supplies*.

Lauri, Inc.
P.O. Box F, Phillips-Avon, ME 04966
Telephone: 207-639-2000
Catalog price: free

Lauri makes dense foam-rubber puzzles and play sets, which are sold in stores and through other mail-order catalogs (see Emily's Toy Box under *Toys*, Herron's Books for Children under *Books*, and The Timberdoodle

Lauri foam-rubber puzzles

under *Educational Supplies*). The company will send a catalog to customers who want to see the complete line, but fills orders by mail only if shoppers can't find what they want in local stores. Parents who own incomplete puzzles and play sets can also order replacement pieces by mail. Send a description of the lost piece with a long self-addressed envelope big enough to hold it, and 50¢.

Learn Me Bookstore
Noncompetitive games. See entry under *Books*.

The Left-Handed Complement
Left-handed playing cards. See entry under *Art Supplies*.

Lucretia's Pieces
Rt. 2, Box 848-MSP, Woodstock, VT 05091
Telephone: 802-457-3877
Catalog price: $1

Custom-made wooden jigsaw puzzles, geared to the ability and special interests of the customer. Lucretia will make a puzzle from a print you request, and will work special picture or letter pieces into the jigsaw design. These are special treats for serious puzzlers—the prices start at about $120.

Marvelous Toy Works
Wooden board games. See entry under *Toys*.

Math Games with Manipulatives
1025 Balboa Street, San Francisco, CA 94118
Telephone: 415-668-4121
Catalog price: long self-addressed stamped envelope

Math games that teach addition, subtraction, multiplication, and division concepts through play. The games come with dice, game boards, and colored rectangular tiles in sizes that correspond to the numbers 1, 5, 10, 50, and 100.

The Metropolitan Museum of Art
Card and board games. See entry under *Gifts*.

Mountain Craft Shop
Traditional Appalachian games and puzzles. See entry under *Toys*.

Museum of Fine Arts, Boston
Ancient board games and new games and puzzles having to do with the fine arts. See entry under *Gifts*.

The Natural Baby Co.
Family Pastimes' games sold at a discount, board games like Catch-a-Mouse and the Royal Game of Goose from Galt Toys, and wooden puzzles from Nashco Products. See entry under *Toys*.

North Star Toys
Wooden puzzles. See entry under *Toys*.

Out of the Woodwork
Wooden name puzzles. See entry under *Furniture*.

Pacific Puzzle Company
378 Guemes Island Road, Anacortes, WA 98221
Telephone: 206-293-7034

Catalog price: free

Wooden puzzles that combine innovation, flair, and quality materials. Number and alphabet puzzles are made of contrasting birch and mahogany plywood, as is a knobbed shape puzzle for younger children. A series of animal puzzles is made up of animal shapes that intertwine and fit against each other—in one a dinosaur's foot nests into another's mouth, while a lizard's tail matches up with the curve of a turtle's back. A line of map puzzles offers children a chance to learn more than the shapes of the 50 states. United States puzzles show major cities, rivers, lakes, and mountains. Puzzles of Africa, Europe, Asia, and North and South America expand horizons further. Three different world puzzles put things in even better perspective.

PlayFair Toys
Games and wooden puzzles. See entry under *Toys*.

The Puzzle People, Inc.
22719 Tree Farm Road, Colfax, CA 95713
Telephone: 916-637-4823

Free brochure

Wooden map, alphabet, and name puzzles, and some cute stand-up puzzles in animal shapes. Map puzzles of the United States, Europe, and several individual states are offered, with the pieces painted in bright colors or stained in contrasting wood tones.

The Puzzle People's alphabet and number puzzle

Sears, Roebuck & Co.
Board games, electronic games, table soccer, and pool tables can all be found in the Sears "Toys" catalog. See entry under *General Catalogs*.

Sensational Beginnings
Knobbed wooden puzzles. See entry under *Toys*.

Shibumi Trading, Ltd.
Japanese Go sets. See entry under *Gifts*.

Sporty's Preferred Living Catalog
Clermont Airport, Batavia, OH 45103-9747
Telephone: 513-732-2411 (800-543-8633 to order)

Catalog price: free

Hundreds of curious gadgets for home, yard, and automobile, from a propane weed torch to a 60-pound mailbox built to withstand the explosion of two M-80's. Equipment for games like darts, croquet, foosball, vol-

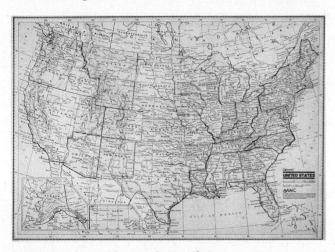

Map puzzle from the Pacific Puzzle Company

leyball, and Ping-Pong always finds space here. Pool alarms, fire extinguishers, space-age tricycles, and the latest barbecue devices are also regular repeaters.

Stave Puzzles
P.O. Box 329, Norwich, VT 05055
Telephone: 802-295-5200

Catalog price: free

Stunningly beautiful but diabolically difficult wooden jigsaw puzzles for those who are obsessed with this pastime. Prices start at about $200 for the simplest 100-piece puzzle and climb up over $3,000 for the largest and most difficult. Factor in therapy and time lost from work, and the actual cost can be much higher.

Sun Sparks
Colored pattern tiles. See entry under *Toys*.

The Sycamore Tree
Games, many based on biblical themes. See entry under *Educational Supplies*.

Tangoes/Rex Games
2001 California Street, Suite 204, San Francisco, CA 94109
Telephone: 415-931-8200

Catalog price: long self-addressed stamped envelope

A set of plastic pattern blocks—tiles in geometric shapes—packaged as a game. Players draw cards that show various designs and try to duplicate the patterns with the tiles. The game can be played alone, with two people, or with teams.

The Timberdoodle
Learning games and Lauri puzzles. See entry under *Educational Supplies*.

Treasured Toys
Simple games and alphabet puzzles. See entry under *Toys*.

Name puzzle from The Puzzle People

Trend Enterprises
Educational Bingo and Lotto games. See entry under *Educational Supplies*.

U.S. Games Systems, Inc.
179 Ludlow Street, Stamford, CT 06902
Telephone: 203-353-8400

Catalog price: $2

Tarot cards and bridge decks in scores of beautiful designs. Some are reproductions of antique cards, others picture works of art or historical figures. A black history deck features portraits of 52 important black Americans; an American history set includes presidents, poets, scientists, and artists. A few children's card games offer simpler entertainment.

World Wide Games
Colchester, CT 06415
Telephone: 203-537-2325 (800-243-9232 to order)

Catalog price: $1

Handcrafted wooden board games and puzzles, joined by a few cardboard bestsellers like Scrabble and Pictionary. Wooden Chinese checkers and chess sets take their place here beside less familiar games, both ancient and new. Adi, Go, and Fox & Geese have been around for centuries; Cathedral and Down Under are recent inventions. More active games like Skittles, Carrom, and table soccer test the coordination of players. Puzzles like Cliff Hanger, Brain Teaser, and Devil's Needle test their patience. Especially for children are oversize domino blocks, French hoops, and pick-up-sticks.

Gardening

Animal Town Game Company
Children's gardening tools. See entry under *Games and Puzzles*.

Brooklyn Botanic Garden
1000 Washington Avenue, Brooklyn, NY 11225-1099
Telephone: 718-622-4433

Free flier describing video

Sells a video-booklet set for child gardeners, *Get Ready, Get Set, Grow: A Kid's Guide to Good Gardening*. The 15-minute videocassette comes with two booklets, one for the young gardener and one for the guiding adult. The video is sold for home use only. Teachers planning to use it in class should ask for special ordering instructions.

W. Atlee Burpee & Co.
Warminster, PA 18974
Telephone: 800-333-5808 (800-362-1991 in PA)

Catalog price: free

The Burpee catalog is well known to gardeners, and almost any of its seeds can be grown by children as well as adults. To make things easier, the company sells a kit called the "Kinder-Garden," a plastic mat with holes marked for planting various vegetables and flowers, and seeds to plant in the holes. The layout helps with spacing and the mat keeps the weeds in check. Children's gardening tools, child-size gardening gloves and aprons, and a miniature wooden wheelbarrow are offered in a separate summer tool and gift catalog.

Henry Field Seed & Nursery Co.
Shenandoah, IA 51602
Telephone: 605-665-9391

Catalog price: free

One of the major garden seed and nursery vendors, Henry Field puts out a wonderfully garish color catalog peppered with such exotic offerings as midget corn (it grows 4-inch ears), yard-long cucumbers, and a flowering crab apple grafted to bloom in five different colors. For children, the catalog offers the "Mighty Mix" seed packet, a mixture of many flower and vegetable seeds in a single packet. It only costs a dime, but children must write the words "Kids' Mighty Mix" on the order form themselves.

Gardener's Supply
128 Intervale Road, Burlington, VT 05401
Telephone: 802-863-1700

Catalog price: free

An excellent tool and supply catalog for grown-up gardeners that offers a small-scale wheelbarrow and a set of children's tools with 30-inch handles.

Gurney Seed & Nursery Co.
Yankton, SD 57079

Catalog price: free

An oversize seed and nursery catalog with thousands of flowers, vegetables, and ornamental plantings, and loads of tools and supplies. The best deal of all is the "Giant Jumble Seed Packet," priced at a mere penny and available only to children. It must be the child's own penny,

A child's shovel and gloves from W. Atlee Burpee & Co.

however, taped to a parent's order. The packet holds a mix of seeds for such easy-to-grow plants as sunflowers, watermelons, peas, and squash.

Plow & Hearth
560 Main Street, Madison, VA 22727
Telephone: 800-527-5247 or 703-948-7010

Catalog price: free

Useful products for garden, yard, and fireplace. Children can join in the gardening fun with a set of four downscaled tools: a leaf rake, a shovel, a garden rake, and a hoe. These can be ordered by the piece or all at once at a small discount. For the cooler months, Plow & Hearth sells an extra-large fireplace screen that extends from the hearth to protect children from burns. It's big enough to fence off a wood stove or to keep toddlers far from the flying sparks and embers of an open fireplace.

Smith & Hawken, Catalog for Gardeners
25 Corte Madera, Mill Valley, CA 94941
Telephone: 415-383-6399

Catalog price: free

High-quality garden tools and outdoor furniture. For children, Smith & Hawken imports scaled-down Bulldog tools from England. The 32-inch spade and fork and the 37-inch rake are probably the sturdiest children's gardening tools on the market. All three feature hardwood handles and rugged solid-socket construction. The fork and spade are built with T-grip handles for ease of manipulation. None of the company's teak garden benches are offered in children's sizes, but a painted Adirondack chair can be had in matching adult and children's models.

T & T Seeds Ltd.
P.O. Box 1710, Winnipeg, MB R3C 3P6, Canada
Telephone: 204-943-8483

Catalog price: $1

One of the major Canadian merchants of flower and vegetable seed, T & T Seeds always tucks an offer somewhere in its catalog for the 10¢ "Kiddies Big Assortment Packet." The oversize packet holds a blend of flower and vegetable seeds. It is not available by itself, only as part of a larger order from mom and dad.

Tryon Toymakers
A hand-painted wooden wheelbarrow. See entry under *Toys*.

Gifts

Abbey Press
Hill Drive, St. Meinrad, IN 47577
Telephone: 812-357-8251

Catalog price: free

A Christian gift catalog that carries a few items for kids. For Christmas 1988, parents could order a magic kit, a crystal-growing set, several puzzles, and a set of Bible storybooks.

American Stationery Company, Inc.
100 Park Avenue, Peru, IN 46970
Telephone: 317-473-4438

Catalog price: free

Personalized stationery, including a set for children with colored balloon decorations. A stationery embosser might make a wonderful gift for a young correspondent—it can be ordered with plates for embossing a name and address, a signature, or a book label ("Library of . . .").

Artisans Cooperative
P.O. Box 2166, West Chester, PA 19380
Telephone: 215-431-3399 (800-521-0231 to order)

Catalog price: $2

Handcrafted toys, dolls, teddy bears, sweaters, and children's quilts. In 1989, offerings included Raggedy Ann and Andy dolls, sock monkeys, a foot-flopping penguin push toy, a dragon pull toy, a wooden rocking horse, sweaters with animal pictures, several bears from the Vermont Teddy Bear Co., and some cute appliquéd bibs.

Bruce Bolind
Boulder, CO 80301
Telephone: 303-443-9688

Catalog price: free

A family gift catalog that doesn't neglect children. Personalized ski hats, wallets, rubber stamps, and hairbrushes are offered for young customers, as well as child-size T-shirts with messages like "Warning: may cause sleepless nights" and "If you think I'm cute you should see Mommy." The company will make china plates and self-adhesive postage-type stamps from your color snapshots. Decorated return-address labels are something of a specialty here—they come in dozens of different designs, from snowflakes and horses to teddy bears and balloons.

Claus & Crew
P.O. Box 455, Churchville, MD 21028

No catalog

Send $2 with a child's name and address, and Claus & Crew will mail back a personalized letter from Santa, typed in script on a word processor and signed in green ink.

The Disney Catalog
475 Oberlin Avenue S., Lakewood, NJ 08701-6989
Telephone: 201-905-0111

Catalog price: $2

Mickey Mouse, Donald Duck, and assorted Disney pals swarm through the pages of this catalog, peering out from T-shirts and sweat shirts, watches, slippers, sheets, bathrobes, infant coveralls, and soft dolls. A silver baby's cup with Mickey's face embossed on the side would make an unusual baby gift. A Goofy or Dumbo doll might be a hit with an older child. The catalog sells videos of many Disney films, collections of Disney cartoons, and a number of book-and-cassette sets.

Walt Disney World
Mail Order Dept., P.O. Box 10,070, Lake Buena Vista, FL 32830-0070
Telephone: 407-824-4718

Catalog price: free

Mickey Mouse T-shirts, party supplies, jewelry, watches, baby dishes, bibs, and other trinkets, and a menagerie of Disney characters as soft dolls. Mickey dominates the doll stage—his 27-inch-tall cuddler is bigger than any of the others—but he shares the spotlight with dolls of Pluto, Jiminy Cricket, Lady, Bambi, Dumbo, the Seven Dwarfs, and a host of companions.

Walter Drake & Sons
68 Drake Building, Colorado Springs, CO 80940
Telephone: 719-596-3854

Catalog price: free

A color catalog of inexpensive gifts that's big on personalized items. Order your child's name on pencils, balloons, bath towels, tote bags, bookplates, and stationery, or have a photograph made into a small jigsaw puzzle. The catalog offers stickers, glow-in-the-dark stars, dinosaur erasers, and other presents that won't break the bank.

Engravables
2626 Main Street, Buffalo, NY 14214
Telephone: 716-833-6892

Catalog price: $1, refunded with purchase

Engraved signs and personalized gifts. Metal enamel bookmarks in the shape of pencils can be engraved with a child's name, as can picture frames and Christmas ornaments. A brass plaque might be appropriate for the lid of a toy chest or the door to a child's room.

Family Tree Originals
P.O. Box 838, Moorpark, CA 93020
Telephone: 805-529-4774

Free brochure

Personalized blue-and-white porcelain plates to commemorate births, weddings, anniversaries, housewarmings, and other occasions. On the birth plate, the name, date, and place of birth are scripted around the rim amid a border of flowers and ferns. A nursery is pictured in the center, with a stork stepping in the window. The time of birth is recorded on the mantel clock, the birth weight on an old-fashioned scale beside the cradle.

Family Tree Originals' birth plate

Roberta Fortune's Almanac
150 Chestnut Street, San Francisco, CA 94111-1004
Telephone: 800-331-2300 or 2232

Catalog price: $2

Fun and stylish gifts for grown-ups (Aalto vases and Michael Graves teakettles), with a few pages of presents for children. The pedal-powered Kettcar, an astronaut suit, a Mickey Mouse watch, a stuffed Dumbo 42 inches tall, and an 11-foot inflatable pool monster have all appeared in recent catalogs. They may not be in the current catalog, but they give an idea of the sort of thing to expect.

Great Days Publishing, Inc.
207 W. Mission Street, Santa Barbara, CA 93101
Telephone: 800-325-2282 (800-325-2285 in CA)

Free brochure

Personalized birthday scrolls that summarize the news on any day from the past hundred years. The recipient's name is penned in calligraphy at the top; the news, weather, and sports follow, with mention of the most popular books, movies, and music of the day.

Hawk Meadow of New England
Rt. 1, Box 182B, Perkinsville, VT 05151
Telephone: 802-263-9400

Catalog price: $3

Beautiful handcrafts, most of them made by New England artists. Past catalogs have held such gems as a sterling silver baby cup with either an elephant's head or a swan's neck for the handle (for about $900), the Arcobaleno rainbow blocks, a toy wooden sailboat, porcelain dolls, and an exquisitely carved hobbyhorse.

Heir Affair
625 Russell Drive, Meridian, MS 39301
Telephone: 800-332-4347 or 601-484-4323

Catalog price: free

A children's gift catalog with a little of a lot of things, spiced up with a few really extravagant luxuries. Your child may not need a chintz-upholstered chaise lounge ($315) or a scaled-down Yamaha motorcycle with a working electric motor ($575), but it's fun to entertain the notion. More within the realm of possibility are a set of giant domino building blocks, a wooden Chinese checkers board, a chime instrument, and a child's calendar. Heir Affair carries a small selection of car seats, baby swings, strollers, and high chairs.

Holst, Inc.
1118 West Lake, P.O. Box 370, Tawas City, MI 48764
Telephone: 517-362-5664

Catalog price: $1

A gift catalog with a rural theme. Old-fashioned weather vanes are offered in profusion, along with amusing posters of baby farm animals and overall-clad toddlers. A musical birthday-cake stand might make a nice gift, as might a pair of stilts with sneakers on the feet.

Initials+ Collection
150 Spear Street, Suite 525, San Francisco, CA 94105
Telephone: 800-782-4438 or 415-781-1243

Catalog price: free

Everything personalized, from engraved brass door

Zolo from The Museum of Modern Art

knockers to embroidered Christmas stockings. Baby blankets, diaper covers, children's painting smocks, sweaters, even child-size raincoats, can be ordered with names or initials prominently embroidered. Personalized pencils and stationery are of course a regular feature of the catalog, as is an engraved silver baby cup and a silver-plated rattle.

Miles Kimball
41 W. 8th Avenue, Oshkosh, WI 54906

Catalog price: free

Inexpensive gifts and household gadgets, some eminently useful, some completely silly. The children's section is stocked with wooden name puzzles, spillproof mugs, windup toys, place mats, children's music cassettes, plastic dinosaurs, and coloring books. One recent catalog featured giant inflatable crayons and a bubble-blowing squirt gun. Parents might want a pair of flamingo lawn ornaments, a makeup cape, or perhaps a doormat that reads "Oh, no. Not you again!"

The Lighter Side
4514 19th Court E., P.O. Box 25600, Bradenton, FL 34203
Telephone: 813-747-2356

Catalog price: free

A crazy catalog crammed with silly hats and T-shirts, funny games, and novelties like an automatic fish caller and a singing statue of Elvis Presley. Diversions for children in the 1988 edition included magnetized blocks, wooden name and map puzzles, a coloring tablecloth, a device for blowing 6-foot soap bubbles, and a plastic rocket that shoots 300 feet, powered by water and air pressure.

Linmar Specialties, Inc.
P.O. Box 402, Woodmere, NY 11598

Catalog price: $1

Personalized porcelain gifts. Linmar will inscribe names, dates, and messages on white porcelain picture frames, plates, Christmas ornaments, children's dishes, and other curios.

The Metropolitan Museum of Art
Special Service Office, Middle Village, NY 11381
Telephone: 718-326-7050

Catalog price: free (request children's catalog)

The Metropolitan Museum of Art, well known for its Christmas catalog, mails a separate children's gift catalog in the fall, filled with exceptional illustrated books and with toys and games relating to the museum's collection. The catalog offers several of the museum's own books, including a wonderful art activity book, a child's journal and photograph album, a songbook illustrated with artwork from the museum, and a lovely baby journal adapted from an 80-year-old French book. New books from other publishers, classic tales in classic editions, card and board games, and toys like a magnetic block set, a plastic knight-and-castle set, and a fill-it-yourself kaleidoscope make up the rest of the collection. An additional catalog of cards, posters, and gifts arrives in the spring, rich with potential for decorating the walls of a child's room.

Museum Books Mail-Order
San Francisco Museum of Modern Art,
P.O. Box 182203, Chattanooga, TN 37422
Telephone: 800-447-1454

Catalog price: free

Children's books chosen for their artwork or innovative design, along with such art-related toys as a Colorshapes set, a soft foam block set, and a book of masks to assemble and wear. The selection changes from year to year, so write and see what's new this season.

Museum of Fine Arts, Boston
Catalogue Sales Department, P.O. Box 1044, Boston, MA 02120
Telephone: 617-427-7791 (800-225-5592 to order)

Catalog price: free

Jewelry, books, prints, and gifts derived from the museum's collection. A section of several pages is devoted to gifts for children: illustrated books, an English tin whistle, a box of marbles, a set of nesting kittens, magnetic blocks, card and board games, and a number of other entertaining and beautiful diversions.

The Museum of Modern Art
Mail Order Department, 11 W. 53rd Street, New York, NY 10019-5401
Telephone: 212-708-9888 (800-447-6662 to order)

Catalog price: $3

A showcase of modern design. The Museum of Modern Art's catalog brings the most elegantly styled place mats, ice-cream scoops, chairs, clocks, calculators, vases, and cookware together into a mail-order market where everything looks sleek and up-to-date. Children aren't a top priority here, but they're not ignored. Recent editions have featured Bauhaus blocks, Colorforms, paper mobiles, and art-supply sets.

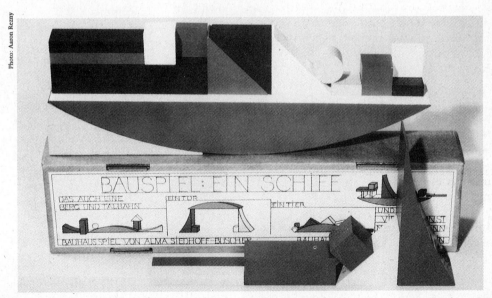

Bauhaus blocks from The Museum of Modern Art

Old Village Shop
Unique Merchandise Mart, Hanover, PA 17333-0008
Telephone: 717-633-3333

Catalog price: free

Inexpensive gifts and housewares, like a plastic car cover, a television pole, and a four-tier illuminated punch fountain. A brief children's section features such things as wooden alphabet puzzles, an inflatable nap mat, doll furniture, and a basketball-net hamper.

The Paragon
Tom Harvey Road, Westerly, RI 02891
Telephone: 401-596-0134

Catalog price: free

A mid-priced gift catalog that's aimed primarily at grown-ups. Children's offerings have included a toy periscope, a Beatrix Potter music box, a blunt-nosed dart set, and a combination piano-xylophone that stands 5 inches tall. The newest edition will likely have a completely different collection.

Perfect Presents by Suzy, Ltd.
270 W. 38th Street, New York, NY 10018
Telephone: 212-302-9177

Free brochure

Gift baskets for baby showers, anniversaries, and other occasions.

Potpourri
120 N. Meadows Road, Medfield, MA 02052
Telephone: 617-359-5440

Catalog price: free

A mid-priced gift catalog stocked with presents in the $10 to $100 range. A handful of items for children move in and out hastily, rarely lingering from one seasonal catalog to the next. Past editions have featured a 5-foot remote-controlled blimp, a set of nesting matrushka dolls, wooden name puzzles, and a pair of Mickey Mouse slippers.

Santa's Elves
Santa Claus, IN 47579

No catalog

Santa's elves, under the direction of postmaster Mary Ann Long, answer children's letters to Santa during the Christmas season. The notes are handwritten, mailed in an attractive envelope, and canceled with the Santa Claus, Indiana, postmark. Ms. Long has set no fixed fee for the letters, but requests donations.

Shibumi Trading, Ltd.
P.O. Box 1-F, Eugene, OR 97440
Telephone: 800-843-2565 or 503-683-1331

Catalog price: free

Traditional Japanese goods, from iron garden lanterns to silk kimonos. Japanese folk toys, tatami mats, rice-paper lanterns, a brass gong, and a beautiful Go set are offered, along with woodblock prints, origami guides, coloring books, illustrated stories, and children's *geta* (sandals). Parents with a taste for things Japanese may be tempted by the elegant cookware and tableware, the adult-size robes, the fabric gift wrap, the calligraphy sets, and the books on Japanese art and culture.

Japanese folk toys from Shibumi Trading, Ltd.

Smithsonian Institution
Dept. 0006, Washington, DC 20073-0006
Telephone: 202-357-1826 (703-455-1700 to order)

Catalog price: free

Toys and gifts having to do with the broad-ranging collections of the Smithsonian's many museums. The Air and Space Museum has contributed glow-in-the-dark sheets, astronaut costumes, and freeze-dried ice cream. From the National Museum of American History come Louisville Slugger baseball bats, baseball uniforms, Mickey Mouse wind socks, a box of marbles, and replicas of old-fashioned teddy bears.

Stocking Fillas
3229 Hubbard Road, Landover, MD 20785
Telephone: 800-638-8886

Catalog price: free

A one-stop mail-order source for those inexpensive favors and stocking gifts that give a happy boost to parties and Christmas. Now that toy stores have drifted away from the cheapest gifts and 5 and 10's have all but disappeared, this catalog fills an important need. Where else can shoppers find magic trees, kazoos, jumping beans, tinsel wigs, windup toys, Styrofoam gliders, glow-in-the dark dinosaurs, animal nose masks, sparkler wheels, and collapsible cups all in one place? Add trendy toys like animal bracelets and California Raisins mugs (or whatever is hot this season) and party supplies like giant balloons, magic candles, and party hats and you have a low-cost formula for a good time. Most of the items sell for less than $2, many for as little as 20¢.

Taylor Gifts
355 E. Conestoga Road, P.O. Box 8500, Wayne, PA 19093-8500
Telephone: 215-789-7007

Catalog price: free

Gifts that run the gamut from personalized doormats to expensive novelty telephones. For the children on your list the catalog sells dolls, battery-powered toys, sacks and boxes for organizing toys, and such things as magnetic marbles and dollhouse kits. A pair of fake antlers might dress up a pet in style for the holidays.

Tiffany & Co.
801 Jefferson Road, Parsippany, NJ 07054-9957
Telephone: 201-428-0570 (800-526-0649 to order)

Catalog price: free

Famous for jewelry and crystal, Tiffany & Co. offers baby gifts for the silver-spoon set. Browse past the sparkling brooches and pendants, and the catalog opens to sterling-silver cups, rattles, and teething rings. Children's bowl-and-mug sets are another regular feature; our favorite has a mug decorated to look like a red drum and a plate ornamented with 1920's-era toys.

Lillian Vernon
510 S. Fulton Avenue, Mt. Vernon, NY 10550
Telephone: 914-633-6400 or 6300

Catalog price: free

A mid-priced gift catalog that devotes a few pages of each edition to fun and handy things for children. Recent catalogs have presented such items as personalized terry diaper covers, an inflatable wading pool, a nylon lunch bag, wooden puzzles, a vinyl floor mat to spread beneath a high chair, and a watch for children learning to tell time.

Wireless
274 Fillmore Avenue E., St. Paul, MN 55107
Telephone: 800-669-9999

Catalog price: $1

A gift catalog "for fans and friends" of Minnesota Public Radio. Spin-offs and sidelines from *A Prairie Home Companion* occupy sacred ground here: cassettes of Garrison Keilor's Lake Wobegon monologues, Powdermilk Biscuit T-shirts (in children's sizes), and gourmet Lutheran coffee. A children's section comes up with presents like ant farms, balloon-sculpture kits, and Mike and Peggy Seeger's album *American Folk Songs for Children*. Recordings of old-time radio shows offer another gift option.

Health and Safety

Alive Productions, Ltd.
P.O. Box 72, Port Washington, NY 11050
Telephone: 516-767-9235

Free flier

Alive Productions offers two instructional videos, *Breastfeeding: The Art of Mothering* and *Baby Alive*. The first is recommended for mothers who are considering or beginning breastfeeding. *Baby Alive* explains how to deal with such medical emergencies as choking, drowning, poisoning, head injuries, cuts, and burns, and shows how to make your home safer for children.

All Points Products
17029 Devonshire Street, #169, Northridge, CA 91325
Telephone: 818-360-7424

Catalog price: long self-addressed stamped envelope

This firm makes the Baby Beeper, which sounds an alarm when a child falls in water or wanders more than 25 feet away from its parent. The child wears a small pendant on a necklace, the parent a transmitter clipped to a belt or pocket. Be prepared for some alarming literature if you write ("Is your baby's safety worth $29.95?").

Health and Safety

Allergy Control Products
96 Danbury Road, Ridgefield, CT 06877
Telephone: 800-422-DUST or 203-438-9580

Catalog price: free

Air filters, face masks, mattress and pillow covers, dehumidifiers, books, and other products to help those with allergies to house dust and mold spore. An insect venom extractor is offered for those allergic to bee stings.

H. G. Arms Company
1449 37th Street, Brooklyn, NY 11218
Telephone: 718-436-2711

Free flier

Sells a steel box with a key lock for securing medicines from curious children. The box fits on a shelf in any standard medicine chest.

Baby Safety Specialists, Inc.
2139 N. University Drive, Suite 196, Coral Springs, FL 33071-9966
Telephone: 800-537-3412 or 305-341-9072

Free brochure

Two dozen products designed to make your house safer for babies and young children. Cabinet locks, outlet covers, doorknob covers, bathtub spout covers, refrigerator locks, toilet seat cover straps, corner cushions for low tables, and cord shorteners all help reduce household risks for toddlers. Shopping cart seat belts, a backseat mirror, and a hand-holding strap reduce the danger on trips outside.

Baby Table Bumper Productions
11684 Ventura Boulevard, #208, Studio City, CA 91604
Telephone: 818-763-2335

Free flier

Sells the Thumper Bumper, a thick padded bumper that wraps around the edge of a coffee or end table to soften the impact of inevitable falls when a baby starts to walk.

Best Selection, Inc.
Baby gates and home safety kits. See entry under *Baby Needs*.

The Cherubs Collection
505 S. Vulcan Avenue, Encinitas, CA 92024
Telephone: 619-436-1120

Free flier

Sells a home first-aid kit and safety organizer called the Kids Care Kit. Each kit comes with a checklist of recommended medicines and safety supplies to keep on hand (the supplies themselves aren't included), hospital consent forms to fill out in case your child needs medical attention when you aren't available, medical history forms, a wipe-clean telephone and message board, an emergency first-aid poster, and a booklet of general medical and safety tips. All this comes in a plastic box with room to spare for all of the suggested medicines and supplies.

Child Safety Catalog
KinderKraft Inc., P.O. Box 5433, Arlington, VA 22205
Telephone: 703-841-1902

Catalog price: free

A color catalog of the best and the safest. The inventory spans toilet lid locks, sealable garbage cans, baby gates, safety covers for doorknobs (including a cloth one designed by KinderKraft), Cosco's Luxury Commuter car seat, the Graco Pack N' Play portable crib, and a childproof medicine box.

Comfortably Yours
61 W. Hunter Avenue, Maywood, NJ 07607
Telephone: 201-368-3499

Catalog price: free

A catalog of health and comfort products for older people that generally tosses in a few things for parents. A hinged baby gate made the list in 1989, as did a fire-escape ladder and a waterproof bed sheet with inflatable bolsters on three sides to keep active sleepers in place.

Courier Health Care, Inc.
Sells a childproof medicine box, a lighted scope for ear exams, a digital pediatric scale, and other home medical products. See entry under *Baby Needs*.

CritiCard, Inc.
445 W. Jackson, Centennial Plaza, Naperville, IL 60540-9990
Telephone: 800-331-8801 or 312-357-6866

Free brochure

Makes CritiKid tags that attach your child's identification and medical history to a shoelace. Parents fill out a form that includes emergency contacts and phone numbers, health insurance information, and any history of allergies or hospitalization and send it in with a fee of about $11. CritiCard sends back two microfilm copies of the information sealed in plastic tags to be laced onto the child's shoe. Check with area hospitals to be sure they're aware of the program. If they're not, the tags won't do much good. If they're actively participating, they may be distributing the tags locally at a better price.

Double A Productions, Inc.
7 Evergreen Way, Dept. SP, Stratham, NH 03885
Telephone: 603-778-3010

Free information

Produces a 90-minute video titled *CPR for Everyone!*, with instructions for administering CPR and dealing with emergency choking situations.

F & H Child Safety Co.
P.O. Box 2228, Evansville, IN 47714
Telephone: 812-479-8485

Catalog price: $1

Products to make homes safer for young children and to make life easier for parents. Outlet covers, cord shorteners, cabinet locks, toilet straps, bathtub spout cushions, doorknob covers, and baby harnesses help keep danger out of children's reach. A plastic grip for rinsing diapers, a device for keeping children's shoelaces tied, and a mesh bath bag simplify some of mom's and dad's tasks.

Family Life Products
900 Town Plaza, P.O. Box K, Dennis, MA 02638
Telephone: 508-385-9109

Free brochure

This firm makes bathtub spout covers sporting the faces of friendly animals. Shoppers can choose a turtle, a bear, a dinosaur, an elephant, or several other creatures to cushion the hard angles of the spout. A padded shopping-cart seat works as an infant cradle or a toddler restraint. Table corner pads, Velcro security straps for toilet or refrigerator, outlet covers, and a wrist leash offer protection from childhood risks.

Federal Emergency Management Association
P.O. Box 70274, Washington, DC 20024

Free information kit

Produces a free kit, "Big Bird Gets Ready for Hurricanes," to help families cope with severe storms. The kit includes a booklet, a board game, and a recording of the song "Hurricane Blues." The assemblage is designed to calm children's fears by teaching them what to expect and what to do if a storm strikes. The booklet also gives practical advice to parents on helping children through an emergency. Other kits are in the works for tornados, earthquakes, and floods.

Growstick Co.
1447 N.W. 191st, Seattle, WA 98177
Telephone: 206-546-6403 or 782-8475

Free brochure

Brightly painted wooden measuring sticks for charting a child's growth to 6 feet and beyond.

Healthy Alternatives, Inc.
P.O. Box 3234, Reston, VA 22090
Telephone: 703-430-6650

Free brochure

Sells a video, *Tender Touch: A Guide to Infant Massage*, which explains the techniques and benefits of baby massage.

Hear You Are, Inc.
4 Musconetcong Avenue, Stanhope, NJ 07874
Telephone: 201-347-7662

Catalog price: free

An extensive catalog of assistive technology devices for the hearing impaired, including personal amplification systems, telephone devices, smoke alarms, and children's games.

JMK Enterprises
14507 Yelm Highway S.E., Yelm, WA 98597
Telephone: 206-458-4492

Free flier

Disposable toothbrushes for travel, for guests, or for hygienic use when a child has a viral infection. The toothpaste is already loaded in the handle; it squeezes out onto the bristles when the handle is compressed.

Julia & Brandon
Childproofing aids. See entry under *Toys*.

Kaplan School Supply Corp.
See entry under *Educational Supplies*.

King-Aire Products
P.O. Box 126, Fortville, IN 46040
Telephone: 317-485-7771

Catalog price: free

Sells a line of air purifiers, water filters, ventilators, humidifiers, and dehumidifiers. The air cleaners filter dust, pollen, smoke fumes, and other impurities from the air and may be a great help to those with asthma and allergies. The smallest model, suitable for cars and other small spaces, costs about $200. The largest can handle a room of over 20,000 cubic feet and costs more than $1,000.

Kotton Koala
17386 Highway 9, Boulder Creek, CA 95006
Telephone: 408-338-2653

Long self-addressed stamped envelope for brochure

Makes a cotton headrest for children that lets them nap in car booster seats without slumping over. The cotton corduroy cover comes off for machine washing.

Lifesaver Charities
P.O. Box 2533, Garden Grove, CA 92640
Telephone: 714-530-7100

Catalog price: long self-addressed stamped envelope

Identification tags to be sewn into children's clothing or shoes. Each tag has space for writing in the child's name, emergency telephone numbers, insurance and allergy information, and a parental consent to emergency medical treatment. Two tags are sent free with the information packet. More can be purchased in bulk for the cost of postage.

MJ & Kids, Inc.
Maple Avenue, P.O. Box 14, Durham, CT 06422
Telephone: 203-349-9406

Free brochure

Offers a first-aid kit for parents traveling with babies. It comes stocked with Tylenol drops, syrup of ipecac, sunscreen, a thermometer, bandages, and a dozen other emergency aids—even coins for a pay phone.

Moonflower Birthing Supply
Homeopathic remedies. See entry under *Baby Needs*.

Naturepath
A home ear scope, a digital thermometer, vitamins, herbal remedies, and Weleda's natural soaps and skin care products. See entry under *Baby Needs*.

J. C. Penney Company, Inc.
Baby gates and first-aid kits. See entry under *General Catalogs*.

Perfectly Safe
7245 Whipple Avenue N.W., Dept. A1, North Canton, OH 44720
Telephone: 216-494-4366

Catalog price: $1

Safety devices for young children and supplies and furnishings for babies. Childproof latches seal drawers, cabinets, appliances, and toilet seats against inquisitive toddlers; gates block dangerous stairways; and an array of devices keep children from toying with electric outlets and cords. Alarms are sold that ring when a child opens a door, adds too much water to the tub, or falls into the pool. A telephone attachment has picture buttons that automatically dial the numbers of parents, grandparents, neighbors, or other emergency help when pressed. Less alarming are the many convenience products for young children: a mat to spread under the high chair, a portable playpen, a baby bounce seat with a wide restraining belt, an infant hammock to stretch across the crib, an electric wipe warmer, and a sheepskin stroller cover. Perfectly Safe sells a stable tricycle from Germany that has a wide wheel base, a hand brake, and a push bar for parents. A bicycle seat (for carrying children on a parent's bicycle) has a padded restraining bar and foot wells that keep children's feet away from spinning wheels. Two helmets are offered to protect bicycle-riding children as young as one year old.

Plow & Hearth
An extra-large fireplace screen which can be used to distance toddlers from sparks and embers or to fence off a wood stove. See entry under *Gardening*.

Preventive Dental Care, Inc.
1147 E. Broadway, #34, Glendale, CA 91205

Catalog price: $2

Dental supplies for the mouths of babies, children, and adults. The company's Rub-a-Dent cleaner for baby teeth is a cloth fingertip cover to be worn by a parent and rubbed over a child's first teeth. Toothbrushes are sold in all manner of sizes and styles—small ones for children, bigger ones for grown-ups, special ones for cleaning around back teeth and braces, and double-headed ones for people with dexterity problems. Floss, picks, pastes, and rinses complete the arsenal. To help make teeth-cleaning fun, the company sells dental timers, mugs, and coloring books for children. A tooth-shaped pillow holds offerings for the tooth fairy. A toothy night-light smiles cheerfully in the dark.

Rainbows & Lollipops, Inc.
Home safety devices. See entry under *Baby Needs*.

The Right Start Catalog
Ear scopes, toddler toothbrushes, digital thermometers, and first-aid and child safety kits. See entry under *Baby Needs*.

SelfCare Catalog
P.O. Box 130, Mandeville, LA 70448
Telephone: 800-345-3371 or 504-892-8032

Catalog price: free

Gadgets for keeping fit and healthy, tools for self-diagnosis, and comfortable clothes for sleep and exercise. Parents of young children will find an ear scope, a travel potty seat, a Strolee car seat, an Olga nursing bra, MagMag's training cup system, and a running stroller. A complete family first-aid kit in a fishing-tackle box looks like an important item. A portable refrigerator that plugs into a car lighter allows families to bring healthy food on long trips. Air filters, blood-pressure monitors, scales, and dental tools may also be of interest to parents.

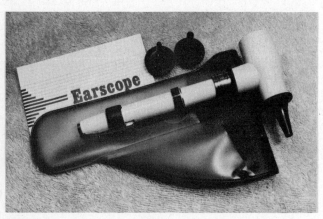

An ear scope from the SelfCare Catalog

Solutions

P.O. Box 6878, Portland, OR 97228-6878
Telephone: 800-342-9988

Catalog price: free

Ingenious devices that solve problems around the house and yard. For families with children, the catalog has offered an ear scope, a bed-wetting cure that sounds an alarm at the first drop of moisture, a direct-dial telephone with space for a photograph on each button, and for car trips, a child's lap desk with built-in storage bin.

Sporty's Preferred Living Catalog

Pool alarms and fire extinguishers. See entry under *Games and Puzzles*.

T'Owl Productions

209 E. Jay Street, Ithaca, NY 14850
Telephone: 607-277-0571

Free information

Terry-cloth rabbit heads, called Boo Boo Bunnies, to hold ice for soothing bumps and bruises from childhood accidents.

Tots World Company

P.O. Box 148, Blue Bell, PA 19422
Telephone: 215-643-3366

Free flier

Sells the Shockblocker, a snap-on outlet cover that keeps children from trouble but provides easy access for adults.

Towards Life Catalogue

Homeopathic remedies and natural skin creams. See entry under *Bath Accessories*.

Westags, Inc.

P.O. Box C, Flourtown, PA 19031
Telephone: 800-232-2873 or 215-233-5141

Free brochure

Makes SneakerMate plastic identification tags, which lace into a child's shoe. Each tag has room for six lines of type, which can be used to give a child's name and address, emergency contacts, allergies or any other information that may be important if a child gets into trouble when you're not around. The letters are embossed on stiff plastic and can't be worn or washed off. Choose red, blue, pink, or black tags to match your child's shoes.

Wet-No-More

Travis Industries, 100 Main-Summer, Coos Bay, OR 97420
Telephone: 800-4-DRY-BED or 503-269-6900

Catalog price: free

A bed-wetting cure that works by sounding an alarm when a child begins to urinate. The alarm and batteries are held in a lightweight foam belt. The sensor is covered by an absorbent fabric sleeve.

The Zaadi Company

836 Chelmsford Street, Lowell, MA 01851
Telephone: 617-453-6508

Free brochure

This company has created a doll to help children cope with and understand medical conditions and surgical procedures. The soft fabric Zaadi doll peels open to show internal organs, major veins and arteries, muscles, and bones. A removable appendix attaches with Velcro; a spare kidney can be used to explain transplants. Other features include sleeping and sad masks, bandages, and an overnight bag. Priced at about $250, the dolls are sold primarily to doctors and hospitals.

SneakerMate identification tags from Westags, Inc.

Hobby Supplies

The American Clockmaker
P.O. Box 326, Clintonville, WI 54929
Telephone: 715-823-5101

Catalog price: $2.50

Kits for making all sorts of handsome clocks. Several full-height grandfather clocks come with tubular brass chimes. Mantel clocks and wall clocks come in cherry, walnut, oak, and other woods. Novelty clocks present options for children. In 1989 the catalog offered a dozen German cuckoo clocks with elaborate carving and lots of activity on the hour.

The American Needlewoman
P.O. Box 6472, Fort Worth, TX 76115
Telephone: 800-433-2231

Catalog price: $1

Patterns and supplies for needlepoint, embroidery, crochet, quilting, and hooked rug projects. The 64-page color catalog is crammed full of idea books, kits, and basic supplies. New parents might want to order a plain quilt and bib ready for embroidery, or one of the many wall-hanging designs. Young needleworkers will find plenty of simple projects to get started on.

Barker Enterprises, Inc.
15106 10th Avenue, S.W., Seattle, WA 98166
Telephone: 206-244-1870

Catalog price: $2

Forty pages of candle molds, dipping vats, waxes, wicks, dyes, scents, decals, and how-to books. Home craftspeople can make simple dipped candles, or use molds to make precisely shaped tapers and dozens of curious figures—from Santas and turkeys to skulls and Buddhas.

Berman Leathercraft
25 Melcher Street, Boston, MA 02210-1599
Telephone: 617-426-0870

Catalog price: $3

Leather-craft supplies, including leather, skins, kits, tools, and books. Belt buckles, snaps, rivets, and other hardware are offered in an abundance of sizes and styles. At the back of the catalog are a few pages of Indian-craft supplies for making war bonnets, drums, moccasins, and beadwork.

Cane & Basket Supply Co.
1283 S. Cochran, Los Angeles, CA 90019
Telephone: 213-939-9644

Catalog price: $2

Cane webbing, chair cane, fiber rush, sea grass, and other materials for making woven chair seats; and reeds, hoops, handles, and rims for basketry. Chair kits (including a child's rocker), tools, finishing oils, and books round out the selection.

Chaselle, Inc.
9645 Gerwig Lane, Columbia, MD 21046
Telephone: 800-CHASELLE (800-492-7840 in MD) or 301-381-9611

Catalog price: free (specify "Art & Craft Materials" catalog)

Just about anything a potter could want can be found in this 400-page catalog, along with equipment and supplies for sculpting, woodworking, screen printing, etching, metal enameling, mosaics, stained glass, weaving, and leather craft. See main entry under *Educational Supplies* for a description of Chaselle's other offerings.

Collectors Marketing Corp.
220 12th Avenue, New York, NY 10001
Telephone: 212-563-2585

Free flier

Topps baseball card sets for collectors. Complete 792-card major league sets can be had for any of the past three years, as can, for a bit more money, complete minor league sets. Reproduction sets of great teams of the past are also listed—teams like the 1927 New York Yankees and the 1959 Los Angeles Dodgers. Albums with vinyl pages are sold to house the collection.

Grey Owl Indian Craft Co.
113-15 Springfield Boulevard, Queens Village, NY 11429
Telephone: 718-464-9300

Catalog price: $2

Indian-craft supplies such as beads, skins, feathers, tomahawk heads, pipe stems, and leather-craft kits. The huge catalog also carries a wealth of ready-made merchandise, from full Sioux war bonnets and Kiowa horned headdresses to Pendleton blankets, beaded bracelets, coonskin caps, and Pueblo log drums. Full-size tepees in either Sioux or Cheyenne style could make your yard the gathering spot of the neighborhood. The 60-page list of books and music would make an ample catalog on its own. A rich source for the costume drawer, the creative craftsperson, and anyone interested in Native American culture. (If you're particularly interested in the blankets, send $1 for the full-color blanket catalog.)

Grieger's
900 S. Arroyo Parkway, P.O. Box 93070, Pasadena, CA 91109
Telephone: 800-423-4181 (800-362-7708 in CA)
Catalog price: free
A massive catalog of jewelry-making supplies, full of specialized tweezers and pliers, gold and sterling chains, rings, gems, minerals, and books. Picks, hammers, and gold pans might make surprising presents for budding prospectors.

H. E. Harris and Co., Inc.
P.O. Box 7087, Portsmouth, NH 03801
Telephone: 603-433-0400 (800-822-5556 to order)
Catalog price: $1, earns $5 discount from first order
The largest and one of the oldest (since 1916) catalogs for stamp collectors. H. E. Harris sells U.S. and foreign stamps, along with a complete line of albums, refill pages, and other supplies.

Heath Co.
Benton Harbor, MI 49022
Telephone: 800-444-3284 or 616-982-3200
Catalog price: free
Kits for building almost anything electronic: stereos, televisions, weather monitors, desktop computers, and much more. A great way to save money and learn how things work at the same time.

Herrschners, Inc.
Hoover Road, Stevens Point, WI 54492
Telephone: 715-341-0604
Catalog price: free
Yarn, thread, fabric, needles, and patterns for making afghans, quilts (including some for babies), hooked rugs, tablecloths, and needlepoint pictures. A few kits are sold for making stained glass window decorations, Christmas tree ornaments, doorknob covers, and other household fancies.

Hobby Surplus Sales
Books and supplies for stamp, coin, and baseball-card collectors. See entry under *Model Trains, Planes, Cars, and Boats*.

Industrial Arts Supply Co.
5724 W. 36th Street, Minneapolis, MN 55416-2594
Telephone: 612-920-7393
Catalog price: free
Kits, plans, tools, and supplies for school industrial arts programs and home craftspeople. Those who want to work with plastic and fiberglass will have the most to choose from, including full-size fiberglass canoe molds and molds for casting hundreds of smaller objects in plastic. Candle-making supplies, electronics kits, plans and parts for making wooden toys, supplies for casting tin soldiers, jewelry and leatherworking kits, and equipment for enameling are also offered.

Klockit
P.O. Box 636, Lake Geneva, WI 53147
Telephone: 800-556-2548
Catalog price: free
Clock kits in scores of different designs, from 6-foot grandfather clocks to mantel clocks and small novelty wall clocks. Buy a complete kit or pick out a movement and dial and create your own. Along with the many timepieces, the catalog offers a good selection of parts, plans, and kits for making wooden toys. The oak-and-walnut kits look like a good way to ease into toy making without investing in expensive machinery.

Leclerc Corp.
Rt. 1, Box 356A, Champlain, NY 12919
Telephone: 518-561-7900
Catalog price: free
Weaving and spinning supplies, including more than a dozen different looms. With the catalog you'll receive a list of dealers in your area. The company requests that customers try to obtain what they want through local stores before resorting to mail order.

Mason & Sullivan Co.
586 Higgins Crowell Road, W. Yarmouth, MA 02673
Telephone: 508-778-0475
Catalog price: free
Kits and supplies for home clock builders. Most of the kits are available either precut or with the plans and rough lumber, and each is coded to indicate the woodworking skill required. The simplest kits need only a screwdriver, a clamp, and some glue to assemble. The more difficult projects call for a good set of power tools and some solid woodworking experience. The color catalog shows old-fashioned mantel clocks, wall clocks, grandfather clocks, and weather instruments. Dozens of faces and movements are offered separately.

Nasco
Equipment and supplies for leatherworkers, jewelry makers, weavers, needleworkers, and other craft hobbyists. See entry under *Art Supplies*.

Pacifica Crafts Equipment
P.O. Box 1407, Ferndale, WA 98248
Telephone: 206-398-7722 or 384-1504

Free brochure

Electric potter's wheels. The basic model costs a bit over $500, a little less if you buy it in kit form. Accessories and additional power can drive the price up twice as far. Call if you're in a hurry—we waited four months for an answer to our letter.

H. H. Perkins Co.
10 S. Bradley Road, Woodbridge, CT 06525
Telephone: 203-389-9501

Catalog price: $1

Basket-weaving and chair-caning kits and supplies. H. H. Perkins sells cane, reed, ash splint, and rush, along with tools, finishes, basket handles, and hoops. Those who want to work from kits can make a doll cradle, a dozen different baskets, two children's chairs, and a number of stools. A good collection of books offers instruction and inspiration.

Pourette Mfg. Co.
6910 Roosevelt Way, N.E., P.O. Box 15220, Seattle, WA 98115
Telephone: 206-525-4488

Catalog price: $2, deductible from first order

A huge catalog of candle-making supplies, offering hundreds of molds and a full line of waxes, wicks, dyes, scents, and candle-decorating equipment. From standard round molds the list moves into avant-garde geometric spires, pyramids, and star-shaped towers, then leaps to ghosts, animals, tiki gods, sports equipment, and holiday fare. Beeswax sheets, candle holders and bases, and instruction books complete the selection. A separate, smaller catalog offers soap-making materials and molds.

Relco Industries
P.O. Box 920839, Houston, TX 77292-0839
Telephone: 713-682-2728

Catalog price: free

Metal detectors for treasure hunters. Prices start at about $100 and work their way up past $300, depending on the features needed.

Sax Arts & Crafts
Materials and equipment for calligraphy, printmaking, ceramics, stone carving, woodworking, jewelry making, weaving, candle making, leather craft, and stained glass. See entry under *Art Supplies*.

School Products Co.
1201 Broadway, New York, NY 10001
Telephone: 212-679-3516

Catalog price: $2

A complete catalog of weaving supplies, including table and floor looms, spinning wheels, warping mills, tapestry beaters, shuttles, hooks, dyes, and yarns. A reference section carries books and videos on weaving and knitting.

Scott Publishing Co.
P.O. Box 828, Sidney, OH 45635

Catalog price: free

Information and supplies for stamp collectors. Scott publishes price guides, checklists, albums, and album pages, and sells protective mounts in many sizes, adhesive hinges, and glassine interleaves. When you write, ask for the name of the nearest dealer, as Scott prefers that customers patronize local merchants before resorting to the mail.

Shillcraft
8899 Kelso Drive, Baltimore, MD 21221
Telephone: 301-682-3064

Catalog price: $1

Latch-hook kits, yarns, and frames for making rugs and hangings. A few cross-stitch and crochet patterns are offered, including some for baby clothes and blankets.

Tandy Leather Co.
P.O. Box 791, Fort Worth, TX 76101

Catalog price: $2.50

A leather-crafter's supermarket. Tandy claims to be the world's largest supplier of leather and vendor of the most extensive line of leather-craft kits and supplies. We don't doubt the claim. The 100-page catalog is jammed with tools, buckles, bolo slides, dyes, conditioners, and kits for making everything from handbags to hoslters. For the basic raw material, Tandy offers rawhides, full-grain vegetable-tanned cowhides, pigskins, goatskins, elk skins, python skins, splits, and various tanned hides for special uses. Monthly sale catalogs give deals on selected kits and tools.

Whittemore-Durgin Glass Co.
P.O. Box 2065, Hanover, MA 02339
Telephone: 617-871-1743

Catalog price: $1

Patterns, tools, glass, and supplies for workers in stained glass. The 100-page catalog arrives as a packet of loose sheets to be punched and put in a binder. A guide at the beginning directs shoppers to products appropriate to their level of expertise.

Kites, Boomerangs, and Juggling Supplies

AG Industries
Paper airplanes. See entry under *Teddy Bears*.

Big City Kite Co.
1201 Lexington Avenue, New York, NY 10028
Telephone: 212-472-2623

Catalog price: $2, deductible from first order

A big color catalog presents scores of colorful and exotic kites. Simple diamond kites can be had with dinosaur, balloon, or flamingo appliqués; dragon kites are offered in the form of sharks and electric eels; box kites come in octagon and star shapes; maneuverable stunt kites are available in a wild array of bright patterns. Japanese carp wind socks are sold for Boys' Day (celebrated May 5 in Japan by hanging a carp wind sock out for every boy in the family), along with dozens of other wind socks in less traditional designs.

Boomerang Man
1806 N. 3rd Street, Room SP, Monroe, LA 71201-4222
Telephone: 318-325-8157

Catalog price: free

You guessed it: boomerangs. But you've surely never seen so many different kinds. Traditional Australian boomerangs are offered in plastic or maple plywood, and the list launches from there into French graphite boomerangs, Colorado boomerangs, Texas boomerangs, cross-stick and tri-blade boomerangs, and night-flying boomerangs designed to be used with cyalume light sticks. The spring catalog lists scores of boomerangs and books and directs shoppers to boomerang contests and boomerang clubs. Just fill out the "order-rang" and they'll be on their way to you.

High Fly Kite Co.
30 West End Avenue, Haddonfield, NJ 08033
Telephone: 609-429-6260

Catalog price: free

Stunt kites, Japanese fighting kites, airfoil kites, wind socks, and ornamental box kites in a fabulous array of shapes and sizes. The simplest stunt kites sell for about $15. A *very* large airfoil kite (250 square feet) sells for over $800. Those who want to make their own can shop from a long list of parts and supplies.

Hyperkites
1819 Fifth Avenue, San Diego, CA 92101
Telephone: 619-231-4977

Catalog price: free

Maneuverable kites in several styles. Hyperkites encourages customers to buy the kites at local kite stores, but will sell by mail to people who don't have a store nearby. The company's ghost kites ("They come in any color you want, as long as it's white!!") are said to be ideal for children. They'll fly in a 3-mph breeze and stand up to a heavy wind without pulling hard on the line. More exotic and powerful kites are offered for advanced fliers.

Into the Wind/Kites
1408 Pearl Street, Boulder, CO 80302
Telephone: 303-449-5356

Catalog price: free

A big color catalog of kites, kite-making supplies, boomerangs, gliders, and other flying diversions. Single-

Legs kite sold by Into the Wind/Kites

string diamond, delta, and dragon kites come in beautiful colors at affordable prices. Double-string stunt kites cost a bit more. Spectacular five-foot snowflake and star kites and six-foot parafoils top out the price list at almost $200.

Klutz Flying Apparatus Catalogue
2121 Staunton Court, Palo Alto, CA 94306
Telephone: 415-857-0888

Catalog price: free

Juggling supplies, books, amusements, and oddities from the publisher of *Juggling for the Complete Klutz*. Need a pink flamingo? Or a pair of boxer shorts covered with glow-in-the-dark ants? The Klutz Flying Apparatus Catalog has both, along with Tom Kuhn's wooden yo-yos, a unicycle, a device for blowing 8-foot soap bubbles, and the famous two-potato clock (it runs on leftover vegetables). Klutz Press's book list includes *Kids Songs: A Holler-Along Handbook*, *Kids Cooking: A Very Slightly Messy Manual*, and how-to books for learning to master lariats, yo-yos, and harmonicas.

The Left-Handed Complement
Left-handed boomerangs. See entry under *Art Supplies*.

Nantucket Kiteman
P.O. Box 508, Nantucket, MA 02554
Telephone: 508-228-2297

Catalog price: long self-addressed stamped envelope

Two delta kites and an eight-pointed snowflake kite, all made by hand on Nantucket Island. The larger delta kite has a 72-inch wingspan and is recommended for teens and adults. The smaller version, with a 52-inch wingspan, can be handled by children over six.

Magic Tricks and Novelties

Abracadabra Magic Shop
P.O. Box 714, Dept. C-534, Middlesex, NJ 08846-0714
Telephone: 201-805-0200

Catalog price: $1 for "Fun Catalog;" $3.95 for "Giant Magic Catalog"

Abracadabra's "Fun Catalog" carries dozens of inexpensive devices for magic tricks and practical jokes. A floating eyeball might make an impression on a special guest; hot and sour candy, an exploding toilet seat, and a squirting lighter offer equally malevolent laughs. Magic accessories include the Zombie Floating Ball, the Five-in-One Miracle Wand, marked cards, a finger chopper, and the required silks, capes, and top hats. Most of the gags sell for less than $2. The magic tricks cost between $1 and $25.

The "Giant Magic Catalog" includes more advanced magic tricks and none of the gags, with prices ranging from about $5 to $100. Since it comes with a copy of the "Fun Catalog," it displays everything Abracadabra has to offer. The magic tricks in both catalogs are geared to the beginning and intermediate illusionist, and no prior knowledge is assumed in the instructions. If a customer can't figure out how to perform a trick from the printed material sent with it, the company is happy to explain by telephone.

Ash's Magic Catalog
4955 N. Western, Chicago, IL 60625
Telephone: 312-271-4030

Catalog price: $2

A hundred pages of magic supplies, marked cards, books, and gags largely aimed at the beginning or intermediate illusionist. Experienced tricksters can pull a skunk from a hat, push a cigarette through a quarter, or produce a glass of water from under a handkerchief using hardware from this catalog. Younger kids can enjoy the magic light bulb that lights in the hand without wires, the coin holder that makes pennies disappear, or the magnetic mummy that rises from its coffin. Pranksters will find squirting cameras, hot-pepper gum, rubber hot dogs, and X-ray glasses.

Dover Publications, Inc.
Magic books. See entry under *Books*.

The Fun House
P.O. Box 1225, Newark, NJ 07101

Catalog price: 25¢

Jokes, tricks, and novelties for thrill seekers and practical jokers. Hot-pepper candy is sure to be a hit with visiting relatives. A squirting toilet seat, fake cockroaches, salt-water taffy with a core of real salt, and the nail-through-the-finger trick all add up to a barrel of laughs for someone, and possibly a pile of headaches for mom and dad. Amazing Live Sea Monkeys offer fun without the annoyance, and the perpetual drinking bird is practically soothing. Parents might want to order the two-headed penny themselves, for deciding arguments over bedtime and television rights.

Hank Lee's Magic Factory
125 Lincoln Street, Boston, MA 02111
Telephone: 617-482-8749 or 8750

Catalog price: $6

For those ready to plunge into magic in a big way, Hank Lee puts out a 350-page monster of a catalog, crammed with everything from a $1 hand buzzer to a $1,200 professional prop for putting a sword through the neck of an audience volunteer. Beginners will find hundreds of gags and simple tricks. Those with more expertise (and more money) can shop for sophisticated coin and card tricks, elaborate stage pieces, and humorous gadgets for livening up a performance. A separate "book book" comes with the main catalog; it lists some 600 volumes of additional tricks and illusions.

Archie McPhee & Company
P.O. Box 30852, Seattle, WA 98103
Telephone: 206-547-2467

Catalog price: $3

"Remarkable bargains of the curious sort," runs the headline inside the Archie McPhee catalog. To "curious" we might add adjectives like weird, amazing, and hilarious. Somehow this company gets its hands on overstock items from the 1950s, bizarre shipments from the Far East, and the novelties we've all been searching for but didn't know it. A recent catalog peddled 30-year-old Howdy Doody key chains, simulated grass hula skirts, glow-in-the-dark cockroaches, Troll dolls, lawn flamingos, Mr. T wigs, X-ray glasses, fake foreign money, windup toys, and sacks of fake snow. A pair of alligator shoes is made of soft vinyl to look like real alligators—18 inches from snout to tail. A secret harmonica is disguised as a lobster claw. And a potato gun "shoots harmless potato pellets." Most amazing of all is the fact that much of this stuff is genuinely cheap—most of it sells for less than $5, some for considerably less. If life at your house needs an emotional lift, write to Archie McPhee.

D. Robbins and Co., Inc.
70 Washington Street, Dept. RM, Brooklyn, NY 11201

Catalog price: $1

Hundreds of magic tricks for beginners and more experienced illusionists, many for as little as $1. Turn pennies into dimes, make water disappear, and suspend a light bulb in thin air. In addition to hardware, the catalog offers a number of instructive books.

Louis Tannen, Inc.
6 W. 32nd Street, New York, NY 10001-3808
Telephone: 212-239-8383

Catalog price: $10.50

Louis Tannen's catalog takes the form of an 800-page hardcover book. An index at the back helps readers find such tricks as "Sawing a Lady in Half" and the "New Burning Alive Illusion." On a level more appropriate for young magicians are dozens of gadgets that make objects appear and disappear, trick cards and coins, light bulbs that light in the hand, and glasses that turn water into silk. A big library of books offers the key to many more tricks performed with sleight of hand rather than fancy hardware. This is a catalog for magicians, not for pranksters. It contains many simple tricks that beginners will enjoy, but steers clear of practical jokes.

Maps

J. L. Hammett Co.
See entry under *Educational Supplies*.

Hoover Brothers, Inc.
Maps and globes. See entry under *Educational Supplies*.

Rand McNally & Co.
P.O. Box 7600, Chicago, IL 60680
Telephone: 312-673-9100
Catalog price: free

A big catalog of maps and globes. The large classroom map sets on spring rollers are probably a bit much for the average home, but simple wall maps, laminated desk maps, and globes should have a place in every house. And what parent would let a child grow up without an atlas on the shelf?

U.S. Geological Survey
Distribution Branch, Box 25286, Federal Center, Denver, CO 80225
Free price list and ordering information

The U.S. Geological Survey publishes excellent topographical maps of all regions of the country, along with thematic maps that show details like coal fields, fold and thrust belts, and oil and gas investigations. Write for a catalog and index of the maps for your state.

Maternity and Nursing Clothes

Babe too! Patterns
3457 E. K4 Highway, Assaria, KS 67416
Telephone: 913-667-5125
Catalog price: long self-addressed stamped envelope

Patterns for making nursing wear. The line includes dresses, blouses, nightgowns, and running clothes.

Beegotten Creations, Inc.
P.O. Box 1800, Spring Valley, NY 10977
Telephone: 800-722-3390
Free brochure

T-shirts, sweatshirts, and caps for families expecting a new baby. A little bee picture is used in emblems on the clothes to complete such messages as "Father to bee," "Sister to bee," "Brother to bee," and "It won't bee over till the fat lady thins."

Bosom Buddies
P.O. Box 6138, Kingston, NY 12401
Telephone: 914-338-2038
Free brochure

Several styles of nursing bras in 100 percent cotton or nylon lace. One underwire style comes in sizes F, G, and H. A cotton sports bra looks like the answer for mothers who exercise. Washable nursing pads, shoulder-strap cushions, and bra extenders top off the list.

Colten Creations
Potter, 54C Burk Drive, Silver Bay, MN 55614
Telephone: 218-226-3716
Long self-addressed stamped envelope for brochure

Patterns for nursing clothes. At this writing the choices include a nightgown, two dresses, and a jumpsuit.

Maternity and Nursing Clothes

Courier Health Care, Inc.

Sells a prenatal pillow, a scale with a wall-mounted remote readout, and breast pumps, nursing bras, and nursing nightgowns. See entry under *Baby Needs*.

Creations by Paula Brown
13211 N.W. 39th Avenue, Vancouver, WA 98685
50¢ for brochure and fabric samples

Nursing dresses in a half-dozen styles and a pair of nursing overalls, all designed by Ms. Brown. The brochure shows sketches of each piece; attached fabric swatches let you see exactly what it can be made of. Cotton and poly-cotton calico prints make up the bulk of the sample pile; stripes, corduroy, and denim are also available.

Daddy's Tees
P.O. Box 160214, Miami, FL 33116
Telephone: 800-541-7202 or 305-271-2073
Free flier

Gift T-shirts for expectant fathers, printed with the message "Daddy to Be." For grandparents-in-waiting the firm sells matching shirts reading "Grandpa to Be" and "Grandma to Be."

Decent Exposures
2202 N.E. 115th, Dept. 196, Seattle, WA 98125
Telephone: 206-364-4540
Free brochure

Makes a simple cotton knit bra without snaps or fasteners in an attractive palette of colors. For nursing, it can be easily lifted up and down. Sizes from 32AA to 42H. Other sizes can be custom made at a small additional charge.

Designer Series
P.O. Box 736, N. Hollywood, CA 91609
Telephone: 818-763-7315
50¢ for brochure

Mary Jane nursing bras in a dozen different styles, from a soft cotton sleep bra to a sturdier sports bra. Sizes range from 32A to 46G. For pregnant women the company offers a tummy sling that looks like a pair of oversized underpants with wide suspenders.

Feeling Fine Programs

Books and videos on fitness and health, including a few on pregnancy, childbirth, and breastfeeding. See entry under *Videos*.

5th Avenue Maternity
P.O. Box 21826, Seattle, WA 98111-3826
Telephone: 800-426-3569 or 206-343-7046
Catalog price: $2

Fashionable maternity wear for work and play in a series of seasonal color catalogs. Sweat suits, cotton pants, and polo shirts look like comfortable casual wear; a flowered party dress might fit the bill for a special occasion. Bathing suits, nursing bras, hosiery, underpants, and camisoles are sold to complete the wardrobe. Watch for the pre-inventory sale in July.

Garnet Hill

Cotton nursing clothes. See entry under *Children's Clothes*.

La Leche League International
P.O. Box 1209, Franklin Park, IL 60131-8209
Telephone: 312-455-7730
Catalog price: free

An organization devoted to teaching parents about breastfeeding, nutrition, and childbirth, La Leche League puts out a catalog of books, information packets, and breastfeeding supplies that's available to members and non-members alike. More than a dozen books and booklets on breastfeeding form the core of the list, with the League's own *The Womanly Art of Breastfeeding* prominently featured. Books and pamphlets on childbirth, child rearing, postnatal exercise, and other parenting subjects fill out the rest of the bookstore. The hardware section presents several different breast pumps and milk storage systems, along with breast shields and shells.

Blouse and pants from 5th Avenue Maternity

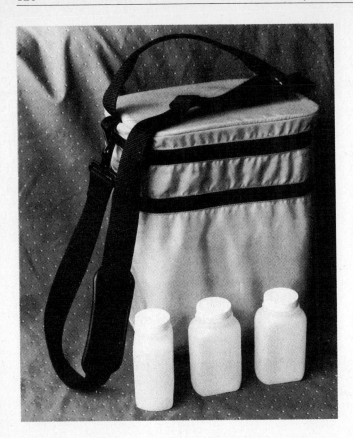

The Lait-Ette Company's Lac-Tote milk storage system

Lait-Ette Company
183 Florence Avenue, Oakland, CA 94618
Telephone: 415-655-5110

Free brochure

Sells the Lac-Tote storage system for chilling and transporting expressed milk, a product designed for working mothers who have to spend time away from their babies. The nylon bag has two compartments: one holds the breast pump and other accessories, the other chills the milk. According to the brochure, the bottles are cooled "nearly to freezing" in less than an hour and kept below 40° for up to 24 hours.

Elizabeth Lee Designs
P.O. Box 696, Tabiona, UT 84072
Telephone: 801-454-3378

Long self-addressed envelope for brochure

Patterns for nursing clothes, including three dresses and a nightgown. One of the dresses can be worn as a maternity outfit.

Lopuco Ltd.
1615 Old Annapolis Road, Woodbine, MD 21797
Telephone: 800-634-7867 or 301-489-4949

Free brochure

Sells the Loyd-B breast pump, a hand-operated pump that Lopuco claims works in half the time of more expensive electric models and requires the use of only one hand.

Los Angeles Birthing Institute
Pregnancy, birth, and parenting books. See entry under *Books*.

Maternity Modes, Inc.
4950 W. Main Street, Skokie, IL 60077
Telephone: 312-677-0099

Catalog price: $1

Maternity clothes, from casual sundresses, T-shirts, and pants to more formal outfits of silk, rayon, and cotton lace. Twenty-five sets and dresses are pictured in the color catalog, many suitable for wear in a professional office.

Mother Nurture Breastfeeding Apparel & Patterns
103 Woodland Drive, Dept. MOSP, Pittsburgh, PA 15228-1715
Telephone: 412-344-5940

Catalog price: $2

Nursing clothes and clothes for early pregnancy. Sewers can order patterns; those without the time or inclination can buy them ready-made. The catalog concentrates on clothes appropriate for work, but also offers casual wear and nightgowns. The catalog shows drawings of the more than 30 outfits sold. Breast shields, pads, and cups are offered from a number of manufacturers. Just one breast pump has been chosen: the battery-powered device from Gentle Expressions.

Motherhood
1330 Colorado Avenue, Santa Monica, CA 90404-2142
Telephone: 800-227-1903 or 213-450-1011

Catalog price: free

Two color catalogs, one of maternity clothes and the other of nursing fashions. Motherhood has 300 stores across the U.S., and the company encourages shoppers to patronize the shops before resorting to the mail. The maternity catalog, in seasonal editions, carries blouses, cotton and knit pants, shorts, collared T-shirts, overalls, jeans, sundresses and dresses for cooler weather, nightgowns, slips, bras, panties, and pantyhose. The nursing catalog sells several styles of nightgowns and robes, a half-dozen nursing bras (including one in black lace), and several shirts and nursing tops.

Mother's Place
6836 Engle Road, P.O. Box 94512, Cleveland, OH 44101-4512
Telephone: 800-444-6864

Catalog price: free

Inexpensive maternity clothes, many of the outfits formal enough for office wear. All of the dresses were priced under $30 in 1989. Blouses, shirts, sweaters, skirts, sweatshirts, and nightshirts were in the $10 to $20 range. Maternity panties, pantyhose, and slips complete the preterm selection. A nursing bra is sold for the months after.

Mother's Wear
1738-S Topanga Skyline, Topanga, CA 90290
Telephone: 800-322-2320 or 213-455-1426

Free brochure

Nursing bras in several styles, and a cotton knit nursing gown with a bed jacket. All the bras are made with cotton cups. Sizes range from 34C to 40E.

Mothers Work
1309 Noble Street, 5th Floor, Philadelphia, PA 19123
Telephone: 215-625-9259

Catalog price: $3

Maternity clothes for working women. Navy suits are sold for the conservative office, plaid jumpers and bright paisley dresses for those who work with looser dress codes. Fabrics include wool gabardine, rayon, chandul crepe, silk, cotton, denim, and acrylic knit. Forty outfits are offered for the sharply dressed worker, a few for evening wear, and a few more for weekends. The catalog is a bit more costly than some, but it comes with a thick wad of fabric swatches.

Dresses from Mothers Work

Motherwear
P.O. Box 114-SP, Northampton, MA 01061
Telephone: 413-586-3488

Catalog price: free

Nursing clothes, baby necessities, and parenting books. The clothes for mother include blouses, a jumper, a sweatshirt, warm- and cool-weather nightgowns, and a high-neck dress. Maternity panties and pantyhose from Mary Jane fill a need in the weeks before the big event. Nikky diaper covers, flannel diapers, cotton newborn hats, infant shoes, and the Baby Matey carrier answer during the months after. The bookshelf is stocked with titles on childbirth, breastfeeding, childhood sleep problems, and other parenting concerns.

Holly Nicolas Nursing Collection
P.O. Box 7121, Orange, CA 92613-7121
Telephone: 714-639-5933

Catalog price: long self-addressed envelope with two first-class stamps

Nursing dresses, blouses, and sleep sets. The brochure describes each of the ten outfits without pictures, using fabric swatches to show the color choices. Some are formal enough for office wear; others are designed for more relaxed environments.

Page Boy Maternity
8918 Governors Row, Dallas, TX 75247
Telephone: 800-225-3103 or 214-951-0055

Catalog price: $2.50

Sophisticated maternity wear in a series of seasonal color catalogs. Cotton knit sundresses, T-shirts, and pants present a note of informality, as does a three-piece sweater suit. But the weight of the inventory leans to more formal attire. A black velvet dress would pass at an inauguration ball. A silk floral tunic over a black dress would shine in any office, as would dozens of other outfits in rayon, cotton, linen, and lace. Dresses in floral prints, bright fuchsia, and pastel lace will answer for special daytime occasions. Prices are on a par with the high styling.

J. C. Penney Company, Inc.
Maternity clothes, nursing bras, and breast pumps. See entry under *General Catalogs*.

Reborn Maternity
564 Columbus Avenue, New York, NY 10024
Telephone: 212-362-6965

Catalog price: $2

A glossy color catalog of fashionable maternity wear, some for the office, some for evening events, and some for more casual occasions. A rayon jacket and skirt with a white cotton blouse combine for a conservative office look; a couple of dresses and two-piece outfits are just

An informal outfit from Page Boy Maternity

as respectable. Knit pant-and-top sets offer a little more comfort. A striped jumpsuit looks like a handy piece for weekend errands.

ReCreations Maternity
P.O. Box 191038, Columbus, OH 43209
Telephone: 800-621-2547 or 614-236-1109

Catalog price: $3

Stylish maternity wear made with comfort in mind. Sharp-looking cotton knit pants, tops, skirts, jumpers, and jumpsuits come in a rack of colors for mixing. Solid-color dresses of linen, silk, and rayon offer a more formal alternative; sundresses and other relaxed warm-weather outfits offer choices for time away from work.

Sears, Roebuck & Co.
A full line of maternity clothes and underwear is offered in the main "Style" catalog, along with nursing bras. See entry under *General Catalogs*.

Tracy Creations
P.O. Box 189641, Farmington Hills, MI 48018

Catalog price: long self-addressed stamped envelope

Makes the Nursing Drape, an oversized bib for mothers to wear while nursing babies in public. It fits around the neck with a Velcro closure and drapes discreetly over the child.

Model Trains, Planes, Cars, and Boats

AG Industries
A radio-controlled model sailboat. See entry under *Teddy Bears*.

Ambient Shapes, Inc.
P.O. Box 5069, 1015 2nd Street, Hickory, NC 28601-5069
Telephone: 800-438-2244 or 704-324-5222

Free flier

Detailed 1/18-scale models of several classic sports cars. Made of metal with soft rubber tires, these cars have almost everything but working motors. The steering wheels work; hoods, passenger doors, and trunks open to reveal minutely reproduced engine compartments, instrument panels, and spare tires. "Suitable for drivers three years old and up," claims the flier.

America's Hobby Center, Inc.
146 W. 22nd Street, New York, NY 10011-2466
Telephone: 212-675-8922

$6 for three years of catalogs

Periodic tabloid-format catalogs crammed with model trains and accessories (mostly HO and N gauge); radio-controlled cars, planes, and boats; sailplanes; and scores of motors, parts, and tools.

Atlas Tool Co.
378 Florence Avenue, Hillside, NJ 07205

Catalog price: long self-addressed envelope with two first-class stamps (request "All Scales Catalog")

HO-, O-, and N-scale model trains, with track, bridges, and trackside buildings (HO only). The slim 32-page catalog offers the most to HO- and N-scale engineers, and just a limited selection of cars and track for O modelers. Those setting up their first train set should send a long self-addressed stamped envelope for the brochure "Beginning Modeling: Track Basics."

Bluejacket Ship Crafters
P.O. Box 18, Castine, ME 04421-0018

Catalog price: $2

Kits and supplies for building wooden ship models. The color catalog shows 14 intricately detailed models which are sold as kits, from the U.S.S. *Constitution* to a 1940s tanker. Modelers who prefer to start from scratch can make up an order from the warehouse of miniature fittings and finely milled lumber. Those who aren't ashamed of shortcuts can order ladders, lifeboats, and figureheads ready-made.

CARS
31566 Railroad Canyon Road, Suite 1-505, Canyon Lake, CA 92380

Catalog price: $2

Hundreds of model cars and trucks in N, O, Standard, and HO scales, with a few ships, airplanes, and horse-drawn carts thrown in for variety. The catalog seems to be split about evenly between kits and finished miniatures.

Childcraft
Miniature racing-car sets. See entry under *Toys*.

Circus Today
3132 S. Highland Drive, Las Vegas, NV 89109
Telephone: 800-782-0022

Catalog price: free

Dozens of radio-controlled model plane and helicopter kits, with a few cars and boats for balance. A full line of radios, engines, accessories, and tools looks like just about anything the ground controller could ask for.

Doug's Hobby Shop
24 Industrial Park Drive, Waldorf, MD 20602
Telephone: 800-445-4544 or 301-843-7774

Catalog price: free

Radio-controlled boats, cars, planes, and helicopters in a manageable catalog that is neither overwhelming nor skimpy. Roughly 50 models are offered in all, along with a range of radios, motors, and other parts.

Doug's Train World
6565 University Avenue, Des Moines, IA 50311
Telephone: 515-274-4424

Catalog price: long self-addressed stamped envelope

Lionel trains and accessories in 027, O, L, and S gauges. A few of the engines and cars are pictured in the catalog, but dozens more are simply listed by name. Several sets are offered by name alone, without further explanation of what they contain. For those who don't know their way around the Lionel train yard without the help of pictures, Doug suggests a letter to Lionel for a full color catalog or a trip to a local hobby shop for a look at the store copy.

Estes Industries
P.O. Box 227, 1295 H Street, Penrose, CO 81240
Telephone: 719-372-6565

Catalog price: free

The first name in model rockets and rocketry supplies. Estes' 70-page catalog includes performance rockets that can soar up to 2,600 feet as well as detailed flying replicas of the Saturn V rocket, the Space Shuttle, and a Mars lander.

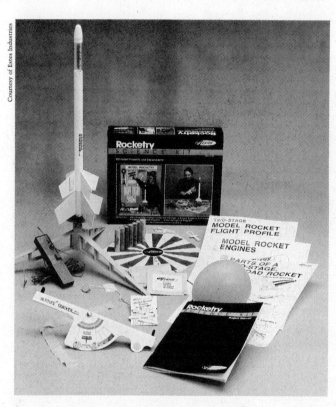

Rocketry Science™ Kit from Estes Industries

H.O. Center of the World
P.O. Box 348, Marion, CT 06444
Telephone: 203-628-8948

Catalog price: $2

Cars, parts, and accessories for HO-scale slot racing. The catalog lists over 200 cars and trucks, among them lighted vehicles, flamethrowers, and a fire engine with flashing lights, a clanging bell, and an extension ladder. Track comes in corkscrews, loops, banked curves, and crossings. Accessories include lap counters, controllers, and road hazards.

Hobby Shack
18480 Bandilier Circle, Fountain Valley, CA 92728-8610
Telephone: 800-854-8471 or 714-963-9881

Catalog price: free

Hundreds of radio-controlled model planes, cars, and boats, sold finished and in kits. Those who want to soup up a model or build from scratch can order from the expansive list of components and accessories. Watch for catalog specials and occasional sales.

Hobby Surplus Sales
287 Main Street, P.O. Box 2170J, New Britain, CT 06050
Telephone: 800-233-0872 or 203-223-0600

Catalog price: $2

A fat catalog packed with model trains, cars, planes, boats, and rockets. Model railroaders will have the most to choose from—more than half of the catalog is devoted to Lionel, American Flyer, HO-, and N-scale trains and accessories. The parts list alone takes up nine pages in tiny type. Engines, cars, track, buildings, scenery, signs, decals, and books fill 60 more. The back of the catalog holds an array of steam engines, model cars, planes, boats, and rockets, plus supplies for collectors of coins, stamps, and baseball cards. Even the dollhouse gets some attention—more than a dozen sets of miniature room furnishings are offered in kit form.

Indoor Model Supply
P.O. Box 5311, Salem, OR 97304
Telephone: 503-370-6350

Catalog price: $2

Kits and supplies for building ultra-light model planes. A few simple gliders are made of balsa wood. More elaborate propeller planes are covered with condensed paper or microlite film and powered by wound-rubber motors. These are quality kits with precision-cut parts, not the ordinary dime-store fliers.

An offshore rescue boat by Robbe Model Sport, Inc.

Lionel Service
50911 W. Boulevard, Mt. Clemens, MI 48045
Telephone: 313-949-4100

$1 for parts price list

Lionel sells replacement parts to individual customers by mail, but not complete trains or sets. Collectors may also want to write for the full catalog, which can be used as a reference when shopping at local hobby stores.

Northeastern Scale Models Inc.
Scale lumber for modelers, including wooden train parts. See entry under *Dollhouses and Miniatures*.

Polk's Modelcraft Hobbies
346 Bergen Avenue, Jersey City, NJ 07304
Telephone: 201-332-8100

Catalog price: $2; free flier

We had some trouble getting hold of this catalog, as it's not always in stock. It finally arrived a year after our initial request. A much smaller flier is free and is mailed without delay. Its six pages display some of Polk's most popular model kits and radio-controlled vehicles. The full catalog offers scores of radio-controlled planes, boats, and cars with all the necessary motors and parts for serious hobbyists, as well as scenery, track, and buildings for model railroaders and kits and parts for wooden-ship modelers.

Race Prep
20115 Nordhoff Street, Chatsworth, CA 91311
Telephone: 818-709-6800

Catalog price: $2

Radio-controlled model cars and parts from many different makers. Race Prep is the official distributor of AYK Racing USA and sells its own line of Race Prep accessories. Kyosho, Tamiya, and Team Associated are among the other manufacturers. The list has no pictures or descriptions, so venture here only if you know your radio controls well.

Rail Scene
T-shirts, sweatshirts, and hats printed with railroad insignia. See entry under *Children's Clothes*.

Robbe Model Sport, Inc.
180 Township Line Road, Belle Mead, NJ 08502
Telephone: 201-359-2115

Catalog price: $6, $4 deductible from order of $50 or more

More than 450 radio-controlled model planes, cars, and boats dramatically photographed in a big color catalog. All are sold as "highly prefabricated" kits designed for the novice. Motors, radios, batteries, switches, propellers, and other parts are sold for more ambitious modelers building from scratch.

Walthers' stock car from Terminal Hobby Shop

F. A. O. Schwarz
Marklin electric trains. See entry under *Toys*.

Scientific Models, Inc.
Wooden ship models. See entry under *Dollhouses and Miniatures*.

Sears, Roebuck & Co.
Model trains, slot cars, and radio-controlled vehicles are listed in the "Toys" catalog. See entry under *General Catalogs*.

Sig Manufacturing Co.
401-7 S. Front Street, Montezuma, IA 50171
Telephone: 515-623-5154
Catalog price: $3

Sig manufactures an extensive line of radio-controlled model planes and rubber-band-driven balsa fliers, along with all the necessary motors, adhesives, fuels, coverings, tools, parts, and balsa pieces. The thick catalog peddles all of the company's own merchandise plus models, radios, engines, and tools from dozens of other makers. If you're ever in Montezuma, stop in for a factory tour any weekday between 7:00 A.M. and 3:30 P.M.

Standard Hobby Supply
P.O. Box 801, Dept. TC, Mahwah, NJ 07430
Telephone: 201-825-2211 (800-223-1355 to order)
Catalog price: $2

Radio-controlled model planes, cars, and boats, sold with separate radios, motors, replacement parts, tools, and supplies. Science and hobby equipment makes up an important sideline here. Telescopes, rock polishers, chemistry sets, and see-through models of combustion engines are offered for those who want to dabble in science. Serious model builders can shop among the vices, saws, airbrush sets, drills, and lathes.

Terminal Hobby Shop
5619 W. Florist Avenue, Milwaukee, WI 53218
Telephone: 800-347-1147 or 414-527-0770
Catalog price: see below

The retail mail-order division of Wm. K. Walthers, Inc., Terminal Hobby Shop publishes three massive model-railroad catalogs. "The World of HO Scale," available for $13.98, weighs in at almost 800 pages. Trains, track, landscaping miniatures, and supplies from nearly 300 manufacturers cram its pages, along with color photographs of exceptional layouts and dioramas. "The World of Large Scale," priced at $9.98, offers trains and accessories from such makers as LGB, Bachmann, Delton, Kalamazoo, American Standard Car, Lionel, Pola, and Walthers. "The World of N & Z Scale" costs $9.98 and runs to a hefty 300 pages.

Tower Hobbies
P.O. Box 778, Champaign, IL 61824-0778
Telephone: 800-637-6050 or 217-398-1100
Catalog price: $3

Radio-controlled planes, cars, and boats presented in a 300-page color catalog with hefty bimonthly updates. Match a stern-wheel riverboat against a Dumas Mach 3.5 V-bottom race boat, or a Fokker triplane against a Mirage III jet fighter. The catalog has hundreds of kits to choose from, and all can be modified with extra radio and engine accessories. Balsa wood, covering materials, adhesives, paints, tools, books, and parts are sold for modelers who work from scratch. Radio-free fliers can choose from an array of gliders and rockets.

Toy Designs
Wound-rubber-powered aircraft kits. See entry under *Woodworking*.

Ye Olde Huff N Puff, Mfg.
P.O. Box 53, Pennsylvania Furnace, PA 16865
Telephone: 814-692-8334
Catalog price: see below

Kits for making model train cars and railside structures in several scales. The HO and HOn3 list, which costs $1, is the biggest of the set, with dozens of specialty cars for hauling pickles, livestock, oil, and coal. Water tanks, engine houses, country stores, and firehouses can be built at whistle-stops. The O-scale list, also $1, offers boxcars, cabooses, and refrigerator cars in a slightly smaller range of choices. The S and On3 lists amount to just two pages each, and can be had for a self-addressed stamped envelope.

Music

ABC School Supply
Records and musical instruments. See entry under *Educational Supplies*.

Alcazar Records
P.O. Box 429, S. Main Street, Waterbury, VT 05676
Telephone: 800-541-9904 or 802-244-8657
Catalog price: free (request children's music catalog)
Hundreds of records, tapes, and compact disks from independent producers, presented in a big descriptive catalog. Artists include Raffi, Sally Rogers, Cathy Fink, Mister Rogers, Bill Harley, Esther Nelson, Tom Paxton, Mike and Peggy Seeger, Riders in the Sky, Jim Valley, Linda Arnold, and many others. Alcazar also sells the Wee Sing books and cassettes, a number of Sesame Street music cassettes, and a nice selection of story recordings, songbooks, and children's videos. Since the company sells music for adults as well as kids, specify the children's music catalog when you write.

Banbury Cross
1408 Kerper Street, Philadelphia, PA 19111
Telephone: 215-745-5121
Catalog price: Long self-addressed stamped envelope
Musical instruments for young children. Finger cymbals, maracas, triangles, cluster bells, handle castanets, rhythm sticks, wrist bells, and a hand drum can all be managed by the youngest musicians and tolerated by patient parents. A bit louder are a pair of 7-inch cymbals and a tom-tom. Several small song bells (like miniature xylophones) and a recorder are sold for children ready to master melodies.

Better Beginnings Catalog
Tapes of children's music. See entry under *Books*.

Chaselle, Inc.
9645 Gerwig Lane, Columbia, MD 21046
Telephone: 800-CHASELLE (800-492-7840 in MD) or 301-381-9611
Catalog price: free (specify "Pre-School & Elementary School Materials" catalog)
The Chaselle catalog offers percussion instruments for younger children—finger cymbals, tambourines, hand bells, maracas, and drums—as well as a plastic recorder, a wooden ukelele, a child's xylophone, and an Autoharp. See main entry under *Educational Supplies* for a description of the company's other offerings.

Children's Book & Music Center
2500 Santa Monica Boulevard, Santa Monica, CA 90404
Telephone: 800-443-1856 or 213-829-0215
Catalog price: free
A delightful and comprehensive 80-page catalog of children's books, records, tapes, videos, and musical instruments. From classic recordings by Pete Seeger, Woody Guthrie, Burl Ives, and The Limelighters to more recent albums by Cathy Fink, Fred Penner, Raffi, Joe Scruggs, Peter Alsop, and Bonnie Phipps, this list has something for every young music lover. And if parents have trouble making choices from the catalog's written descriptions, they can dial a special number (using a touch-tone phone: 213-385-5312) to hear samples of both new recordings and old favorites. Durable record and cassette players are offered, along with headphones and a big assortment of unusual rhythm and melody instruments. For a description of the book and video offerings, see separate entries under *Books* and *Videos*.

Children's Music House
109 E. Main Street, Clinton, CT 06413
Telephone: 203-669-9521
Catalog price: free
Over 500 children's recordings on record and cassette. Stories, folk tales, and children's poetry recordings start

A musical gift basket from Banbury Cross

out the list, including *Pinocchio* read by Cyril Ritchard, Rudyard Kipling's *Just So Stories* narrated by Boris Karloff, *The Wizard of Oz* performed by Ray Bolger, and Carl Sandburg reading his own poems. Sound tracks to popular films and musicals take up another chunk of the catalog, from *Annie* to *You're a Good Man, Charlie Brown*. Folk songs, songs in French and German, and collections by such artists as Raffi, Rosenshontz, Tickle Tune Typhoon, and Hap Palmer round out the list. A rich inventory that is certainly worth exploring.

Children's Recordings
P.O. Box 1343, Eugene, OR 97440
Telephone: 503-485-1634
Catalog price: free

Stories and music for children on records and cassettes. The catalog offers recordings by Raffi, Fred Penner, Cathy Fink, Kevin Roth, Tom Glazer, Hap Palmer, Malvina Reynolds, and dozens of other artists. Story recordings include Claire Bloom reading *The Tale of Peter Rabbit* and *Black Beauty*, Cyril Ritchard's performance of *Pinocchio*, and Jack Nicholson's versions of *The Elephant's Child* and *How the Camel Got His Hump*.

Chinaberry Book Service
Children's music cassettes. See entry under *Books*.

Constructive Playthings
Percussion instruments, plus records and cassettes for young children. See entry under *Educational Supplies*.

Cumberland General Store
Dulcimers, mandolins, banjos, fiddles, and harmonicas. See entry under *Toys*.

Discovery Music
4130 Greenbush Avenue, Sherman Oaks, CA 91423
Telephone: 800-451-5175 or 818-905-9794
Free flier

Discovery Music produces the popular *Lullaby Magic* and *Morning Magic* tapes, which blend older melodies from Mozart and Brahms with more contemporary tunes from the Beatles and Paul Simon. On one side the lyrics are sung quite beautifully by Joanie Bartels. The other side plays just the instrumental track so parents can do the singing. These and three more tapes, *Lullaby Magic II*, *Travelin' Magic*, and *Sillytime Magic*, are available directly from the source. (They can also be found in most of the other children's music catalogs.)

Early Learning Centre
Cassettes and musical instruments. See entry under *Toys*.

Educational Activities, Inc.
Records and cassettes for young children. See entry under *Educational Supplies*.

Educational Record Center, Inc.
Building 400, Suite 400, 1575 Northside Drive, N.W., Atlanta, GA 30318-4298
Telephone: 404-352-8282 (800-438-1637 to order)
Catalog price: $1

More than 500 records, cassettes, videos, and filmstrips in a teacher's catalog that parents will love. The most popular children's singers are here—Raffi, Fred Penner, Cathy Fink, Joe Scruggs, Mister Rogers, Tom Glazer, Ella Jenkins, Greg and Steve, Rosenshontz—along with such other attractions as Sousa marches, the sound tracks to *The Wizard of Oz* and *Mary Poppins*, classical recordings like the *William Tell Overture*, *Nutcracker Suite*, and *Carmen*, Christmas collections, and "greatest hits" collections from the 1930s, '40s, and '50s. To complement its wide-ranging musical offerings the company sells an equally diverse selection of spoken recordings, from Claire Bloom reading Beatrix Potter to Martin Luther King's greatest speeches and Vincent Price performing ghost stories.

Educational Teaching Aids
Percussion instruments and records for young children. See entry under *Educational Supplies*.

Elderly Instruments
1100 N. Washington, P.O. Box 14210, Lansing, MI 48901
Telephone: 517-372-7890
Catalog price: free (specify which catalogs)

Elderly Instruments puts out three big catalogs. One lists instruments and accessories; another records, cassettes, and compact disks; and the third songbooks and videotapes. Guitars lead off the instrument catalog and set the tone for things to follow. Folk and bluegrass music are the special loves here, and the instruments run to banjos, mandolins, dulcimers, violins, Autoharps, and concertinas, accompanied by several pages of picks, pegs, capos, and straps. Children can flip farther back in

Bambina musical toys, sold by Elderly Instruments and Music for Little People

the catalog to find harmonicas, kazoos, jaw harps, nose flutes, slide whistles, and plastic percussion instruments in brightly colored animal shapes. A musical saw, though it sounds like a toy, could be a dangerous instrument in the wrong hands.

The book and video catalog lists hundreds of song and instruction books, including many for children. Several guitar instruction books are offered for young musicians, and more than 50 songbooks. The catalog of recordings is the densest of all, with thousands of listings from bagpipe to zydeco. Younger children can concentrate on the children's section, with recordings by Raffi, Utah Phillips, Pete Seeger, Sally Rogers, Ella Jenkins, and dozens of other artists. Older children will want to wander through the rest of the catalog, into cowboy songs, reggae, country, and classical recordings.

A Gentle Wind
P.O. Box 3103, Albany, NY 12203
Telephone: 518-436-0391

Catalog price: free

An independent children's recording label with some big awards under its belt. Rachel Buchman's *Hello Everybody!*, a collection of songs for toddlers, won a 1987 Parents' Choice Award. Troubadour's *Are We Almost There?* and Doug Lipman's *Tell It With Me* have also won awards in recent years. Shoppers can choose from new music or traditional songs freshly interpreted. In the spoken-word department, A Gentle Wind offers original stories, traditional tales, and classics like Kipling's *Just So Stories*. All are available on cassette only, at reasonable prices.

Grey Owl Indian Craft Co.
Recordings of American Indian music. See entry under *Hobby Supplies*.

Growing Child
Recordings and simple instruments. See entry under *Toys*.

Harps of Lorien
610 N. Star Route, Questa, NM 87556
Telephone: 505-586-1307

Catalog price: free

Harps and other musical instruments, many of them made specifically for children. A baby harp with 11 widely separated strings would make a lovely first instrument for a young child. A somewhat larger version is offered for older children, and a full-size model for adults. All three harps are tuned to the pentatonic scale. Among the other instruments suited for young children are wooden recorders, panpipes, Sioux and Pueblo drums, and a percussion set (12 instruments for less than $30). Older players can choose from bagpipes, dulcimers, banjos, violins, psalteries, and Tibetan ritual instruments.

John Holt's Book & Music Store
Violins down to $1/16$-scale, recorders, pitch pipes, and pianicas. See entry under *Books*.

Hoover Brothers, Inc.
Musical instruments, records, and tapes. See entry under *Educational Supplies*.

House of Oldies
35 Carmine Street, New York, NY 10014
Telephone: 212-243-0500

Catalog price: $8 (also send want list)

The House of Oldies has more than a million 45s and LPs dating back to the early 1950s. If you want to res-

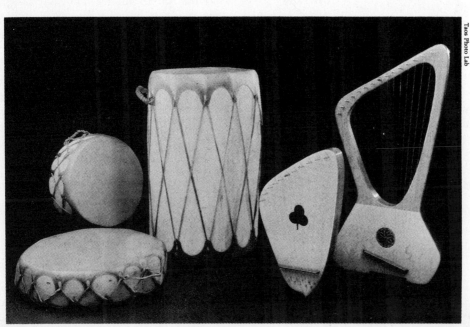

Harps and drums from Harps of Lorien

urrect a favorite record from your past, the company may be able to help. The inventory is made up primarily of popular music, with just a limited selection of children's recordings. Allan Sherman is well represented, as are the early rock 'n' rollers.

M. P. Hughes Dulcimer Co.
4419 W. Colfax Avenue, Denver, CO 80204
Telephone: 303-572-3753

Free brochure

Dulcimers, harps, lyres, banjos, mandolins, and balalaikas sold as kits and finished instruments. The kits are an inexpensive way to buy beginner's instruments, and a wonderful way to learn how they're made.

Kaplan School Supply Corp.
See entry under *Educational Supplies*.

The Kids on the Block, Inc.
9385-C Gerwig Lane, Columbia, MD 21046
Telephone: 800-368-KIDS or 301-290-9095

Catalog price: free

Offers a record (or cassette) of music and stories that deal with "sensitive issues such as disabilities, cultural differences, and divorce." The selection is distilled from the performances of The Kids on the Block puppet troupe.

Kimbo Educational
P.O. Box 477, Long Branch, NJ 07740
Telephone: 800-631-2187 or 201-229-4949

Catalog price: free

Musical teaching aids on records and cassettes. Choose from nursery rhymes, baby and children's games, dance music, music and encouragement for aerobics, and activity recordings like the "Do-It-Yourself Kids Circus" and "Walk Like the Animals." Popular children's performers like Raffi, Rosenschontz, Eric Nagler, and Tickle Tune Typhoon are also sold. The entire list runs to more than 100 titles.

Learn Me Bookstore
Records and cassettes. See entry under *Books*.

Little Angels
P.O. Box 1749, Pacifica, CA 94044

Free brochure

Little Angels has produced a cassette of music-box lullabies and children's songs that is intended to calm newborns and to help young children wind down for bed.

Little Ears
P.O. Box 56168, Tucson, AZ 85703
Telephone: 602-888-2830

Catalog price: free

The folks at Little Ears have chosen what they consider to be the 50 best children's tapes, and they offer them for sale at 10 to 20 percent off the suggested retail prices. We have few arguments with their selection. The choices skip from sing-alongs like Bob McGrath's *If You're Happy and You Know It*, to country recordings from Cathy Fink and Riders in the Sky, to stories narrated by Tammy Grimes, Jack Nicholson, and Julie Harris. Add a little Raffi, some lullaby and wake-up tapes, some classical recordings for children, and some Bible songs, and the mix is complete. The catalog is free and the prices are good, so how can you lose?

Living Sound Productions
P.O. Box 5486, Balboa Island, CA 92662
Telephone: 213-547-4679

Catalog price: free

Relaxation tapes for children and adults. The tapes combine nature sounds, music, guided imagery, and relaxation techniques. Two have been designed for children who are especially active at bed and nap time; two for adults who have trouble getting to sleep at night.

Marlboro Records
845 Marlboro Spring Road, Kennett Square, PA 19348
Telephone: 215-793-1432

Free brochure

Children's music by Kevin Roth. *Parents* magazine judged these recordings "everything children's music should be: funny, warm, sweet, entertaining, a little bit silly, and very comforting." A lullaby album is joined by three collections of daytime songs.

The Mind's Eye
Books and music on cassette. See entry under *Books*.

Music Box World
P.O. Box 7577, Rego Park, NY 11374
Telephone: 718-626-8153

Catalog price: $2

Dozens of music boxes crafted of wood, porcelain, and other materials. Choose the box you want, then select a tune from the list of almost 1,000 possibilities. If you're tired of hearing music boxes tinkle out "It's a Small World" and "Jingle Bells," try ordering one that plays "When You Wish Upon a Star" or "Your Cheating Heart." Only the Minnie and Mickey Mouse boxes are stuck with tunes from the factory. He plays "Everything Is Beautiful." She plays "Some Enchanted Evening."

Music for Little People
P.O. Box 1460, Redway, CA 95560
Telephone: 800-346-4445

Catalog price: free

A beautiful color catalog, illustrated with colored-pencil artwork, that brings together children's recordings and a fascinating array of musical instruments. The records and tapes include Wee Sing, Peter Alsop, Raffi, Pete Seeger, Rosenschontz, John McCutcheon, Riders in the

The Suzuki Method of Music Education

Shinichi Suzuki traveled from Japan to Germany in 1924 to study classical violin. While embarking on the arduous task of mastering the German language, he marveled at the ease with which young children learn the complexities of their mother tongue. Though such an amazing feat of learning is accepted as natural by most people, to Suzuki it suggested that young children hold an incredible potential to master difficult subjects.

When he returned to Japan to teach violin, Suzuki began to apply some of his new theories. Children learn language, he reasoned, because they are surrounded by it in their environment. When they begin to speak, their parents repeat their first words back to them, both to encourage them and to let them hear the sounds spoken clearly. Suzuki also noted that children seemed most adept at absorbing from and conforming to the atmosphere around them while very young. He decided that all children, not just those deemed gifted or musical, could learn to play the violin if it was presented to them as naturally as was their mother tongue.

Suzuki first outfitted his young pupils with instruments of a size they could handle and showed them basic fingering and bowing techniques. Then he played fine examples of violin music and encouraged them to imitate him. He required that daily home practice be supervised by an interested family member who also attended the lessons, and asked parents to play violin with their children at home. As the lessons progressed he organized groups so that children could play music with their peers. Only later, after his students had mastered playing by imitation, did he begin to teach reading from notation.

Through the late 1940s and early 1950s, the method began to gain in popularity in Japan, and Suzuki's pupils were able to stage some remarkable concerts. At a national demonstration in Tokyo in 1955, Suzuki conducted 1,200 young violinists, aged 4 to 15, in a program that included Vivaldi's Concerto in A minor and selections from Bach and Handel. The more advanced students played movements from Bach's Concerto in A minor and Mozart's Violin Concerto in A major. Through accounts and films of these concerts, music teachers in the West began to take note of Suzuki's teaching methods. In

Photo: Arthur Montzka

1964, when he brought a group of children to perform at the Music Educator's National Conference in Philadelphia, he won scores of converts. Suzuki made several return visits to lead workshops for music educators in the United States, and other teachers have arranged for annual performance tours to show the proficiency of young Japanese students.

Today thousands of children in the United States and Canada study music by the Suzuki method, and the principles have been adapted to the teaching of other instruments. Shar Products Company (see entry opposite) sells scaled-down violins and cellos—both the original Nagoya-Suzuki instruments and some from other makers—along with adjustable stools for young players and a full line of books and accessories. John Holt's Book & Music Store (see entry under *Books*) sells small-scale violins.

Sky, and Tickle Tune Typhoon. Among the instruments are half- and three-quarter-size guitars, a ukulele, Peruvian panpipes, a step xylophone, animal-shaped rhythm instruments, harmonicas, kazoos, Pueblo and Irish drums, a West African mbira, recorders, and a lap harp. The catalog also carries a good collection of recorded stories and children's videos. Among them: Joe Hayes's traditional tales of the American Southwest, Jay O'Callahan and Naomi Baltuck's original yarns, and a library of classic tales from The Mind's Eye.

Music in Motion
109 Spanish Village, #645-D, Dallas, TX 75248
Telephone: 800-445-0649 or 214-231-0403

Catalog price: $3

Recordings, musical instruments, songbooks, and gifts for music lovers of all ages. For children the company offers the Wee Sing book and cassette collection; a number of books on composers, instruments, dance, and music history; games relating to music; and a selection of easy-to-master instruments. Parents may be interested in the brass candlesticks shaped like trumpets and French horns, or the toilet seat decorated with a piano keyboard pattern.

Nasco
Records, cassettes, and musical instruments for young children. See entry under *Educational Supplies*.

The Natural Baby Co.
Alcazar Records' catalog of children's music. See entry under *Toys*.

PlayFair Toys
Music cassettes and simple musical instruments. See entry under *Toys*.

Recorded Books
Books on cassette. See entry under *Books*.

Richman Cotton Company
Percussion instruments and children's music tapes. See entry under *Children's Clothes*.

The San Francisco Music Box Co.
Mail Order Department, P.O. Box 7817, San Francisco, CA 94120-7817
Telephone: 800-227-2190 or 415-653-3022

Catalog price: free

A fat color catalog devoted exclusively to music boxes. The familiar wooden boxes are here, along with musical snow globes, chiming teddy bears, night-lights, and jack-in-the-boxes that play tunes when they pop. Many of the instruments come with a choice of music, and the company has set up a toll-free dial-a-tune service to let shoppers hear the songs before they order.

Sears, Roebuck & Co.
Musical instruments and record and tape players can be found in Sears's "Toys" catalog. See entry under *General Catalogs*.

Shar Products Company
2465 South Industrial Highway, P.O. Box 1411, Ann Arbor, MI 48106
Telephone: 800-248-SHAR (800-482-1086 in MI) or 313-665-7711

Catalog price: free

Instruments, music, and accessories for classical string players. Suzuki violins and cellos are sold for small players, the violins in sizes down to $1/16$ scale and the cellos down to $1/10$ size. Parents who can't afford the original Nagoya-Suzuki instruments can try less expensive Twinkle and Saito makes. Much more expensive instruments are also sold, from such makers as David Burgess, Roman Teller, and Nicholas Deluane. Violas by A. R. Seidel are sold in lengths from 13 to $16½$ inches. An adjustable Suzuki cello stool makes smaller players comfortable; cases and bows fit all sizes of instruments. The accessory list continues on to strings, mutes, rosin, chin rests, shoulder pads, metronomes, music stands, bridges, pegs, and repair tools. A long list of sheet music and student books offers something for players at all levels.

Sweet Baby Dreams
220-F Dela Vina Avenue, Monterey, CA 93940
Telephone: 408-659-3259

Free brochure

Soothing lullabies and instrumental music for calming babies, children, and adults when it's time to wind down for sleep. *Slumberland* combines flute, piano, and synthesizer music in original calming compositions. *Sweet Baby Dreams* adds the sound of a heartbeat to the music, to help crying babies relax. *Lullabies from Around the World* is a collection of traditional bedtime melodies performed in gentle instrumental arrangements. A dozen other tapes create a range of soothing atmospheres.

The Sycamore Tree
Cassettes and songbooks, including a number of Christian hymns. See entry under *Educational Supplies*.

Troubadour, Inc.
11 Spring Valley Road, West Roxbury, MA 02132
Telephone: 617-327-8965 or 734-1416

Catalog price: free

A Gentle Wind (see above) sells two of Troubadour's albums, but parents who want the entire opus can go direct to the source. Troubadour offers an album of songs composed by children, two other music and poetry collections, a songbook, and two books of poetry for parents.

Outdoor Gear

Laura Ashley by Post
Dress wool coats. See entry under *Children's Clothes*.

Baby Furs by Scandinavian Origins
Sheepskin baby buntings and fur child-carrier liners. See entry under *Baby Needs*.

Eddie Bauer
Cotton sweaters and insulated boots. See entry under *Children's Clothes*.

L. L. Bean, Inc.
Freeport, ME 04033
Telephone: 800-221-4221

Catalog price: free

L. L. Bean sells sleeping bags, tents, and other outdoor gear for family vacations, but no outdoor clothing in children's sizes. The spring 1989 catalog offered a backpack child carrier; the fall and winter list presents such things as the Baby Bag snowsuit and skis for toddlers.

Birth & Beginnings
Baby Bag snowsuits and PolarPlus jackets from American Widgeon. See entry under *Baby Needs*.

Cabela's
812 13th Avenue, Sidney, NE 69160
Telephone: 800-237-4444 or 8888

Catalog price: free

A fat color catalog of outdoor gear and hunting and fishing supplies. There's little specifically for children—some camouflage outfits, some insulated boots, and a couple of sleds—but the wealth of tents, sleeping bags, and other outdoor hardware should merit a look by families who who like to camp, hunt, or fish.

Campmor
810 Route 17 N., Box 997-F, Paramus, NJ 07653-0997
Telephone: 800-526-4784 or 201-445-5000

Catalog price: free

A hundred pages crammed with outdoor gear and camping equipment. Campmor has made at least some effort to outfit younger campers. Four models of sleeping bags come in children's sizes, one of them a mummy bag good down to 5° F. Polypropylene and Thermax long underwear is offered in children's sizes (from 6–8 to 14–16), as are hiking and ski socks, Sorel insulated boots, a water-repellent parka-and-pants set, and a number of cold-weather caps. The company sells the Kelty Ridgeway child carrier (a frame backpack), the Snugli II baby carrier, Acorn slipper socks in small sizes, the Baby Bag snowsuit, and a PolarPlus bunting bag. Loads of tents, cooking sets, compasses, knives, lanterns, and other supplies are offered for trip leaders.

The Company Store
Down parkas, coats, and vests. See entry under *Sheets, Blankets, and Sheepskins*.

Down coat, hat, and mittens from The Company Store

French Creek Sheep & Wool Company
Elverson, PA 19520
Telephone: 215-286-5700

Catalog price: $2

Sells a shearling baby bunting that looks like a toasty way to get an infant through winter errands. The basic sack is rectangular with a zipper up the front. A car seat model has slits for the center belt. Parents will have to resist the pull of a catalog full of adult-size shearling jackets and sweaters in order to get to the bunting.

Frostline Kits
2512 W. Independent Avenue, Grand Junction, CO 81505-7200
Telephone: 800-KITS-USA or 303-241-0155

Catalog price: free

Kits for making jackets, down parkas, rain gear, sleeping bags, tents, backpacks, and other items of protection from the elements. Many of the garments can be ordered in children's sizes. The catalog also offers a whole line of baby-supply kits, including a diaper bag with changing pad, a down crib comforter, a baby carrier, and three different buntings. Many of the listings can also be purchased ready-made.

Gander Mountain, Inc.
P.O. Box 248, Wilmot, WI 53192
Telephone: 414-862-2331

Catalog price: free

A mail-order department store for hunters, crammed with an amazing variety of camouflage and cold-weather wear, decoys, guns, scopes, hunting bows, sleeping bags, knives, and other paraphernalia. Very little of the merchandise is specifically for children, but a page of boots, gloves, jackets, shooting shirts, and camouflage suits should provide the necessities for young hunters (size 8 and up).

Good Gear for Little People
Washington, ME 04574
Telephone: 207-845-2211 or 2233

Catalog price: free

All the necessities and lots of extra fun for the younger members of the family (clothes to size 8). The company sells a great selection of winter wear: the Baby Bag infant snowsuits with optional linings; toddler and child-size snowsuits from American Widgeon, Wyoming Woolens, and Chuck Roast; balaclavas, mittens, and booties made of PolarPlus; and a toasty wool cardigan with matching leggings. Diaper bags, baby carriers, a hook-on high chair, the Gerry baby monitor, and a toddler-size table-and-chair set are sold for new parents. Parents of older children can consider life preservers, a bicycle helmet, a bicycle-mounted child carrier, an artist's smock, and a good collection of activity books.

Les Petits
Winter coats, snowsuits, and boots imported from France. See entry under *Children's Clothes*.

Mountaintop Industries
P.O. Box 66, Eden, VT 05652
Telephone: 802-253-8488

Catalog price: $1

Fishing tackle and wildlife T-shirts and sweatshirts. Of the 20 shirt designs offered, only two—a picture of a charging grizzly bear and one of mallards landing on lily pond—come in children's sizes. A youth's small (6–8) is the smallest shirt available. Thirty pages of hooks, lures, and other fishing gear can outfit anglers of any size.

Munchkin Outfitters
Day packs and a child-sized frame backpack. See entry under *Children's Clothes*.

Patagonia Mail Order, Inc.
Synchilla and Capilene jackets that are at once light and very warm. See entry under *Children's Clothes*.

Patagonia's Synchilla jacket

J. C. Penney Company, Inc.
Tents, sleeping bags, and winter coats. See entry under *General Catalogs*.

Petit Pizzazz
American Widgeon outerwear. See entry under *Children's Clothes*.

The R. Duck Company
Nylon rain suits and ponchos. See entry under *Children's Clothes*.

Ramsey Outdoor
226 Route 17, P.O. Box 1689, Paramus, NJ 07653-1689
Telephone: 201-261-5000

Catalog price: free

Tents, sleeping bags, and day packs. Outdoor wear takes up most of the catalog, but none of it comes in children's sizes.

Recreational Equipment, Inc.
P.O. Box 88126, Tukwilla, WA 98138-0125
Telephone: 800-426-4840 or 206-575-3287

Catalog price: free

Outdoor gear at excellent prices. REI (as the business is known to its friends) is a 50-year-old consumer cooperative with more than 2 million members. Children can be outfitted with Thermax long underwear, Sorel boots, PolarPlus jackets and infant suits, down and polyfill parkas, and a lightweight snowsuit with Thinsulite insulation. A diaper bag has the efficient look of mountaineering equipment. Parents will find luggage, tents, sleeping bags, rainwear, boots and shoes, flannel shirts, Capilene long underwear, and all manner of cold-weather outerwear. Flannel sheets are sold in tasteful patterns for grown-ups and in wild dinosaur patterns for younger tastes. Anyone can shop from the REI catalog and take advantage of its great prices; members get a refund at the end of the year based on their total purchases.

Sears, Roebuck & Co.
Tents and sleeping bags, sold in the "Toys" catalog. See entry under *General Catalogs*.

Stowe Canoe & Snowshoe Co.
Box 207 River Road, Stowe, VT 05672-9968
Telephone: 802-253-7398

Free brochure

Manufacturer of Tubbs snowshoes in several traditional designs. The Junior Alaska Trapper is a long, narrow shoe made to accommodate children and lighter adults. Snowshoe fanatics can order a complete set of snowshoe furniture.

Tackle-Craft
P.O. Box 280, Chippewa Falls, WI 54729-0280
Telephone: 715-723-3645

Catalog price: free

Tools, hooks, jig molds, feathers, and other supplies for fisherpeople who tie their own flies. A stomach pump lets you find out exactly what flies the fish are biting. A long list of exotic feathers, fur from such creatures as badgers and wolves, paints, and threads let you create the perfect lure for the situation.

Vermont Bird Company
P.O. Box 1008, Rutland, VT 05701-1008
Telephone: 802-775-6754

Catalog price: $2 ($5 for international orders)

Cashmere knits for those who can afford the best. Hats, mittens, scarves, and mufflers come in children's sizes, but at grown-up prices. A tiny hat for newborns is packaged in a gift box with a wooden rattle. The children's mittens come with a connecting string to protect the investment from loss. For the extra money, customers get the softest winter wear imaginable, a choice of 22 colors, and detailed attention to sizing.

Cashmere for children from the Vermont Bird Company

Voyageur's
P.O. Box 409, Gardner, KS 66030
Telephone: 913-764-7755

Catalog price: free

Life jackets, waterproof bags, flotation bags, repair kits, and other canoeing and kayaking equipment.

Wilderness Press
2440 Bancroft Way, Berkeley, CA 94704
Telephone: 415-843-8080

Catalog price: free

Maps and guidebooks for hikers and canoeists. The list concentrates on California trails, with ventures up the Pacific Crest Trail into Washington, across the Pacific to Hawaii, and east as far as Minnesota's Boundary Waters Canoe Area. Hikers with young children may be interested in *Backpacking with Babies and Small Children* and *Sharing Nature with Children*.

Parents' Magazines

General Parenting and Family

The Brown University Child Behavior and Development Letter
Manisses Communications Group, Inc., P.O. Box 3357, Wayland Square, Providence, RI 02906-0357
Telephone: 401-831-6020
Annual subscription: $67, US$72 in Canada
Sample issue: free

"Monthly reports on the problems of children and adults growing up." This eight-page journal is thick with news of new research and articles on such subjects as a nursery school's program that teaches children to resolve conflicts peacefully, breastfeeding, practical advice about summer camp, and how school can help children of divorced parents. The reports are more thorough than those found in most magazines and generally end with suggestions for further reading. The small-type format is geared to professionals in the field of child development—no pictures or advertisements break up the pages—but parents able to get past the forbidding format (and the high subscription price) will be rewarded with a wealth of information.

Child
P.O. Box 3176, Harlan, IA 51593-0367
Telephone: 800-772-0222
Annual subscription: $12, $21 outside mainland U.S.
Sample issue: $3

An excellent bimonthly magazine published by The New York Times Magazine Group. An average issue bulks out to about 150 pages, stuffed with a dozen articles on such topics as how children catch colds, how birth order affects a child's personality, how to raise a child with a positive attitude, an analysis of disposable diaper brands, and tips for traveling with young children. Three or four photographic fashion features show up in each edition; as do regular columns on health, medicine, money management, manners, and learning; and reviews of records, books, videos, television shows, and toys. The articles generally go into satisfying depth, and as a barometer of the latest toy and children's fashion trends, *Child* is hard to beat. Many of the articles are illustrated with photographs of unusual playthings, items of furniture, room decorations, and other items; a tiny-type shopping guide at the back shows where it all can be bought.

Christian Parenting
P.O. Box 3850, Sisters, OR 97759
Telephone: 503-549-8261
Annual subscription: $14.97, $18.97 outside U.S.
Sample issue: $2.95, $3.95 outside U.S.

A bimonthly parenting magazine written from a Christian perspective. Regular columns on family health, single parenting, marital relations, and fathering are supplemented by feature articles on such topics as children in the kitchen, allergies, raising a Christian child, and balancing career and family. Eight regular parenting advice columns are broken down by children's ages, from "Infant" to "High School."

Communique
A newsletter for interracial families. See entry for Interracial Family Alliance under *Associations, For Parents and Families*.

Compassionate Friends Newsletter
The Compassionate Friends, Inc., P.O. Box 3696, Oak Brook, IL 60522-3696
Telephone: 312-323-5010
Annual subscription: $10, US$12 in Canada
Sample issue: free

A quarterly newsletter for parents whose children have died. See entry under *Associations, For Parents and Families*.

Dads Only
Family Development Foundation, P.O. Box 340, Julian, CA 92036
Telephone: 619-471-6629
Annual subscription: $24
Sample issue: free

A subscription to *Dads Only* brings six bimonthly newsletters and two "Dad Talk" audiocassettes. The newsletters hold articles like "Building Your Wife's Self Esteem," "Respect or Rebellion" (on managing children), and one on sports equipment appropriate for younger children. Sandwiched between the feature stories are news briefs, book reviews, and ideas for activities. On each hour-long cassette, editor Paul Lewis interviews a noted specialist on marriage, family, child development, or men's issues.

Exchange

A quarterly newsletter for working mothers. See entry for Working Mothers Network under *Associations, For Parents and Families*.

Family Reader

P.O. Box 534, Onalaska, WI 54650

Annual subscription: $15, US$20 in Canada
Sample issue: $3, US$4 in Canada

A bimonthly magazine that culls articles from more than 40 alternative parenting publications, including *Mothering, Growing Without Schooling, Home Education Magazine,* and *Nurturing Today*. A great way to keep informed without spending a fortune on subscriptions.

Focus on the Family

801 Corporate Center Drive, Pomona, CA 91768
Telephone: 714-620-8500

Annual subscription: free
Sample issue: free

A magazine from a Christian ministry "dedicated to the preservation of the home." (Founder James C. Dobson also broadcasts his evangelical message on radio.) The sample issue we saw dealt with parent/teen relations, including a long article on "Helping Your Teen Say No to Sex" and an excerpt from the book *Ten Mistakes Parents Make With Teenagers*. The ministry also publishes two magazines for children: *Focus on the Family Clubhouse* and *Focus on the Family Clubhouse Jr*. Both are almost entirely devoted to scriptural teachings.

ForParents

Interpersonal Communication Services, Inc., 3011 Schoolview Road, Eden, NY 14057
Telephone: 716-992-3316

Annual subscription: $12.95, US$16.95 in Canada
Sample issue: $2, US$3 in Canada

A newsletter that "provides parents of school-age children with advice, resources, and skills for fostering positive family communication and moral development." Regular columns offer "together time" activities, sources of further reading, put-downs to avoid, and exercises for practicing communication skills. Five issues are mailed each year.

Gifted Children Monthly

P.O. Box 115, Sewell, NJ 08080
Telephone: 609-582-0277

Annual subscription: $25, US$35 in Canada
Sample issue: $1

A publication for parents of children "with great promise" between the ages of 4 and 12. Articles cover such subjects as helping an underachiever get back on track, computers and school success, study-skill secrets, and personality patterns in gifted children. A four-page pull-out section for children contains puzzles, tricks, mazes, games, and stories sent in by young readers, a question-and-answer column, and a pen-pal directory. Nine monthly issues are published during the school year.

The Home Business Advocate

195 Markville Road, Unionville, ON L3R 4V8, Canada
Telephone: 416-477-3641

Annual subscription: $25
Sample issue: not available

A monthly newsletter for people who run home businesses and for people considering the idea. Profiles of parent-run businesses are a regular feature, along with practical advice and sources for further information.

Loving Fathers

c/o George Marx, 403B Eagle Heights, Madison, WI 53705
Telephone: 608-238-0526

Annual subscription: $5, US$7 in Canada
Sample issue: free

A quarterly newsletter with writings on fatherhood. Readers share thoughts and experiences in letters, and the editors write on topics like parental leave activism.

Marriage & Family

St. Meinrad, IN 47577
Telephone: 812-357-8011

Annual subscription: $14.95, US$18.95 in Canada
Sample issue: $1, US$1.50 in Canada

A monthly magazine published by the Benedictine monks of St. Meinrad Archabbey. Articles deal with issues such as alcoholism in a marriage, bringing up a child in a religious environment, and public attitudes toward adopted children.

Mother

Argus Consumer Publications Ltd., 12/18 Paul Street, London EC2 4JS, England

Annual subscription: £16 (or £24 for airmail postage)
Sample issue: £2

A glossy parenting publication from England, for those who want to expand their horizons. Content is similar to American women's and parent's magazines, with articles on such things as post-partum depression, coping with a child's asthma, and dieting. Knitting and sewing patterns, recipes, and equipment reviews also find a place. Published monthly.

Mothering

P.O. Box 1690, Santa Fe, NM 87504
Telephone: 505-984-8116

Annual subscription: $18, US$21 in Canada
Sample issue: $1, US$2 in Canada

An alternative parenting quarterly that focuses on the usual parent-magazine subjects from a different angle. Article titles like "Camping with Toddlers" and "Teenage Skin Care" might be found in mainstream publications, but here the writers discuss the problems of children and pit toilets and the use of herbal skin treatments. Articles on natural childbirth (by Sheila Kitzinger), underwater birth, sex education for adolescents, home-based businesses, home schooling, and grief after miscarriage have all appeared in recent issues.

Mother's Choice
1531 Haywood Road, Hendersonville, NC 28739
Telephone: 704-692-6405

Annual subscription: $12
Sample issue: free

A bimonthly newsletter for mothers who choose to stay at home. Articles in a recent issue dealt with taking the step from the workplace to full-time motherhood, locating a pen pal, the emotional strain of tight family finances, and a woman's nutritional needs.

Nurturing Today: For Self and Family Growth
The Fathers' Exchange, 187 Caselli Avenue, San Francisco, CA 94114
Telephone: 415-861-0847

Annual subscription: $16, US$20 in Canada
Sample issue: $3.50, US$4.50 in Canada

A quarterly magazine of parenting and early childhood education. A section is set aside in each issue for a discussion of fathering issues, and another for a review of nurturing programs and workshops across the country. Articles, editorials, book and film reviews, and resource guides fill the pages. Checks for samples and subscription should be made payable to The Fathers' Exchange.

Parenting for Peace & Justice
4144 Lindell Boulevard, Suite 122, St. Louis, MO 63108
Telephone: 314-533-4445

Annual subscription: $15
Sample issue: 50¢

A newsletter for parents who want to raise their children into a better world. Articles give advice on how to combat sexism and racism at home and at large, work for peace as a family, find alternatives to television violence, and bring up children who care about others.

Parenting Magazine
501 Second Street, San Francisco, CA 94107
Telephone: 800-525-0643, 303-447-9330, or 415-546-7575

Annual subscription: $18, US$24 in Canada
Sample issue: $2, US$3 in Canada

A hefty bimonthly magazine for parents, with articles long enough to delve into serious issues as well as practical matters. A recent report gave a good analysis of the limits of prenatal tests and the moral dilemma they present. Other stories dicuss such day-to-day matters as meting out praise to a child, setting behavior limits, discipline, buying a home, treating a cold, and hosting a child's tea party. Book, software, and video reviews keep parents informed of what's good and what's not. Regular columns discuss education, finances, and children's health; a recipe column dishes out ideas for meals, snacks, and parties. Send requests for sample issues, with payment, to Suite 110 at the address above.

Parents
Argus Consumer Publications Ltd., 12/18 Paul Street, London EC2 4JS, England

Annual subscription: £16.50 (or £25 for airmail postage)
Sample issue: £2

A monthly parenting magazine that offers a British perspective on the usual lineup of recipes, decorating tips, rainy-day activities, and sickroom advice. Feature articles treat topics like "AIDS and Birth" and "How Good Are Disposable Nappies [diapers]?"

Parents Magazine
P.O. Box 3055, Harlan, IA 51593-2119
Telephone: 800-727-3682

Annual subscription: $20, US$27 in Canada
Sample issue: not available

One of the most popular family magazines, the monthly issues of *Parents* stretch to more than 200 pages. Articles in each issue address the needs, questions, and interests of parents in every stage of their children's growth, from pregnancy, birth, and infancy through the teenage years. And to make things easier for readers, the table of contents starts out with a breakdown of articles by children's ages. Feature articles probe subjects like child care (the March 1989 issue ran an exclusive one-page letter from President Bush on his child-care policies), building family traditions and rituals, how to help preschoolers learn, and how to steer children clear of alcohol and drug abuse. Many of the articles offer first-hand advice from parents who've dealt with tough situations. Readers participate more intimately through regular polls on parenting opinions, the results of which are tabulated and printed, and through question-and-answer columns. Discounts are almost always offered from the published subscription price, so write for current rates before sending a check.

Parents Magazine Newsletter
Sent to members of the Parents Magazine Read Aloud Book Club; see entry under *Book Clubs*.

Priority Parenting
P.O. Box 1793, Warsaw, IN 46580-1793
Telephone: 219-453-3864

Annual subscription: $14, US$16 in Canada
Sample issue: $2, US$2.50 in Canada

A monthly newsletter "designed to encourage and educate parents who believe in quality, natural child care."

Each monthly issue tackles a different subject—war toys, miscarriage, home schooling, immunizations—through the writings of parents, educators, and other experts. The firsthand accounts are enlightening, and they're backed up by book and tape reviews, announcements of upcoming events, and helpful resource lists that connect readers to support groups, newsletters, and hard-to-find publications. Tamar Orr, the newsletter's editor, also offers a book called *Not On the Newsstands: A Parent's Resource Book*, that guides readers to more than 100 alternative publications and newsletters. It's only $7.50 if ordered along with a subscription to *Priority Parenting*. Order a six-month subscription for half the annual price if you want to give the newsletter a test run.

Shop Talk
5737 64th Street, Lubbock, TX 79424

Annual subscription: $24, US$29 in Canada
Sample issue: $4

A monthly newsletter of shared parenting experiences. Readers' own stories are combined in theme issues on such topics as sibling relationships, miscarriage, and being a mother. The tales make great reading, often provide practical solutions to troubling problems, and offer a wonderful network of support.

The Single Parent
Parents Without Partners, Inc., 8807 Colesville Road, Silver Spring, MD 20910
Telephone: 301-588-9354

Annual subscription: $15, US$18 in Canada
Sample issue: $1

A bimonthly magazine for single parents that blends practical advice with real-life experiences. Book reviews, legislative news, and schedules of local support groups make the publication even more useful. For information about the activities of Parents Without Partners, see entry under *Associations, For Parents and Families*.

Stepfamilies & Beyond
Listening, Inc., 8716 Pine Avenue, Gary, IN 46403
Telephone: 219-938-6962

Annual subscription: $12, US$13 in Canada
Sample issue: $1.25, US$1.50 in Canada

A four-page monthly newsletter about remarriage. A question-and-answer column deals with specific family problems, while articles probe such issues as coping with visitation, who pays the medical bills, and false charges of child abuse in divorce cases. The back page is given over to a story for children, generally with a message about adapting to a changing family situation.

Stepfamily Bulletin
Stepfamily Association of America, Inc., 602 E. Joppa Road, Baltimore, MD 21204
Telephone: 301-823-7570

Annual subscription: $14, US$16 in Canada
Sample issue: $4.50, US$5.50 in Canada

A quarterly journal for stepfamilies. A benefit of membership in the Stepfamily Association of America (see entry under *Associations, For Parents and Families*), it is also available by regular subscription. Articles share firsthand experiences, suggest ways to resolve conflicts, and offer advice from school and family counselors. A short story for children in each issue addresses some of the emotional problems that come with remarriage. Book reviews alert readers to helpful resources.

Today's Christian Woman
465 Gundersen Drive, Carol Stream, IL 60188
Telephone: 312-260-6181

Annual subscription: $14.95, US$18.95 in Canada
Sample issue: not available

A bimonthly Christian woman's magazine with articles on parenting, marriage, and faith; recipes; children's book reviews; and a question-and-answer column that addresses child-raising and marriage problems.

Welcome Home
Mothers at Home, P.O. Box 2208, Merrifield, VA 22116
Telephone: 703-352-2292

Annual subscription: $15, US$18 in Canada
Sample issue: $2

A monthly journal for mothers who choose to stay home with their children. Articles by readers recount real-life experiences, offer support for other mothers, and provide practical advice on such things as clothes shopping with teenagers. An October issue presented ideas for Halloween costumes to make at home. A special issue dealt with the emotional pain of miscarriage. In addition to the journal, Mothers at Home publishes a book, *What's a Smart Woman Like You Doing at Home?* Mothers at Home is not currently a membership organization, but *Welcome Home* subscribers in some areas have organized local support groups.

Work & Family Life
6211 W. Howard Street, Chicago, IL 60648
Telephone: 800-727-7243 or 312-647-6860

Annual subscription: $36 in U.S. and Canada
Sample issue: free

A monthly color newsletter published by the Bank Street College of Education that deals with topics relating to marriage, parenting, child care, elder care, and work. A regular parenting column gives suggestions for constructive play activities and other ideas to make life with children more fulfilling. Another discusses problems encountered at work and ways they can be resolved. One called "Quick Cuisine" serves up healthy recipes that can be prepared with a minimum of fuss. Health and finance columns give sound advice from experts.

Working Mother
Customer Service, P.O. Box 53861, Boulder, CO 80322-2886
Telephone: 800-525-0643 or 212-551-9500
Annual subscription: $12.95, US$18.95 in Canada
Sample issue: not available

A glossy monthly magazine for mothers who work. The publication covers the usual parents' and women's magazine territory—reader surveys; articles on child rearing, fashion, and food; columns on health and money—with some additional material added about work and child care. Though a sample issue is technically not available, interested readers can write for a trial subscription and pay or cancel after receiving the first copy.

For New Parents

American Baby Magazine
P.O. Box 53093, Boulder, CO 80322-3093
Telephone: 800-525-0643
Annual subscription: $12
Sample issue: $2 (see below)

A glossy monthly magazine for parents of infants and toddlers and for couples about to have a baby. Articles deal with such subjects as bathing and feeding a baby and exercise and diet for a new mother. Regular columns answer questions about child health, nutrition, and behavior; short news items report on medical studies, new products, and other items of interest to parents. Send sample-copy requests (with $2) to American Baby, 475 Park Avenue S., New York, NY 10016.

Baby Talk
Parenting Unlimited, Inc., 636 Avenue of the Americas, New York, NY 10011
Telephone: 212-989-8181
Annual subscription: $11.95
Sample issue: not available

A monthly baby magazine that's been going strong for more than 50 years. Recent issues have printed articles on such subjects as screening a sitter, toddler nutrition, vision problems in young children, books for babies, car safety, the development of locomotion skills, and coping with postpartum depression. Regular columns deal with women's health issues, breastfeeding, and news in the field of children's health. Each 60-page issue holds a lot to read.

Chatelaine's New Mother
Maclean Hunter Building, 777 Bay Street, Toronto, ON M5W 1A7, Canada
Telephone: 416-596-5422
Annual subscription: not available
Sample issue: $4 in U.S., free in Canada

A biannual magazine for new parents, distributed free of charge at Canadian hospitals. The publication is designed as one-time reading matter, with no continuing subscriptions. That means that each issue carries a great deal of information on a range of parenting topics, from jaundice and ear infections to car safety and helping siblings accept a new baby. While the title suggests a bias toward women, the magazine actually includes a number of articles on fathering.

Mother & Baby
Argus Consumer Publications Ltd., 12/18 Paul Street, London EC2 4JS, England
Annual subscription: £16.50 (or £25 for airmail postage)
Sample issue: £2

A monthly baby magazine from England that hasn't been influenced by the progress of feminism (fathers are rarely mentioned here). Health, finances, child development, recipes, and child-rearing tips all get their share of attention, with feature articles on such things as buying a car and coping with miscarriage.

Mothers Today
P.O. Box 56, Wynnewood, PA 19096
Telephone: 212-481-9030
Annual subscription: $10 for two years, $19.80 outside U.S.
Sample issue: free

A bimonthly magazine for expectant and new parents, with comparison shopping guides to baby equipment, book reviews, and regular columns on health (both mother's and child's), child development, and new baby products. Articles cover such subjects as children's furniture design, the pitfalls of working at home, and home finances.

For Parents of Young Children

Building Blocks
3893 Brindlewood, Elgin, IL 60120
Telephone: 312-742-1013
Annual subscription: $10, US$13 in Canada (specify "Family Edition")
Sample issue: $2

A monthly newspaper for parents of young children, sprouted from the editors' parenting classes more than a decade ago. Each issue is rich with ideas for things that parents and children can do together, including dozens of craft projects, suggestions for seasonal activities, and all sorts of games, from finger play to simple board games. One page is given over to things to make and activities for very young children, such as a large "peek-a-boo" box that a child can crawl into and ideas for puzzle play with measuring cups, Jello molds, or a coffee percolator. A "Child Care Edition" adds a section with ideas for group play and costs an additional $5 per year.

Growing Child
Dunn & Hargitt, Inc., 22 N. 2nd Street, Box 620,
Lafayette, IN 47902
Telephone: 317-423-2624

Annual subscription: $19.95, US$25.95 in Canada
Sample issue: free

An eight-page monthly newsletter geared to the age of your child (from birth to six years). Articles deal with stages of child development, with age-appropriate activities in each issue. *Growing Child* also puts out a mail-order catalog (see entry under *Toys*).

The Parent Connection
700 Grove Street, Worcester, MA 01605
Telephone: 508-852-5658

Annual subscription: $15, US$18 in Canada
Sample issue: $1.50 with a long self-addressed stamped envelope, US$2.50 in Canada

A monthly newsletter for parents of young children published by a nonprofit family resource program and support group. Parents in the Worcester area can take advantage of the toy lending library, the speakers, and group activities. Others can read the newsletter, which prints articles on things like listening to sounds with children, dental care, and ways to play with clay and finger paints. A monthly calendar offers something fun to do every day, from "Go to the library and check out a record" to "Bake some muffins and share them with a friend."

Sesame Street Magazine Parents' Guide
See entry for *Sesame Street Magazine* under *Children's Magazines, For Young Children*.

ADOPTION

OURS
A bimonthly magazine for adoptive and prospective adoptive familes. See entry for Adoptive Families of America, Inc., under *Associations, Adoption*.

BOOK AND MEDIA REVIEWS

The Bulletin of the Center for Children's Books
University of Chicago Press, Journals Div., P.O. Box 37005, Chicago, IL 60637
Telephone: 312-962-7600

Annual subscription: $24, US$27 in Canada
Sample issue: $2.50

A journal of children's book reviews. More than 50 new titles are given critical appraisal in each monthly issue (11 come with a subscription, due to a break in August).

Children's Book News
The Canadian Children's Book Centre, 229 College Street, 5th Floor, Toronto, ON M5T 1R4, Canada
Telephone: 416-597-1331

Annual subscription: $20 in U.S., free in Canada
Sample issue: self-addressed stamped envelope in Canada

A quarterly journal filled with news of Canadian children's books. Reviews of new titles take up about half of the journal's 16 pages. Activities for children, author interviews, and publishing news fill the rest.

Children's Video Report
145 W. 96th Street, Suite 7C, New York,
NY 10025-6403
Telephone: 212-227-8347

Annual subscription: $35, US$43 in Canada
Sample issue: long self-addressed envelope with two first-class stamps

A bimonthly journal that reviews children's videos. Each issue casts its critical eye on more than 20 new tapes, offering judgments that can save readers time and money. Written by child development and media experts, the reviews also draw on the opinions of young viewers.

The Five Owls
2004 Sheridan Avenue S., Minneapolis,
MN 55405-2354
Telephone: 612-377-2004

Annual subscription: $18, US$23 in Canada
Sample issue: $3, US$3.75 in Canada

A bimonthly book review journal written for both parents and children's book professionals. Each 16-page issue starts out with an in-depth article on an area of children's literature—songbooks or fantasy stories, for example—followed by a list of recommended books, each briefly analyzed. Exceptional new books on all subjects receive careful scrutiny in the second half of the magazine, with reproductions of sample illustrations. Space is also given to interviews with such children's authors and illustrators as Robert McCloskey, Chris Van Allsburg, and Jean Fritz.

The Horn Book Magazine
31 St. James Avenue, Boston, MA 02116-4167
Telephone: 617-482-5198

Annual subscription: $28, US$32 in Canada
Sample issue: free

The most comprehensive of the children's literature journals, *The Horn Book Magazine* reviews approximately 70 new children's books in each bimonthly issue, and supplements that useful guide with extensive articles on many aspects of the children's book world. A study of E. B. White's manuscripts for *Stuart Little*, for example, revealed how the story evolved through edits and revisions. Comments by educators, librarians, publishers, writers, and artists shine light on the field from many sides.

Parents' Choice
Parents' Choice Foundation, P.O. Box 185, Waban, MA 02168
Telephone: 617-965-5913

Annual subscription: $15
Sample issue: $1.50

A quarterly tabloid-format review of children's media, including books, television, movies, videos, recordings, toys, games, computer programs, and rock 'n' roll. The foundation's annual awards, published in a special issue of the magazine, are widely respected and often proudly cited in advertisements for the winning products. Both the regular reviews and the awards draw on the expertise of nationally recognized critics and the insights of creative artists in the various media. The magazine supplements its well-written and useful reviews with interviews and articles.

CHILDBIRTH

Birth Notes
A newsletter promoting home birth. See entry for Association for Childbirth at Home, International under *Associations, Childbirth*.

Childbirth Alternatives Quarterly
Janet Isaacs Ashford, 327 Glenmont Drive, Solana Beach, CA 92075
Telephone: 619-481-7065

Annual subscription: $20
Sample issue: $5

A 20-page journal that blends articles, reviews, and news updates on subjects relating to natural childbirth. Among the topics covered in the past are the growth of midwifery, the medical safety of home birth, and the history of traditional childbearing practices. Ms. Ashford also sells books, slides, and cards through her Childbirth Resources catalog (see entry under *Childbirth*).

The Compleat Mother
c/o Tamra B. Orr, P.O. Box 1793, Warsaw, IN 46580
or P.O. Box 399, Mildmay, ON N0G 2J0, Canada
Telephone: 219-453-3864, 519-855-4449 in Canada

Annual subscription: $10
Sample issue: $3 in U.S., Can$3.50 in Canada

A quarterly magazine with a focus on pregnancy, natural childbirth, and breastfeeding. First-person accounts of births, breastfeeding problems and solutions, and other parenting stories fill most of each issue.

NAPSAC News
A quarterly journal of childbirth and infant care. See entry for NAPSAC International under *Associations, Childbirth*.

Special Delivery
Informed Homebirth, P.O. Box 3675, Ann Arbor, MI 48106
Telephone: 313-662-6857

Annual subscription: $12, US$15 in Canada
Sample issue: $2, US$3 in Canada

A quarterly newsletter for midwives and for parents interested in natural childbirth options.

EDUCATION AND CHILD CARE

Child and Youth Care Quarterly
Human Sciences Press, Inc., 72 Fifth Avenue, New York, NY 10011
Telephone: 212-243-6000

Annual subscription: $36, $43 outside U.S.
Sample issue: free

A scholarly journal, published quarterly, for professionals in the fields of day care and residential youth care.

Child Care Information Exchange
P.O. Box 2890, Redmond, WA 98073
Telephone: 206-883-9394

Annual subscription: $35, US$40 in Canada
Sample issue: not available

A bimonthly magazine aimed at directors of day-care centers.

Childhood, The Waldorf Perspective
Nancy Aldrich, Rt. 2, Box 2765J, Westford, VT 05494

Annual subscription: $20, US$23 in Canada
Sample issue: $5

A quarterly journal about Waldorf education, a holistic approach to schooling based on the writings of Rudolph Steiner. The longest article in each issue deals with the practical application of the Waldorf philosophy to the teaching of school-age children. Craft projects, ideas for celebrating festivals and holidays, and book reviews are also regular features.

Childhood Education
See entry for Association for Childhood Education International under *Associations, Education and Child Care*.

Children & Teens Today
Atcom Publishing, 2315 Broadway, Suite 300, New York, NY 10024-4397
Telephone: 212-873-5900
Annual subscription: $55
Sample issue: $5

A monthly newsletter "for child and adolescent specialists." The eight-page issues summarize news stories, court decisions, surveys, scientific studies, and other pertinent bits of information to keep readers abreast of current trends. A reference or contact name is included at the end of each brief for readers who need to know more.

Children Today
Office of Human Development Services, U.S. Dept. of Health and Human Services, 200 Independence Avenue S.W., Rm. 356-G, Washington, DC 20201
Annual subscription: $7.50, US$9.40 in Canada
Sample issue: $2, US$2.50 in Canada

A bimonthly magazine that delves into a wide range of child development topics, from latchkey children in the library to visitation rights for grandparents. Though aimed at professionals, the magazine is not scholarly and many of the articles deal with problems faced by parents. Send subscription orders to: Superintendent of Documents, Government Printing Office, Washington, DC 20402-9371.

Child's Play
See entry for Canadian Alliance of Home Schoolers under *Associations, Education and Child Care*.

Day Care & Early Education
Human Sciences Press, Inc., 233 Spring Street, New York, NY 10013-1578
Telephone: 212-620-8000
Annual subscription: $19, US$26 in Canada
Sample issue: free

A quarterly journal for day-care and early education professionals, with articles that might be of interest to parents. A recent issue included articles on "Encouraging Science through Playful Discovery," the development of literacy in children, and "Questioning Techniques to Promote Thinking." Regular review columns analyze new books and videos, and a story is provided to read aloud.

Dimensions
A journal of early childhood development and education. See entry for Southern Association of Children Under Six under *Associations, Education and Child Care*.

Family Day Care Bulletin
The Children's Foundation, 815 15th Street N.W., Suite 928, Washington, DC 20005
Telephone: 202-347-3300
Annual subscription: $10, US$12 in Canada
Sample issue: free, US50¢ in Canada

A quarterly newsletter for family day-care providers, with practical advice, legislative updates, and news of conferences and publications. The Children's Foundation also publishes a number of directories, bulletins, and papers relating to family day care, including an annual directory of family day-care associations and support groups.

Family Day Caring
450 N. Syndicate, Suite 5, St. Paul, MN 55104
Telephone: 612-488-7284
Annual subscription: $12 in U.S. and Canada
Sample issue: $2

A bimonthly magazine for family day-care providers. Regular columns deal with health and safety issues, the relationship between providers and parents, guidance and child development, and business management. A feature article goes into more depth on a subject such as creativity or teaching styles.

First Teacher
P.O. Box 6781, Syracuse, NY 13217
Telephone: 800-341-1522 or 303-367-4400, ext. 791
Annual subscription: $17.95 in U.S. and Canada
Sample issue: $5

A monthly "workshop" in tabloid form for people who work with young children. Several pages of projects turn everyday materials into learning activities. Articles explore a different theme each month. A recent issue on "Sharing and Caring" included a dozen reports from classroom teachers on ways to promote sharing, many with instructions for games and activities. An article on caring for classroom pets presented practical advice with a list of suggested children's books on the subject.

Gifted Child Quarterly
National Association for Gifted Children, 4175 Lovell Road, Suite 140, Circle Pines, MN 55014
Telephone: 612-784-3475
Annual subscription: $45, $50 outside U.S.
Sample issue: not available

A scholarly journal for educators and psychologists who work with gifted children, and for parents who are interested in learning more about the problems facing and opportunities open to their children. Article titles in a recent issue included "Curricular Decision-Making for the Education of Young Gifted Children," "Precocious Reading Ability: What Does It Mean?," and "A Cross-Sectional Developmental Study of the Social Relations of Students Who Enter College Early."

The Gifted Child Today
350 Weinacker Avenue, P.O. Box 6448, Mobile, AL 36660-0448
Telephone: 205-478-4700

Annual subscription: $29.97, $35.97 outside U.S.
Sample issue: $5

A practical magazine for educators who work with "gifted, creative, and talented children and youth," stocked with "how-to" articles, advice from experts, curriculum-material and software reviews, and an annual directory of summer programs. Though aimed at teachers, the bimonthly magazine might be of interest to parents involved in their children's education.

Growing Without Schooling
2269 Massachusetts Avenue, Cambridge, MA 02140
Telephone: 617-864-3100

Annual subscription: $20 in U.S. and Canada
Sample issue: $3.50

A bimonthly newsletter founded in 1977 by John Holt, *Growing Without Schooling* continues to spread his message of alternative, creative, "interest-initiated" education, with an emphasis on home schooling. Articles on such subjects as "Questioning Standardized Tests" are supplemented by book reviews, interviews with educators and home-schoolers, news items, and a regular column titled "Watching Children Learn."

Home Education Magazine
P.O. Box 1083, Tonasket, WA 98855
Telephone: 509-486-1351

Annual subscription: $24, US$35 in Canada
Sample issue: $4.50, US$6 in Canada

A bimonthly for home-schoolers. Book and curriculum reviews, children's activities, news briefs, and an extensive section of letters from readers form the skeleton of each issue. Interviews with educators and articles on educational methods and philosophies provide the flesh. Many of the articles end with a list of sources.

The Learning Edge
Clonlara HBEP, 1289 Jewett, Ann Arbor, MI 48104
Telephone: 313-769-4515

Annual subscription: $6, US$7 in Canada
Sample issue: $2, US$3 in Canada

The newsletter of the Clonlara Home Based Education Program, published five times a year for parents involved in home schooling. Letters from parents, creative writing from students, teaching tips, projects, and book reviews fill out each issue.

Lollipops
P.O. Box 299, Carthage, IL 62321-0299
Telephone: 217-357-3981

Annual subscription: $15
Sample issue: free

A magazine for preschool and early childhood educators that is rich with craft projects, activities, stories, and teaching ideas, many of them designed to be photocopied from the magazine.

National Nanny Newsletter
976 W. Foothill Boulevard, Suite 591, Claremont, CA 91711
Telephone: 714-622-6303

Annual subscription: $20
Sample issue: $5

A quarterly newsletter for in-home child-care providers.

Pre-K Today
Scholastic Inc., P.O. Box 2038, Mahopac, NY 10541

Annual subscription: $19.95
Sample issue: free

A widely read magazine for child-care workers and preschool teachers, loaded with practical articles, expert advice, activity ideas, and media reviews.

Preschool Perspectives
P.O. Box 7525, Bend, OR 97708
Telephone: 503-382-4657

Annual subscription: $24, US$28 in Canada
Sample issue: $2.50, US$3 in Canada

A monthly newsletter (with a break for July and August) written for teachers who work with children between the ages of three and six. The eight-page issues deal with health and safety concerns, social development, parent involvement, the teaching of science and math, and much more. Activity ideas and recommended children's books complement the articles. Most of the pieces are written from firsthand experience by working teachers. Both writing and design are top-notch.

Shining Star
P.O. Box 299, Carthage, IL 62321
Telephone: 217-357-3981

Annual subscription: $16.95 in U.S. and Canada
Sample issue: $4.50, Can$6.25 in Canada

A quarterly magazine of reproducible puzzles, games, and stories "for Christian educators and parents," with lesson guides and advice on how to teach from the material. Aimed at children aged 5 to 13.

Skole
A biennial journal of alternative schooling. See entry for National Coalition of Alternative Community Schools under *Associations, Education and Child Care.*

Texas Child Care Quarterly
4029 Capital of Texas Highway S., Suite 102, Austin, TX 78704
Telephone: 512-450-4167
Annual subscription: $8
Sample issue: $2.50

A publication of the Texas Department of Human Services that offers information for child-care providers everywhere. Aside from a couple of pages of local news, very little in the magazine is specific to Texas. A recent issue held long articles on the ins and outs of bottle feeding, the efficient use of space in the family day home, and on the selection of child-care staff. Several pages of art and craft projects and group activities relating to astronauts and the Moon offered some fascinating learning and entertainment. A review page focused on children's books about the Moon and outer space. A list of additional books, pamphlets, curriculum modules, and audiovisual aids for child-care providers can be had by writing the Distribution Coordinator, Media Services Division 206E, Texas Department of Human Services, P.O. Box 2960, Austin, TX 78769 and requesting the catalog of "Child Development Program Materials."

Totline
Warren Publishing House, Inc., P.O. Box 2255, Everett, WA 98203
Telephone: 206-485-3335
Annual subscription: $15 in U.S. and Canada
Sample issue: $1 in U.S. and Canada

Each bimonthly issue of *Totline* brings 24 pages of activity ideas, games, and craft projects for teachers (and parents) who work with preschool children. These could range from instructions for building ant farms and simple percussion instruments from discarded containers to new lyrics for familiar tunes and ways to make an art museum stimulating for a preschooler.

Young Children
National Association for the Education of Young Children, 1834 Connecticut Avenue N.W., Washington, DC 20009
Telephone: 202-232-8777
Annual subscription: $25, US$35 in Canada
Sample issue: free

A bimonthly journal of early childhood education, aimed primarily at educators and child-care workers. Each issue holds advice on dealing with children (on avoiding adversarial discipline, for example), reports on new research, legislative news, book reviews, and several in-depth articles on aspects of teaching young children.

HEALTH AND SAFETY

Child Health Alert
P.O. Box 338, Newton Highlands, MA 02161
Annual subscription: $19, US$23 in Canada
Sample issue: free

"A monthly survey of current developments affecting child health," published in the form of a six-page newsletter. Each issue goes into some detail on a half-dozen medical topics. Subscribers in past months have learned about the link between stomach problems and fruit juices in young children, ways of dealing with night terrors, the dangers of treating chicken pox with Caladryl lotion, and the hazards of bicycle-mounted child seats.

Family Safety & Health
National Safety Council, 444 N. Michigan Avenue, Chicago, IL 60611-3991
Telephone: 312-527-4800
Annual subscription: $8.10 in U.S. and Canada
Sample issue: not available

A quarterly magazine of health, nutrition, fitness, and safety. Subscribers in recent months have learned about the safe use of space heaters and cosmetics, winter driving techniques, the benefits of a high-carbohydrate diet, how to stay fit during pregnancy, how to fall-proof a home for the elderly, dieting to beat middle-aged spread, cross-country skiing for exercise, and how to deal with a child addicted to television. A dozen shorter health and safety tips add even more information to each issue, and a letter column prints stories from readers about accidents and ways they could have been prevented.

Parents' Pediatric Report
77 Ives Street, Providence, RI 02906
Annual subscription: $35
Sample issue: $3

An eight-page newsletter, mailed monthly, which strives to bring up-to-date information about children's health to child-care providers. Though aimed at professionals, the newsletter may also be of interest to parents who can afford a subscription. In fact some articles, such as one on the dangers of snowmobile use by children, seem more pertinent to the home than to the child-care center. Each issue provides in-depth discussion of several childhood diseases, including symptoms, risks of communication, and treatment methods. Articles on such childhood problems as crossbites and intoeing offer additional information.

Pediatrics for Parents
P.O. Box 1069, Bangor, ME 04401
Telephone: 207-942-6212
Annual subscription: $15, US$18 in Canada
Sample issue: $2

Medical advice for parents, in a ten-page monthly newsletter. Reports warn of common household dangers (such as the risk of suffocation from balloons), announce product recalls, give nutrition advice, and offer simple tips for home medical care (a way to make a fingertip bandage, for example, or to test whether an inhaler is empty). Many of the articles end with suggestions for further reading, and all cite their sources. Book reviews and news briefs alert parents to the latest medical findings and opinions.

REGIONAL AND CITY PUBLICATIONS
(Listed alphabetically by state and province)

Bay Area Parent
Kids Kids Kids Publications, 455 Los Gatos Boulevard, Suite 103, Los Gatos, CA 95032
Telephone: 408-358-1414
Annual subscription: $12
Sample issue: $1.50

A thick monthly "newsmagazine" (a format akin to that of *Rolling Stone*) for residents of Santa Clara County and the greater Bay Area. Each 64-page issue deals with subjects like abduction, premenstrual syndrom, area schools, and how to choose a preschool. Regular columns separate out information for parents of babies, young children, school-age children, and teens, as well as offering craft projects, restaurant and movie reviews, and tips for working parents. The calendar section spreads over more than 15 pages.

L.A. Parent Magazine
443 E. Irving Drive, P.O. Box 3204, Burbank, CA 91504
Telephone: 818-846-0400
Annual subscription: $12
Sample issue: $2

A hefty monthly newspaper for Southern California families, each issue spanning almost 100 pages. Much of that bulk is filled with ads for local stores, schools, and other family services, but extensive articles and a lengthy calendar section give the publication substance. An annual childbirth issue presents expert advice on the before and after, plus a massive comparative chart of services offered by area birthing facilities.

S. F. Peninsula Parent
P.O. Box 89, Millbrae, CA 94030
Telephone: 415-342-9203
Annual subscription: $15
Sample issue: $2

A monthly newspaper for San Francisco and the Peninsula, stocked with parenting articles, reports on schools and camps, ideas for activities and day trips, media and fashion reviews, and a big calendar of exhibits, story hours, concerts, festivals, and other family events.

San Diego Family Press
P.O. Box 23960, San Diego, CA 92123
Telephone: 619-541-1162
Annual subscription: $12
Sample issue: $2

A monthly four-color magazine for parents in San Diego County. A big calendar section lists seasonal events, performances, library happenings, and other family activities. Articles explore parenting topics in more detail, backed up by media reviews and features on day trips and places to visit.

Wet Set Gazette
P.O. Box 603, Azusa, CA 91702
Telephone: 818-332-7958
Annual subscription: $10
Sample issue: free

A monthly tabloid for parents of young children in the Los Angeles metropolitan area, published by Dy-Dee Diaper Service. Each 20-page issue is divided into sections for expectant parents and new parents so readers can get right at the material that interests them. Directories of childbirth educators, parenting classes, and parent support groups are combined with book reviews and informational articles.

Denver Parent
818 E. 19th Avenue, Denver, CO 80218
Telephone: 303-832-7822
Annual subscription: $14
Sample issue: $2.50

A newspaper for parents in the Denver area and for Colorado residents as far away as Fort Collins, Colorado Springs, Greeley, and Pueblo. Feature reports deal with issues like teen pregnancy, college choices, teaching children to manage money, and childhood dental care. Movie, book, and software reviews find a regular spot, as does a monthly calendar of events.

Parents Plus
716 N. Tejon, Colorado Springs, CO 80903
Telephone: 719-633-1778
Annual subscription: $5
Sample issue: 5 × 9 self-addressed stamped envelope

A parents' tabloid for residents of Colorado Springs and southern Colorado. Bimonthly issues combine reviews of restaurants, museum exhibits, recordings, books, videos, and toys with news of support groups, drop-in clinics, and shopping center babysitting services, as well as offering a calendar of events.

Connecticut Parent
8 Glastonbury Avenue, Rocky Hill, CT 06067
Telephone: 203-721-7455
Annual subscription: $18
Sample issue: free

A monthly tabloid for Connecticut residents. A theme-issue approach lets the newspaper explore a subject in some detail. A recent one on college plans ran articles on the application process, saving for college costs, and finding financial aid. A regular question-and-answer column caters to parents with children of all ages, as do book and software reviews. A calendar spreads across the two center pages for easy posting.

Potomac Children
P.O. Box 39134, Washington, DC 20016
Telephone: 301-656-2133
Annual subscription: $10
Sample issue: $1

A monthly tabloid aimed at parents and children in Washington, D.C., Maryland, and northern Virginia. A pull-out calendar of family events takes up the center spread. The remaining pages are filled with recipes, restaurant reviews, articles on parenting subjects, and advertisements for area stores and schools. An annual summer camp directory fills most of the April issue.

Atlanta Parent
P.O. Box 8506, Atlanta, GA 30306
Telephone: 404-325-1763
Annual subscription: $12
Sample issue: $1

A monthly tabloid for the Atlanta area. Each 32-page issue brings a calendar of events, parenting tips, interviews, and news of places to visit. Copies are distributed free at area stores.

Youth View
1401 W. Paces Ferry Road N.W., Suite A-217, Atlanta, GA 30327
Telephone: 404-231-0562
Annual subscription: $12
Sample issue: $1.50

A parents' newspaper for the state of Georgia, published monthly with a one-issue summer break. A calendar of events spreads across the two center pages, fleshed out with movie reviews, notes about ongoing attractions, a school directory, and ideas for family activities. Regular columns deal with children's athletics, preschool and child care, pets, and children's health. A special spring issue brings a comprehensive directory to summer camps and day programs.

Chicago Parent News Magazine
7001 N. Clark Street, Suite 217, Chicago, IL 60626
Telephone: 312-508-0973
Annual subscription: $12
Sample issue: $1

A 48-page tabloid for Chicago-area parents, published in ten monthly issues. Articles, ideas for things to do, reviews, and a calendar of events fill each edition. Distributed free at area libraries, schools, and stores.

Indy's Child
8888 Keystone Crossing, Suite 1050, Indianapolis, IN 46240
Telephone: 317-843-1494
Annual subscription: $12
Sample issue: free

A monthly parents' newspaper for Indianapolis, with articles on aspects of child rearing, regular columns on health and family activities, and a pull-out calendar of events. The newspaper is distributed free at area daycare centers, doctor's offices, libraries, and stores.

Kansas City Parent
6400 Glenwood, Suite 300, Overland Park, KS 66202
Telephone: 913-262-3635
Annual subscription: $15
Sample issue: $1.50

Kansas City's entry in the regional parent-newspaper field. A pull-out calendar of events takes up the center spread, with a choice of plays, concerts, and exhibits on every day of the month. Articles like "Weight Training and Steroids: Every Parent's Nightmare" and "A Common Sense Approach to Summer Visits" give each monthly issue some reading interest.

Baltimore's Child
11 Dutton Court, Baltimore, MD 21228
Telephone: 301-367-5883
Annual subscription: $10 ($15 if sent by first-class mail)
Sample issue: $2

A hefty monthly newspaper for parents in the Baltimore metropolitan area. Each 40-page issue includes articles of parenting advice, book reviews, a health column, and an extensive calendar section. The March 1989 issue delved into the subject of summer camp and came up with a six-page directory of area camps, printed as a chart comparing facilities and services. That solid information was backed up by two long articles with tips on how to pick the right camp for a particular child.

Parent & Child
7048 Wilson Lane, Bethesda, MD 20817
Telephone: 301-229-2216
Annual subscription: $12.95
Sample issue: $3

A bimonthly magazine for the Washington area, written for parents with children from infants to teens. Feature articles provide practical advice on such concerns as choosing a summer camp and planning a birthday party, often with a list of addresses tacked on at the end. Regular columns answer questions about education and learning, medicine and health, childhood emotional problems, and the work/family balance. A thorough calendar of events is supplemented by longer write-ups on places to visit, day trips, and ideas for family activities. Reviews give the lowdown on movies, television, videos, music, and performances.

Boston Parents Paper
P.O. Box 1777, Boston, MA 02130
Telephone: 617-522-1515

Annual subscription: $9
Sample issue: free at area stores

A monthly for parents in eastern Massachusetts. Each 40-page issue holds more than a dozen articles on such topics as separation anxiety, eating disorders, children's fashion, places to visit, family pets, and day-care issues. Reviews of books, videos, television programs, stores, and toys show up regularly. A long calendar section keeps parents posted on seasonal happenings.

Grand Rapids Parent
40 Pearl Street N.W., Suite 1040, Grand Rapids, MI 49503
Telephone: 616-459-4545

Annual subscription: $12, US$19 in Canada
Sample issue: not available

A glossy monthly magazine for parents in the Grand Rapids area, stocked with articles about school programs, summer camp, and parenting matters, profiles of noted parents, and regular advice on finances and health (the medical column written by T. Berry Brazelton). Family restaurants, places to go, and a calendar of area happenings have their place here, as in all the regional magazines.

Minnesota Parent
100 W. Franklin Avenue, Suite 105, Minneapolis, MN 55404
Telephone: 612-874-1155

Annual subscription: $15
Sample issue: free

A monthly newspaper for the Twin Cities area with parenting articles, suggestions for things to do, and a calendar of events. Issues are distributed free at area stores, schools, and doctor's offices.

Suburban Parent
575 Cranbury Road, Suite B5, East Brunswick, NJ 08816
Telephone: 201-390-0566

Annual subscription: $7
Sample issue: $1

A parenting newspaper for central New Jersey, specifically Middlesex, Mercer, and Monmouth counties. Readers will find a monthly calendar; book, music, video, and movie reviews; financial advice; a fathers' column; articles on things to do and places to go; and longer feature stories on such things as scoliosis, birthday party ideas, and new fashions in children's clothes.

Big Apple Parents' Paper
Buffalo-Bunyip, Inc., 67 Wall Street, Suite 2411, New York, NY 10005
Telephone: 212-254-0853

Annual subscription: $15
Sample issue: not available

A monthly newspaper for the New York metropolitan area. An effort is made to include material of interest to parents with children of all ages, with a concentration in each issue on a different age group. A special edition on preschool and toddlers discussed the future of day care, gave advice on organizing a playgroup, and reported on a talk by Dr. Benjamin Spock. Restaurant and movie reviews are a regular feature, along with a medical column and a calendar of events.

New York Family Magazine
420 E. 79th Street, Suite 9E, New York, NY 10021
Telephone: 212-744-0309

Annual subscription: $30 for two years
Sample issue: $3.50

A glossy magazine with information for parents in Manhattan and the surrounding boroughs. A comprehensive calendar section is backed up with articles about day trips appropriate for families, theater and exhibit reviews, child-care and school information, news of after-school programs, and educational tips for parents. Eight issues are mailed each year. The same publisher puts out *Westchester Family Magazine* for residents of the northern suburbs (see below).

Parentguide News
P.O. Box 1084, New York, NY 10021
Telephone: 212-213-8840

Annual subscription: $11.90
Sample issue: $1.50

A monthly tabloid for the New York metropolitan area. A calendar of events is broken down by borough and county, with regular listings that stretch from Middlesex and Morris counties in New Jersey to Suffolk and Westchester counties in New York, with plenty of action in between. Close to 20 feature articles in each issue discuss topics like coping as a single parent, childhood allergies, exercises for parents and children, and the symptoms of a learning disability. Photographic fashion features serve a double role: as news of what's in style and as a guide to area stores.

WNY Family Magazine
297 Parkside Avenue, P.O. Box 244, Buffalo, NY 14215-0244
Telephone: 716-836-3486

Annual subscription: $11.95
Sample issue: $1.50

A monthly magazine for families living in western New York. Reports on museum exhibits and performances, suggested day trips, and restaurant reviews range over territory from Rochester to Jamestown, with Buffalo as the central focus. A calendar section lists classes, workshops, and meetings for parents, zoo and library events, plays, and concerts for children. Parenting articles supplement the activity guides.

Westchester Family Magazine
420 E. 79th Street, Suite 9E, New York, NY 10021
Telephone: 212-744-0309

Annual subscription: $30 for two years
Sample issue: $3.50

A glossy magazine, published eight times a year, with information for parents living in the northern suburbs of New York City. A calendar section is supported by longer write-ups of suggested day trips and activities, theater and exhibit reviews, child-care and school information, news of after-school programs, and educational tips for parents. The same publisher puts out *New York Family Magazine* (see above) for residents of Manhattan and the boroughs.

All About Kids
University of Cincinnati, Dept. of Pediatrics, 231 Bethesda Avenue, Cincinnati, OH 45267-0541
Telephone: 513-221-4600

Annual subscription: $9.50
Sample issue: available in stores

A monthly parenting newspaper for greater Cincinnati. Articles cover school news, parenting advice, and other subjects. A monthly calendar spreads across the center pages so that it can be pulled out and pinned up.

Birth to Three
3411-1 Willamette Street, Eugene, OR 97405
Telephone: 503-484-4401

Annual subscription: $15
Sample issue: $2

A bimonthly newsletter of "positive parenting" that deals with babies and young children. Children's feelings, fatherhood, toy-making tips, and infant dental care were all explored in articles in a recent issue. *Birth to Three* sponsors a program for teenaged parents and works to raise money for young children in need of expensive medical care. News of area happenings takes up roughly a third of the publication.

Portland Family Calendar
1714 N.W. Overton Street, Portland, OR 97209
Telephone: 503-220-0459

Annual subscription: $10
Sample issue: $1

As its name suggests, a calendar of activities provides the framework on which this magazine is hung. Articles flesh out the monthly issues with information on such subjects as traditional street games, things to do in the snow, and the circumcision controversy. Area stores and newsstands hand out the publication free of charge.

Pittsburgh's Child
P.O. Box 418, Gibsonia, PA 15044
Telephone: 412-443-1891

Annual subscription: $15
Sample issue: free

A monthly magazine in tabloid form for parents, educators, and child-care workers in the Pittsburgh area, geared to children from birth to about age 12. A pull-out "Kaleidoscope" section is written by children for children. For adults the magazine offers a calendar of area events, articles on health, education, and entertainment, and regular reviews of books, records, videos, television, movies, and live performances.

The Baby Connection
Parent Education Center, Subscription Service, Drawer 13320, San Antonio, TX 78209
Telephone: 512-342-4632

Annual subscription: $18, US$28 in Canada
Sample issue: not available

A monthly newspaper for San Antonio parents with articles that might be of interest beyond the local area. The ten pages of each issue hold articles on such topics as a two-minute house cleaning and how a newborn senses and learns to know her parents. Schedules of local events keep parents in touch with what's happening.

Dallas Child
P.O. Box 110760, Carrollton, TX 75011-0760
Telephone: 214-960-8474

Annual subscription: $12
Sample issue: $2

A monthly tabloid for families in the Dallas area. A calendar section lists events and activities. Feature articles guide readers to such seasonal atractions as pick-your-own fruit and vegetable farms, and offer advice on parenting topics.

Our Kids
6804 West Avenue, San Antonio, TX 78213 or P.O. Box 630412, Houston, TX 77263
Telephone: 512-349-6667 or 713-781-7535

Annual subscription: $15
Sample issue: not available

A monthly newspaper that comes out in two regional editions: one for San Antonio, the Hill Country, and South Texas; the other for Houston and East Texas, including Galveston. Both contain extensive calendar sections, area resource guides (to art education opportunities or indoor skating rinks, for example), ideas for day trips and family activities, reviews of books and concerts, columns on baby care and childhood health, as well as general articles. While sample issues aren't available, the subscriptions come with a satisfaction guarantee. It it doesn't meet your expectations, you can cancel at any time and the publisher will refund the unused balance of the subcription money.

Parent Express
346 Pierpont Avenue, Salt Lake City, UT 84101
Telephone: 801-363-1336

Annual subscription: $15
Sample issue: $1.50

A "news and resource" magazine for Utah families. The news is of local performances, legislative activity on family issues, and medical discoveries. As a resource the paper offers a monthly calendar of events, a directory of family services, and articles on topics like "Childbirth Education: A Consumer's Guide" (with a list of classes and programs across the state).

Eastside Parent
P.O. Box 22578, Seattle, WA 98122
Telephone: 206-322-2594

Annual subscription: $15
Sample issue: $1.50

A monthly tabloid for parents living in Bellevue and other Eastside communities. The editorial content duplicates *Seattle's Child* (see entry below), with calendar listings and advertisements tailored to the Eastside.

Northwest Baby
15417 204th Avenue S.E., Renton, WA 98056
Telephone: 206-634-BABY

Annual subscription: $10
Sample issue: free

A monthly newspaper for parents of young children in the greater Seattle and Tacoma areas, with calendar listings for King, Pierce, Snohomish, and Kitsap counties. General articles on such subjects as "Music and the Young Child" and "Nursing Your Baby," a monthly guide to infant development, book and music reviews, news from the state legislature, and other tidbits of information extend the publication to readers across the Pacific Northwest.

Pierce County Parent
P.O. Box 22578, Seattle, WA 98122
Telephone: 206-565-4004

Annual subscription: $12
Sample issue: $1.50

A monthly newspaper for Tacoma parents, published with much of the same editorial content as *Seattle's Child* (see entry below), but with activities, reviews, and calendar listings geared to happenings in Pierce County.

Seattle's Child
P.O. Box 22578, Seattle, WA 98122
Telephone: 206-322-2594

Annual subscription: $15
Sample issue: $1.50

Seattle's Child is one of three linked publications for parents in the Seattle and Tacoma areas. *Eastside Parent* prints the same general articles but tailors its calendar and activity listings to residents of Bellevue and other Eastside communities. *Pierce County Parent* focuses on family activities in the Tacoma area. (See entries above.) All three publications combine activity listings with reviews and longer parenting articles. A pull-out calendar fills the center spread, presenting the events of the month in abbreviated format for the bulletin board. Several pages of descriptive listings offer more information; reviews of performances and exhibits give even more detail. An annual guide to Seattle and Eastside schools, both public and private, is a wonderfully useful resource that comes with the February issue. A directory to camps and summer schools comes with the April issue; a holiday shopping guide is inserted in December.

Kids Toronto
542 Mt. Pleasant Road, Suite 401, Toronto, ON M4S 2M7, Canada
Telephone: 416-481-5696

Annual subscription: Can$12, US$18 in U.S.
Sample issue: $2

A thick monthly tabloid for the Toronto area, about evenly divided between news of local happenings and general parenting articles. A special birthday party issue in March 1989 included a comprehensive directory of local party supply stores, party hosts, and caterers, and lots of tips for successful party planning.

Special Needs

Deaf American
See entry for National Association of the Deaf under *Associations, Special Needs*.

Exceptional Children
The Council for Exceptional Children,
1920 Association Drive, Reston, VA 22091
Telephone: 703-620-3660

Annual subscription: $35, $39.50 outside U.S.
Sample issue: $7

A bimonthly journal for educators in the field of special education. Scholarly in tone, it publishes data-based research and position papers, and policy analysis.

The Exceptional Parent
P.O. Box 3000, Dept. EP, Denville, NJ 07834-9919
Telephone: 617-730-5800

Annual subscription: $16, $21 outside U.S.
Sample issue: free

A magazine for parents of children with disabilities. Practical articles offer information about education, social skills, health care, recreation, sexuality, financial planning, technology, and adaptive equipment; a "Family Forum" column allows readers to share experiences and provide mutual support; and every issue holds stories of how families cope with raising a handicapped child. A resource directory is a regular feature; once a year, in a special issue, it is expanded into a comprehensive guide to assistive technology.

Intensive Caring Unlimited
8 Haycroft Avenue, Springhouse, PA 19477
Telephone: 215-233-4723

Annual subscription: $8, US$10 in Canada
Sample issue: free

A bimonthly newsletter for parents and professional caretakers of children with medical and developmental problems, grieving parents, and those experiencing high-risk pregnancies.

Their World
An annual magazine for parents of children with learning disabilities. See entry for National Center for Learning Disabilities under *Associations, Special Needs*.

Travel

Family Travel Times
Travel With Your Children, 80 Eighth Avenue, New York, NY 10011
Telephone: 212-206-0688

Annual subscription: $35, Can$54 in Canada
Sample issue: $1 in U.S. and Canada (U.S. currency)

A travel newsletter loaded with specific advice for families on where to go, where to stay, how to get there, and how much it will cost. Money-saving tips can be found in every issue—universities that rent furnished townhouses during the summer, new hotels offering attractive start-up rates, and packages that put travelers up in British homes. Every issue explores one or two different types of vacation in some detail. One month it might be family dude ranches and European castles, the next it could be accommodations in the National Parks and what's new at Disneyland and Walt Disney World. A book review column takes a look at new travel books and suggests children's books appropriate to the vacations featured.

Twins and Multiples

Twinline Reporter
See entry for Twinline under *Associations, Twins and Multiples*.

Twins Magazine
P.O. Box 12045, Overland Park, KS 66212
Telephone: 913-722-1090

Annual subscription: $18, US$27 in Canada
Sample issue: $5

A bimonthly magazine for multiples and their parents. The usual family magazine columns have their place here—advice on children's health, photography tips, an editorial on fatherhood, and a report on educational issues—but other problems particular to multiples are given special attention. One article deals with the question of room sharing, another asks "Should Twins be Given Gifts to Share?," and another offers ideas on planning and efficiency (this last written by a mother of triplets).

Party Supplies

The Ark Catalog
Pewter birthday-candle holders. See entry under *Toys*.

The Ark Catalog's pewter birthday-candle holders

Childcraft
Dinosaur piñatas, dress-up costumes, and a changing array of other party supplies. See entry under *Toys*.

Current
Gift wrap, party supplies, and greeting cards. See entry under *Stickers and Rubber Stamps*.

Walt Disney World
Mickey Mouse party supplies. See entry under *Gifts*.

Just for Kids!
See entry under *Toys*.

Maid of Scandinavia
3244 Raleigh Avenue, Minneapolis, MN 55416
Telephone: 800-328-6722 (800-851-1121 in MN)

Catalog price: $1

A 200-page catalog stuffed with cake-decorating and candy-making supplies, ice-cream makers, cotton-candy machines, barbecue tools, taco racks, and other cooking gadgets for hosts and hostesses, along with party supplies like crepe-paper streamers, paper honeycomb balls, party signs, plastic table covers, doilies, paper plates, and piñatas.

Stocking Fillas
Party favors and stocking gifts. See entry under *Gifts*.

Pet Supplies

Gaines Pet Care Booklets
P.O. Box 877, Young America, MN 55399
Telephone: 312-222-8792

Free list

Information booklets and posters for dog and cat owners. Most are free, but a few of the posters and bigger books carry a nominal charge. Among the free booklets are *First Aid for Dogs, Eliminating Behavior Problems in Cats,* and *Introducing Your Dog to Your New Baby. My First Puppy* is written for children. *Gaines Guide to America's Dogs* is a wall poster showing all the AKC-registered breeds.

H. Kauffman & Sons
141 E. 24th Street, New York, NY 10010
Telephone: 800-US-BOOTS or 212-684-6060

Catalog price: $3

The *grande dame* of equestrian suppliers. H. Kauffman & Sons has been a New York City fixture since 1875, and remains the best source for English riding equipment and one of the best for western gear. Horse lovers will find dozens of saddles for hunting, jumping, dressage, and polo, along with pads, bridles, snaffles, crops, farrier's tools, blankets, and stirrups. Riders can dress themselves from the array of hats, coats, jodhpurs, breeches, and boots, many of the styles available in children's sizes.

Massachusetts Society for the Prevention of Cruelty to Animals
Attn.: Betty Stevens, 350 S. Huntington Avenue, Boston, MA 02130
Telephone: 617-522-7400

Catalog price: long self-addressed stamped envelope

Booklets, posters, and teaching aids that deal with the care of animals. Separate pet care booklets are available for dogs, cats, birds, horses, and small mammals.

Orvis
Comforts for dogs. See entry under *General Catalogs*.

Our Best Friends Pet Catalog
79 Albertson Avenue, Albertson, NY 11507
Telephone: 800-852-PETS or 516-742-7400

Catalog price: $2

Supplies and luxuries for dog and cat owners. Rawhide bones direct puppies' teething efforts away from the furniture; plastic-lined absorbent pads protect the floor. Cats can be furnished with scratching posts and climbing towers, bell collars, and automatic refrigerated feeding dishes that pop open as programmed. Pet beds, carriers, pooper scoopers, leashes, feeding mats, grooming supplies, books, and first-aid kits are offered to keep the critters happy and healthy. Extra-tough balls and Frisbees answer the needs of rough-playing larger dogs.

Pedigrees, The Pet Catalog
15 Turner Drive, P.O. Box 110, Spencerport, NY 14559-0110
Telephone: 716-352-1232

Catalog price: free

Supplies and gifts for pet owners. Rawhide bones are sold by the case; doorway gates in a choice of styles; and pet carriers in wicker, nylon, and rigid plastic. Matching dog-and-master sweaters gave us pause, as did a set of doggie booties that protect carpets from wet and soiled paws. Pet doors, leashes, beds, feeding dishes, grooming supplies, and cat scratching posts come in an almost overwhelming array of choices. Pet toys and owners' T-shirts are equally abundant in this 44-page shop. A few pages at the back offer cages, books, and treats for birds, hamsters, fish, ferrets, and other smaller pets.

J. C. Penney Company, Inc.
Cages, pet beds, and carriers. See entry under *General Catalogs*.

Renew America Catalog
Nontoxic flea and tick controls. See entry under *Toys*.

Wilshire Book Co.
12015 Sherman Road, North Hollywood, CA 91605
Telephone: 213-875-1711

Catalog price: free

Books for horse lovers. Reasonably priced paperback editions cover such subjects as first aid, training, horsemanship, dressage, western riding, and buying a horse.

Photographs, Albums, Frames, and Picture Plates

Associated Photo Co.
P.O. Box 14270, Cincinnati, OH 45214
Telephone: 513-421-6620
Free brochure
Photo Christmas cards, birth announcements, calendars, invitations, and change-of-address cards made from your color snapshots.

Bruce Bolind
Photo plates and postage-type stamps. See entry under *Gifts*.

Capri Photo Company
P.O. Box 1381, Long Island City, NY 11101
Catalog price: free
Color processing, from wallet-size prints to posters. Capri can make jigsaw puzzles, buttons, and self-adhesive postage-type stamps from your prints.

Color Lab
8 Burnett Avenue, Maplewood, NJ 07040
Free price list
Duplicate prints made from your photographs. Color Lab can make any size print from wallet to 8 × 10 from any standard-size original (send prints only, no negatives or slides). The company can also turn a photograph into a jigsaw puzzle.

Cumberland Crafters/Kid's Art
Your child's artwork reproduced on T-shirts, sweatshirts, tote bags, calendars, and place mats. See entry under *Children's Clothes*.

Eastman Kodak Co.
343 State Street, Dept. 412L, Rochester, NY 14650
Catalog price: $1 for annual *Index to Kodak Information*
Kodak publishes scores of books and booklets to aid photographers of every stripe. *Photographing Friends and Family* and *Photographing Your Baby* look like good basic guides; *Building a Home Darkroom* should help with the next step. Canadian customers should write to Kodak Canada, Inc., 3500 Eglington Avenue W., Toronto, ON M6M 1V3.

Exposures
475 Oberlin Avenue S., Lakewood, NJ 08701
Telephone: 800-222-4947 or 201-370-8110
Catalog price: free
Beautiful frames and archival scrapbooks for those extraspecial family photographs. Pick from wood inlay, silver, leather, wicker, and jeweled frames. Several are made specifically for baby pictures; others would suit an heirloom photograph of great-great-grandfather. Exposures also sells slide viewers, magnetic frames, archival storage boxes, and paper dolls made from your child's photograph.

Karen Studios
P.O. Box 175, Rye, NY 10580
Free brochure
Ceramic plates, ashtrays, and tiles made from children's drawings. The company can reproduce most pencil, crayon, and simple watercolor pictures, but can't work from finger paints or photographs.

Light Impressions
439 Monroe Avenue, Rochester, NY 14603
Telephone: 800-828-6216 (800-828-9629 in NY) or 716-271-8960
Catalog price: free
Archival supplies for storing and displaying photographs. If you're concerned that your child's pictures

The Light Impressions Photo ★ Archive® album

may not survive her childhood, Light Impressions can furnish you with archival photo albums, scrapbooks, negative and slide files, storage boxes, and framing supplies. The albums are made with acid-free paper, polypropylene sheet protectors, and nondestructive adhesives.

Philip's Foto Co.
262 E. Main Street, Elmsford, NY 10523
Free price list

Photo processing, from wallet-size prints to 8 × 10 enlargements. Photographic Christmas and Chanukah cards are offered in season.

Pictures–U.S.A.
P.O. Box 5135, Chicago, IL 60680
Free price list

Color and black-and-white photographic processing. The company handles all the usual print sizes and can work with unusual film formats, stereo slides, and movie film. Customers can order fresh film with their processing, from Kodak and other manufacturers. Write for the address of the nearest lab.

Portraits, Inc.
Painted and sculpted portraits. See entry under *Posters and Decorations*.

Qualex
3000 Croasdaile Road, Durham, NC 27705
Telephone: 919-383-8535
Free price list

Photographic processing, including film developing, color and black-and-white prints, stereo slides, and movie film. Write for prices and the address of the nearest processing plant.

Reliance Color Labs, Inc.
P.O. Box 159, Stamford, CT 06904-9993
Free price list

Copy prints made from black-and-white or color photographs. No negatives are needed for making wallet-size prints, regular snapshots, or enlargements up to 20 × 28 inches. Prints can also be made into Christmas cards and calendars. Reliance does not work from slides or film.

Small Fry Originals
Div. of Plastics Manufacturing Co., 2700 S. Westmoreland, Dallas, TX 75233
Telephone: 214-330-8671
Free information

Melamine picture plates and plastic mugs decorated with your child's artwork or with a favorite photograph. When you decide to order, the company sends an art kit with markers and 50 pieces of paper. Set your children

A child's artwork reproduced on a plate and a mug by Small Fry Originals

to work and pick out the best results. You're charged for the initial kit and for each plate or mug you order. Photographs can be sent in with your initial order—they require special preparation.

A. Strader Folk Art
Folk-art portraits painted from photographs. See entry under *Posters and Decorations*.

Surmacz Originals
P.O. Box 303-SP, Granger, IN 46530-0303
Telephone: 219-277-6531
Catalog price: self-addressed stamped envelope

This company will custom-make a color photo calendar from a favorite negative, slide, or print. The picture is enlarged to almost 8 × 10 inches and attached to a 12-sheet monthly calendar.

York Color Labs
400 Rayon Drive, Parkersburg, WV 26102
Free price list

York processes color print and slide film, makes reprints and enlargements from prints, slides, or negatives, and sells color film by mail. Write for prices and for the address of the nearest processing center.

Young Rembrandts
P.O. Box 3160, Vero Beach, FL 32964-3160
Catalog price: self-addressed stamped envelope

Children's artwork reproduced on plates, ceramic tiles, trinket boxes, and ashtrays, or embedded in Lucite paperweights. No special papers or materials are required—customers simply send a favorite drawing with their payment. Any number of reproductions can be made from a single work of art.

Posters and Decorations

American Arts & Graphics, Inc.
P.O. Box 75239, Seattle, WA 98125-0239
Telephone: 800-524-3900

Catalog price: $1

A poster catalog that wanders from kittens to cheesecake, with sports cars, jets, unicorns, rock stars, and teddy bears thrown in for good measure. Teens and preteens will find plenty of interest here (and plenty their parents won't want on the walls). There's not much for the younger set. One special feature is an array of wall murals, most of them about 9 × 13 feet. Scenes include Manhattan by night, a tropical beach, the planet Saturn, and a map of the world. The catalog also presents rubber stamps, stickers, calendars, note pads, and gift ideas like battery-powered light-up shoelaces and Mickey Mouse ice trays.

The American Clockmaker
German cuckoo clocks. See entry under *Hobby Supplies*.

American Library Association
ALA Graphics, 50 E. Huron Street, Chicago, IL 60611
Telephone: 800-545-2433 (800-545-2444 in IL, 800-545-2455 in Canada) or 312-944-6780

Catalog price: free

Inexpensive posters aimed at promoting reading and library use. For kids, the association offers a dozen designs featuring Mickey Mouse, Ramona, Curious George, Pinocchio, Miss Piggy, Garfield, Yoda, the Cat in the Hat, Pippi Longstocking, and Maurice Sendak's Wild Things.

Laura Ashley by Post
Wallpaper and fabric. See entry under *Children's Clothes*.

Bobbi Becker—Folk Artist
P.O. Box 292, New Oxford, PA 17350-0292
Telephone: 717-624-4953

Catalog price: $3

Prints and cards with an old-fashioned folk art look. Farm scenes, Santas, and Noah's ark are among the subjects pictured. The 16-page color catalog features ten prints and as many cards.

T. E. Breitenbach
P.O. Box 538A, Dept. S, Altamont, NY 12009
Telephone: 518-861-6054

Catalog price: $1

Several posters of original artwork by T. E. Breitenbach that might be described as a cross between the styles of Hieronymus Bosch and Salvador Dali. Some are probably too strange for children, but a couple look like they'd be big hits. *Catchpenny* is a conglomeration of dozens of nursery rhyme and fairy tale characters in a single picture that is great fun to hunt through. *Proverbidioms* is a similar graphic assemblage of proverbs and cliché phrases. A close look turns up a cat coming out of a bag, a man with his head in the sand, and a woman sweeping dirt under a rug.

T. E. Breitenbach's Catchpenny *poster*

Dover Publications, Inc.
Dinosaur, wildflower, and wild-bird posters. See entry under *Books*.

Fabrications
P.O. Box 67, East Meadow, NY 11554
Telephone: 516-496-8730

$1 for brochure, refunded with order

Personalized wall plaques with pegs for hanging clothes. A sky-blue kite plaque has yellow stars and pegs in three corners. A pink ballet slipper has a single peg at the pointed toe. Bears, planes, trains, ducks, and rainbows can ring the rest of the room.

Hansen Planetarium Publications
1098 S. 200 W., Salt Lake City, UT 84101
Telephone: 801-538-2242 (800-321-2369 to order)

Catalog price: free

Posters, slides, and postcards of space scenes. Planets (including the Earth), stars, the Space Shuttle, and astronauts on the moon are all pictured in spectacular color.

Holst, Inc.
Posters of children and baby farm animals. See entry under *Gifts*.

J & M Designs, Inc.
317-F Highway 70 E., Garner, NC 27529
Telephone: 919-779-2294

Free brochure

Vinyl window shades printed with colorful designs. Among the choices are a clown, a rocking horse, clouds, a rainbow, dinosaurs, a train, and teddy bears.

Just Sew Creative
P.O. Box 6638, South Bend, IN 46660
Telephone: 800-338-3422

Free brochure

Quilted wall hangings. The line includes a tiger, a lion, a bear, an elephant, and a clown, all rendered in a friendly cartoon style.

The Metropolitan Museum of Art
Posters. See entry under *Gifts*.

Modern Homesteader
A mounted jackalope trophy. See entry under *Children's Clothes*.

Moss & Associates
651 Serotina Court, Mt. Pleasant, SC 29464
Telephone: 803-884-5139

Catalog price: $1

Decorative crafts presented in a photocopied brochure. Hand-painted switch plates could add a cute touch to a child's room. One surrounds the switch with a little house, another a swimming duck, and a third a border of flowers. Hand-painted barrettes can be personalized with a child's name.

National Gallery of Art
Publications Office, Washington, DC 20565
Telephone: 202-842-6466

Catalog price: free (request print and poster catalog)

Prints of paintings in the collection of the National Gallery, many for as little as 50¢. The museum also loans (at no charge) videos, films, and slide programs on such subjects as Alexander Calder's mobiles, Henri Matisse's paper cut-outs, the Christmas story in art, painting techniques, Shaker crafts, and Eskimo art. Those interested in this loan program should address their inquiries to "Extension Programs" at the National Gallery of Art, or call 202-737-4215.

The Bear Dance *poster from The New-York Historical Society*

The Nature Company
Nature posters. See entry under *Science and Nature*.

The New-York Historical Society
The Museum Store, 170 Central Park West, New York, NY 10024-5194
Telephone: 212-873-3400

Catalog price: free

Books, posters, and cards having to do with New York City history or with items in the museum's collection. A page of children's books holds antique treasures like *The Slant Book* and a collection of rebuses. Posters include fine reproductions of Audubon's bird lithographs, a beautiful 1876 portrait of the young Dyckman sisters, and a fantastic forest scene titled *The Bear Dance*, with hundreds of bears enjoying an imaginary celebration. The catalog is a good source of unusual holiday cards.

The Peaceable Kingdom Press
2954 Hillegass Avenue, Berkeley, CA 94705
Telephone: 415-654-9989

Catalog price: free

Posters and cards featuring the art of popular children's book illustrators. Maurice Sendak fans can order posters from *In the Night Kitchen* and *Where the Wild Things Are*, or a special poster he created for the International Year of the Child. A *Goodnight Moon* poster is an almost irresistible bedside decoration, as are two made from *The Runaway Bunny*. George and Martha, Babar, Curious George, and Peter Rabbit all rate their own colorful sheets, as do some of the great illustrations by N. C. Wyeth, Chris Van Allsburg, and Thacher Hurd. One beautiful poster lays out all the known species of whales and dolphins in a comparative picture chart. Many of the children's book illustrations show up again in the card rack, joined by Maurice Sendak's Little Bear and the folks from Oz.

Maurice Sendak's Where the Wild Things Are *poster from The Peaceable Kingdom Press*

J. C. Penney Company Inc.
Wallpaper, curtains, and other decorating needs. See entry under *General Catalogs*.

Persnickety
P.O. Box 289, 117B Harry F. Byrd Highway, Sterling, VA 22170
Telephone: 800-336-7777

Catalog price: $3

Curtains, wallpaper, bedcovers, peg boards, picture frames, and other decorating accessories. None of the window treatments or wallcoverings scream "children's room," though of course they'd all look as nice in a child's room as anywhere else. One wallpaper border has a row of Victorian-era children in subtle blue and white. A child-size white wicker rocker comes plain or découpaged with fabric flowers.

The Playmill
Rt. 3, Box 89, Dover-Foxcroft, ME 04426
Telephone: 207-564-7702 or 8122

Catalog price: free

Wooden clothes racks, pull toys, crayon holders, and children's-room lamps. The pieces come in a regular zoo of animal shapes: elephants, crocodiles, horses, whales, bears, puppies, ducks, and dragons, to name a few. Most can be ordered with a simple painted decoration of flow-

Year of the Young Reader poster from The Peaceable Kingdom Press

ers or balloons, or personalized with a child's name. Some of the clothes racks come with a name cut out as part of the design.

Portraits, Inc.
985 Park Avenue, New York, NY 10028
Telephone: 212-879-5560

$1 for brochure

Painted or sculpted portraits. The brochure shows portraits by a few of the gallery's artists, and they are all handsome works. Most of those pictured have been executed in oil on canvas, but the artists also work in pastel, watercolor, pencil, bronze, and marble. Expect the cost of a portrait to fall between $1,000 and $35,000.

Poster Originals, Ltd.
330 Hudson Street, New York, NY 10013
Telephone: 212-620-0522

Catalog price: $10

A regular art book of a catalog. Poster Originals sells hundreds of fine-art posters from museums and galleries around the world and an excellent sampling of art posters made to advertise cultural events. All are illustrated in color in the glossy 100-page catalog. In addition to the wealth of high-quality exhibition broadsides, the company offers 40 circus and zoo posters from Poland, several striking designs from the 1972 Olympics, and a year-by-year lineup of the Georgia O'Keefe posters from the Santa Fe Chamber Music Festival. A few posters made just for children are added to the pile, including three by illustrator Nancy Carlson and one from an exhibit at the Solomon R. Guggenheim Museum.

The Prairie Pedlar
Rt. 2, Lyons, KS 67554
Telephone: 316-897-6631 or 257-2937

$2.50 for brochure

Prints of carousel horses and barnyard animals, and reproductions of antique postcards featuring teddy bears, antique dolls, and Dolly Dingle. A set of handmade marionettes dangle with rope legs and necks. The larger ones are a bit pricey for children, but a little wooden duck might make a nice toy.

Priss Prints, Inc.
3002 Jeremes Landing, Garland, TX 75043
Telephone: 214-278-5600 (800-543-4971 to order)

Free brochure

Giant self-adhesive wall decorations. One set features the main Disney characters, another the Disney babies. Friendly-looking jungle beasts and clowns could perk up a room. A picture alphabet could turn it into a learning experience.

Rowhouse Press
P.O. Box 20531, Dept. SP, New York, NY 10025
Telephone: 212-662-9604

Catalog price: $2.50

Publishes two quilt posters that might be appropriate for children's rooms. One shows a quilted alphabet, the other a patchwork sampler with a hidden puzzle full of optical illusions. Kits for making calico Christmas tree ornaments and cloth fruits and vegetables are also sold, along with a nice set of quilt Christmas cards.

Shaker Workshops
Shaker peg racks and hanging shelves. See entry under *Furniture*.

Shibumi Trading, Ltd.
Tatami mats, rice-paper lanterns, a brass gong, woodblock prints, and other traditional Japanese goods. See entry under *Gifts*.

StenArt, Inc.
P.O. Box 114, Pitman, NJ 08071-0114
Telephone: 609-589-9857

Catalog price: $2, refunded with order

Stencils for decorating walls, floors, curtains, and furniture with country-style borders and patterns. Hearts,

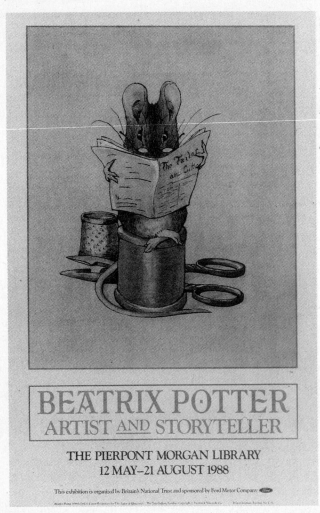

Beatrix Potter poster from Poster Originals, Ltd.

flowers, sheep, teddy bears, rabbits, and geese are among the motifs offered. Made of clear plastic, the stencils can be washed and reused as often as the job requires.

A. Strader Folk Art
100 S. Montgomery Street, Union, OH 45322
Telephone: 513-836-6308

Catalog price: $2

Prints of portraits painted by Ms. Strader in an early 19th-century folk art style. Children are the subjects of many of the portraits, sometimes with favorite pets and toys, sometimes with rural landscapes in the background. Those who want something more personal can commission a painted portrait. Send some unsmiling photographs of the subject, suggestions for belongings, pets, or backgrounds, and Ms. Strader will create a folk art portrait on canvas with the look of an antique.

Think Big!
Oversize versions of everyday objects for dramatic decorating accents. See entry under *Furniture*.

The University Prints
21 East Street, P.O. Box 485, Winchester, MA 01890
Telephone: 617-729-8006

Catalog price: $3; free information packet

University Prints publishes a huge library of inexpensive color and black-and-white prints to be used in the study of art history. The 5 × 8-inch reproductions can be ordered individually, in sets, or in custom-bound collections. The information packet includes prices for some of the most popular sets, such as "Knighthood," "Classic Mythology," "Flemish Art," and "Landscape Painting." The complete catalog, at almost 300 pages, lays out all the choices.

Puppets

Maher Ventriloquist Studios
P.O. Box 420, Littleton, CO 80160
Telephone: 303-798-6830

Catalog price: $1, deductible from first order

Ventriloquist puppets, dialogue books, and home-study courses for learning the basics and the finer points of ventriloquism. The simplest puppets move their heads, mouths, and arms. Professional dummies swivel their eyes, raise their eyebrows, and wink.

Munchkin Outfitters
Hand puppets. See entry under *Children's Clothes*.

The Prairie Pedlar
Marionettes. See entry under *Posters and Decorations*.

Treasured Toys
Plush hand puppets. See entry under *Toys*.

The Zany Zoo Gang
P.O. Box 253, Randallstown, MD 21133
Telephone: 301-655-1912

Self-addressed stamped envelope for brochure

"Handcrafted huggable handpuppets." Gang members include Beauregard Bunny, Pamela Panther, Ellie Elephant, Wendel Worm, Thadius Turtle, Perry Pig, Francis Flamingo, Lars Lion, and Samantha Swan.

Rocking Horses

Back to Basics Toys
A spring-mounted wooden rocking horse. See entry under *Toys*.

The Bartley Collection, Ltd.
See entry under *Furniture*.

Bighorn Sheepskin Company
Sells a sheepskin-covered rocking sheep. See entry under *Sheets, Blankets, and Sheepskins*.

Boston & Winthrop
A hand-painted rocking horse. See entry under *Furniture*.

Cabin North
A pine rocking horse. See entry under *Furniture*.

Chaselle, Inc.
Sells a baby-safe rocking horse with a railing around the seat that keeps small children from toppling. Another tip-proof rocker is offered for children a little older. See entry under *Educational Supplies*.

Gloucester Classics Ltd., Inc.
811 Boylston Street, Boston, MA 02116
Telephone: 800-444-3633 or 617-424-0027

Free brochure

Makes an exquisite rocking boat in the shape of a traditional New England dory, which nestles on rockers carved to simulate frothy waves. This is a quality piece of work—the seat and handle are made of ash, the rockers of clear pine, and the inside faced with mahogany. And as a finely crafted toy, it carries a demanding price tag. For youngsters tired of rocking and ready to roll, the business sells a riding fire truck, a riding school bus, and a deluxe wooden wagon made by Sun Designs. The wagon's side rails can be added or removed by a child driver, and a canopy top can be ordered as an accessory.

Hartline
3 Crafts Road, Gloucester, MA 01930
Telephone: 508-283-1933

Free brochure

Beautiful children's rocking toys in the form of such ocean creatures as a sea otter, a porpoise, a seal, and a killer whale. All rest on exquisitely painted wave rockers. The porpoise and whale spout a spray of white-painted wood, which forms the handgrip. The company either will point you to a store in your area, or, if none is close by, will sell directly by mail.

Ice Wood Designs
An affordable wooden rocking horse. See entry under *Toys*.

Imagination Toys
A wooden rocking horse. See entry under *Toys*.

Laura D's Folk Art Furniture
Rocking reptiles and frogs. See under *Furniture*.

Marvelous Toy Works
Wooden rocking horses. See entry under *Toys*.

The Natural Baby Co.
The Nantucket Rocking Dory. See entry under *Toys*.

The Right Start Catalog
Rocking cows, pigs, planes, and boats. See entry under *Baby Needs*.

The Gloucester Rocker from Gloucester Classics Ltd.

Royal Rocking Horses
1820 Medina, College Station, TX 77840
Telephone: 409-693-5702

Catalog price: free

Spectacular mahogany rocking horses imported from England. The wood is laminated in alternating light and dark layers to produce a striking contoured effect. The creatures are finished with natural horsehair manes and tails, and sealed with varnish and beeswax. They come mounted on graceful bows or on a stand for more controlled movement. Prices start at about $1,000.

F. A. O. Schwarz
See entry under *Toys*.

Toys Unique
595 N. Westgate, Grand Junction, CO 81505
Telephone: 303-242-7092

$1 for brochure (ask for retail price list)

A wild collection of rocking animals that improve on traditional wood construction with plush cushioned seats and padding along the back of the neck, where a child can rest his head. Among the beasts in this rocking zoo are dragons, pelicans, luxuriously maned lions, moose, elephants, longhorn steers, roosters, swans, and dinosaurs. Even the most fearsome of the animals have friendly faces, and all give comfortable rides.

Treasured Toys
A traditional painted wooden rocking horse. See entry under *Toys*.

Tully Toys
4606 Warrenton Road, Vicksburg, MS 39180
Telephone: 601-638-1724

Catalog price: free

A curious assemblage of wooden rocking animals made with comfortable rounded seats and handholds appropriate to the species. Riders of the ostrich and giraffe will have to hang on to the neck, a ram's horns curl back to provide a grip, and a dinosaur's neck plate sticks out just where it's needed. More conventional bar grips are built into the lion, shrimp, armadillo, hippopotamus, and alligator. Prices are reasonable (about $50, plus shipping).

Wisconsin Wood Products
Pat and Rich Rubasch, Rt. 4, Box 202, Viroqua, WI 54665
Telephone: 608-624-5445

Free information

The Rubasches make a lovely maned rocking horse in two sizes—one for toddlers and one for older children. The smaller horse is made of either cherry or black walnut; the larger can be had in walnut or stained pine. Both are made with small safety wheels at the end of the rockers to keep them from tipping over. Extending the barnyard are rocking cows and pigs, painted with a durable enamel finish. The prices here are extremely reasonable, given the quality of materials and workmanship.

Woodmonger
George Baumgardner, 111 N. Siwash, Tonasket, WA 98855
Telephone: 509-485-3414

Free flier

Rocking horses carved from silver pine to show the dramatic grain of the wood. This woodworker really cares about his raw material—all of the wood comes from forests burned by wildfires rather than from live trees. The standing deadwood ages and cures to a natural silver-blue color that won't be found in live-cut timber. The horses stand 2 feet tall from the floor to the tip of the ears, a scale designed for toddlers. One-year-olds can mount with help. By age five, children will be a little big for the steed and ready to move on to a bicycle.

The Woodworkers' Store
Plans, parts, tools, and lumber for making rocking animals. See entry under *Woodworking*.

Treasured Toys' personalized rocking horse

Science and Nature

ABC School Supply
Magnets and other science supplies for young children. See entry under *Educational Supplies*.

American Meteorite Laboratory
P.O. Box 2098, Denver, CO 80201
Free flier

Sells a Meteorite Crater Study Kit ($6 postpaid in 1989), which comes complete with fragments of the Canyon Diablo meteorite and a 65-page booklet, *A Comet Strikes the Earth*. Other books are offered for the more advanced meteorite enthusiast.

American Science Center
601 Linden Place, Evanston, IL 60202
Telephone: 312-475-8440
Catalog price: 50¢

Optical supplies, from advanced microscopes to $2 plastic bug boxes. In between are reading magnifiers, inspection glasses, prisms, fiber-optic strands, filters, and a broad range of unmounted lenses. A child might get a kick out of the firm's reducing glass—a reverse magnifier that makes things look *smaller*. Jerryco, an affiliate at the same address, offers industrial surplus (see entry below).

Ampersand Press
Science games. See entry under *Games and Puzzles*.

Anatomical Chart Co.
8221 N. Kimball Avenue, Skokie, IL 60076
Telephone: 800-621-7500 or 312-764-7171
Catalog price: free

Anatomical posters and models for people in the medical profession. Some of the charts may be of interest to older children with a strong interest in biology, but most of the models are too expensive for parents to consider. The company also sells T-shirts in children's sizes (small, medium, and large) with anatomical designs silk-screened on the front. One shows the digestive system, another the upper body skeleton, and another the muscles of the chest and stomach. Several humorous designs simulate the dress of various hospital workers.

Audubon Workshop
1501 Paddock Drive, Northbrook, IL 60062
Telephone: 312-729-6660
Catalog price: free

A color catalog well stocked with birdhouses, feeders, and seed.

Bird 'n Hand
40 Pearl Street, Framingham, MA 01701
Telephone: 508-879-1552
Catalog price: free

Bird feeders and seed. A window-mounted tray feeder made of clear plastic will give younger children an exciting close-up introduction to wild birds. Various other tube and platform feeders will attract more reticent birds. More than a dozen types of seed are sold.

Bunting Magnetics Co.
500 S. Spencer Avenue, P.O. Box 468, Newton, KS 67114-0468
Telephone: 316-284-2020
Catalog price: free

A serious magnet catalog for the true magnet fanatic. The $40 minimum on orders will send most parents hunting elsewhere, but for high-power alnico and ceramic magnets, Bunting is the place. All listings include precise dimensions and holding force. Many small but powerful magnets can be had for under $3. Others of industrial size are priced at over $300.

Center for Environmental Education
Whale Gifts, 167-2 Elm Street, P.O. Box 810
Old Saybrook, CT 06475-0810
Catalog price: free

Books, videos, T-shirts, toys, games, and gifts relating to ocean wildlife. This is primarily an adult gift catalog, but several pages are given over to stuff for children. If your child likes whales, turtles, penguins, seals, or polar bears, you won't go wrong by writing for this catalog.

Chaselle, Inc.
Science equipment for schools and preschools. See entry under *Educational Supplies*.

Childcraft
Telescopes, microscopes, chemistry sets, magnets, and other science tools for younger children. See entry under *Toys*.

Cloud Chart, Inc.
P.O. Box 21298, Charleston, SC 29413-1298
Telephone: 803-577-5628

Free flier

Offers a series of cloud posters that show photographs of various cloud types and explain the weather they predict.

Constructive Playthings
Science toys for young children. See entry under *Educational Supplies*.

Cuisenaire Company of America, Inc.
Science equipment and supplies. See entry under *Educational Supplies*.

Discovery Corner
Lawrence Hall of Science, University of California, Berkeley, CA 94720
Telephone: 415-624-1016

Catalog price: $1

Toys, gadgets, and gifts from U. C. Berkeley's Lawrence Hall of Science, presented in an attractive color catalog. The emphasis here is on fun. Offerings include a glow-in-the dark stencil kit for turning a bedroom into a planetarium, an inexpensive but optically crisp Bausch & Lomb microscope, giant inflatable dinosaurs, robot kits, Slinkies, and crystal-growing kits.

Dover Publications, Inc.
Dinosaur, wildflower, and wild-bird posters, and science activity books. See entry under *Books*.

Dover Scientific Co.
P.O. Box 6011, Long Island City, NY 11106
Telephone: 718-721-0136

Catalog price: $1

Shells, fossils, minerals, and Indian artifacts described in a typed catalog with a few black-and-white illustrations. Buyers will have to have an idea of what they're looking for, as the descriptions are very brief. A color portfolio of shells and minerals, available for $2, makes a good guide to the collection. Dover also sells a number of science and nature books, including some illustrated guidebooks that will help make sense of the catalog.

Droll Yankees Inc.
Mill Road, Foster, RI 02825
Telephone: 401-647-3324

Catalog price: free

Twenty styles of bird feeders, including several free-standing pole models, a window feeder, and two hummingbird feeders. Most combine a vertical tube with several stations and an open tray below. An entertaining extra here is the collection of Droll Yankee recordings. Two dozen albums fill our ears with bird songs, barnyard noises, swamp choruses, and the sounds of steamboats, tugboats, ocean surf, and the Green Mountain Railroad.

Droll Yankees' Winner window feeder

Duncraft
Penacook, NH 03303
Telephone: 603-224-0200

Catalog price: free

Rest and nourishment for the birds. Duncraft sells more than 40 feeders and a dozen seed mixtures to fill them with. Several different hummingbird feeders are pictured in the color catalog, with a choice of nectar mixtures. Three different window-mounted feeders will bring the activity up close for younger birders. Tube feeders, ball feeders, and tray feeders, to be hung or mounted on a pole—the catalog seems to have something for every bird and location. Houses are offered for wrens, bluebirds, and purple martins.

Edmund Scientific
101 E. Gloucester Pike, Barrington, NJ 08007
Telephone: 609-573-6260 or 547-8880

Catalog price: free

A little of everything for the science-loving child (or parent), from microscopes and electronic components to boomerangs and glow-in-the-dark necklaces. When our catalog arrived we savored its 80 pages as bedtime reading for a week. Among the offerings are magnets, prisms, periscopes, dissecting kits, gyroscopes, plastic dinosaurs, weather instruments, radiometers, magnifiers, lasers, motors, blacklight tubes, stethoscopes, star maps, holograms, steam-powered models, the famous perpetually drinking bird, and a toy boat powered by ice cubes. All are presented in tiny color pictures with clear descriptions.

The Exploratorium's Bubble Hoop

Environments, Inc.
Science toys for younger children. See entry under *Educational Supplies*.

Exploratorium Store, Mail Order Dept.
3601 Lyon Street, San Francisco, CA 94123
Telephone: 415-561-3093

Free price list

The Exploratorium Store generally offers toys relating to science and mathematics in a fun color catalog, but decided to skip the catalog in 1989. All of the merchandise is still available by mail, however. The Bubble Thing and the Exploratorium's own Bubble Hoop both make giant soap bubbles up to 8 feet in diameter. A drawing pendulum creates beautiful patterns with a pen and a swing. Bill Ding stack-and-hook wooden clowns are sold because they demonstrate principles of physics and gravity. And a potato-powered clock draws electric current from fruit and vegetables.

Librarian, Henry Ford Museum & Greenfield Village
P.O. Box 1970, Dearborn, MI 48121
Telephone: 313-271-1620

Free plans

Anyone who takes the time to write can receive a free set of plans for building a working model of Thomas Edison's original tin-foil phonograph. The supplies should be readily available at any hardware store, though real tin foil is suggested as better than aluminum foil, and this must be ordered from a scientific supply company. Just as Edison had to tinker with his invention before it recorded sound, so the modern builder will have to adjust and fiddle before it works just right.

Jack Ford Science Projects
P.O. Drawer 1009, Duluth, GA 30136

Catalog price: $1

Brochures on various scientific subjects, each containing several articles and plans for various experiments. The list holds practical titles like "Testing of Textiles" and "Simple Organic Chemistry" along with such exotic booklets as "E.S.P.," "How to Build a Synthetic Laser Model Pistol," "UFO Analysis," and "What Mysteries Do the Pyramids Hold?"

The Gifted Children's Catalog
Science kits and such nature study aids as an ant farm and a leaf press. See entry under *Toys*.

Hagenow Laboratories Inc.
1302 Washington Street, Manitowoc, WI 54220
Telephone: 414-683-3339

Catalog price: $1.50

Chemicals, glassware, and laboratory apparatus for the older science enthusiast. Among the offerings that might appeal to parents are several inexpensive microscopes, prepared slides, some starter chemical sets, rock and mineral assortments, radiometers, and magnifiers. Hagenow takes a more businesslike and less "fun" approach than Edmund Scientific, but the prices are good, and the catalog is a good source of supplies for junior-high and high-school science projects.

J. L. Hammett Co.
Microscopes, laboratory apparatus, rock collections, science activity kits, anatomical models, magnets, chick incubators, and astronomy charts and models. See entry under *Educational Supplies*.

Hansen Planetarium Publications
Space photographs as posters, slides, and postcards. See entry under *Posters and Decorations.*

Hoover Brothers, Inc.
Science kits. See entry under *Educational Supplies.*

Hyde Bird Feeder Co.
56 Felton Street, P.O. Box 168, Waltham, MA 02254
Telephone: 617-893-6780

Catalog price: free

A color catalog nicely stocked with birdhouses, feeders, books, and other bird-watching supplies. Two window feeders are designed to stay in place even when the weather turns cold (when feeders attached by suction cups tend to fall off). One mounts on the windowsill; the other sticks to the glass with adhesive. Four different hummingbird feeders are sold, 12 hanging tube feeders, and an assortment of suet holders, post feeders, and squirrel baffles.

Jerryco
601 Linden Place, Evanston, IL 60202
Telephone: 312-475-8440

Catalog price: 50¢

An affiliate of American Science Center (see entry above), Jerryco offers industrial and military surplus. The enterprising parent can find cheap glass lenses, magnifying mirrors, electric motors, and other entertaining science trinkets.

Learning Things, Inc.
68A Broadway, P.O. Box 436, Arlington, MA 02174
Telephone: 617-646-0093

Catalog price: free

A catalog aimed at science and math teachers, but one that can be a joy for parents and children as well. Its 48 pages are filled with science and math toys, microscopes, carpentry tools, magnifiers, magnets, star maps, wall charts, stethoscopes, polyhedral dice, rubber-stamp carving kits, sand timers, prisms, books, rock collections, and blocks for learning mathematical and geometrical principles. A specialty is the firm's own student microscope, which is sturdy enough for classroom use and cheap enough to buy for home.

Lick Observatory OP
"Catalog Request," Univ. of California, Santa Cruz, CA 95064
Telephone: 408-429-2201

Catalog price: 50¢

Prints and slides made from astronomical photographs taken at the observatory. Among the choices are a number of views of the moon's surface, some spectacular comets, and an array of nebulas, star clusters, and galaxies.

Lindsay Publications Inc.
P.O. Box 12, Bradley, IL 60915-0012
Telephone: 815-468-3668

Catalog price: $1

Books of electronics projects, many of them reprints of older works. Readers can find out how to build shortwave receivers, Tesla coils, metal detectors, and lasers. A two-volume set titled *Experimental Science* is crammed with tinkering possibilities. Equally fascinating are *Fifty Perpetual Motion Mechanisms*, first published in 1899, and William Corliss's *Handbook of Unusual Natural Phenomena*, a collection of reputable scientific reports of fantastic occurrences.

MMI Corporation
2950 Wyman Parkway, P.O. Box 19907, Baltimore, MD 21211
Telephone: 301-366-1222

Catalog price: free

Materials for studying astronomy and geology, in two separate catalogs. Each lists dozens of globes, maps, and demonstration models and hundreds of slides. The geology booklet offers mineral and fossil sample sets.

Malick's Fossils
5514 Plymouth Road, Baltimore, MD 21214
Telephone: 301-426-2969

Catalog price: $3

Thousands of fossils, from Precambrian algae to Pleistocene mammoth teeth, listed in computer type in a thick and forbidding catalog. A few pages at the back add several hundred archaeological artifacts and a small mound of meteorites to the inventory.

Learning Things' Ultrascope

Science kits from Merrell Scientific/World of Science

Merrell Scientific/World of Science
1665 Buffalo Road, Rochester, NY 14624
Telephone: 716-426-1540

Catalog price: $2

Two hundred pages of "science teaching support materials," which may be as appealing to parents and children as they are to educators. Ant farms, bug boxes, magnifiers, dinosaur models, and Magic Rocks are offered for younger kids. Older investigators may be interested in the erupting volcano kit, or the wide array of microscopes, balances, model rockets, chemistry sets, incubators, anatomical models, balances, and dissecting sets. Merrell also offers a good selection of science and nature books for children of all ages, from coloring books to field guides and reference texts.

Nasco
901 Janesville Avenue, Ft. Atkinson, WI 53538
Telephone: 414-563-2446

Catalog price: free (request "Science" catalog)

Nasco, one of the country's largest educational-supply houses, offers several thick catalogs. (See additional entries under *Art Supplies* and *Educational Supplies*.) The science catalog runs to almost 300 pages and is crammed full of dissection kits (and specimens), laboratory equipment, microscopes, telescopes, magnifiers, animal cages, aquaria, butterfly nets, animal track molds, weather instruments, star maps, scales, mineral collections, rocket models, electric motors, magnets, and books. The firm offers some real values in good, low-priced equipment. A Bausch & Lomb student microscope that we've seen priced elsewhere at over $60 sells here for under $40. Among the more "fun" items in the catalog: blacklight fixtures, Slinkies, a fluorescent mineral collection, and a large flexible Fresnel lens. Among the more unusual items: a skinned mink for dissection and an enlarged model of a human jaw showing several types of tooth decay.

National Audubon Society
950 Third Avenue, New York, NY 10022
Telephone: 212-832-3200

Catalog price: free

Nature books, posters, videos, bird feeders, and other nature items. Many of the books and posters come from Dover Publications (see entry under *Books*), but the Society also sells its own field guides and calendars. Dover's excellent nature coloring books are a good bet for children.

The Nature Company
P.O. Box 2310, Berkeley, CA 94702
Telephone: 800-227-1114

Catalog price: $2

More of a gift catalog than some of the other catalogs in this section, The Nature Company offers pretty and interesting things having to do with the natural world. Kids will be interested in the kaleidoscopes, crystal-growing kits, giant inflatable dinosaurs, animal nose masks, and pop-up science books. The selection changes each season, so expect to find other curiosities not mentioned here.

Opportunities for Learning, Inc.
20417 Nordoff Street, Chatsworth, CA 91311
Telephone: 818-341-2535

Catalog price: free (request "Sourcebook for Science" catalog)

A specialized catalog for grades K through 9 from a general school-supply company (see separate entry under *Educational Supplies*). Activity kits and computer software dominate the 32-page science booklet. Among the kits are magnet, chemistry, and crystal-growing sets. Rock collections, ant farms, bug viewers, dinosaur models, and microscopes might also appeal to young researchers.

PlayFair Toys
Star finders, an insect-collecting kit, a microscope, and some other science and nature-study supplies. See entry under *Toys*.

Renew America Catalog
Solar-powered toys. See entry under *Toys*.

Sears, Roebuck & Co.
Telescopes and microscopes are sold in Sears' "Toys" catalog. See entry under *General Catalogs*.

Small World Toys
P.O. Box 5291, Beverly Hills, CA 90210
Catalog price: free

An original and fascinating collection of science toys for children aged 4 to 12. A pair of glasses is made with rotating dials of color filters, a detective kit comes with solutions to check for starch and luminosity, and a battery-operated land-rover model must be assembled with rubber-band pulleys and belts. Other toys demonstrate principles of magnetism, conductivity, solar energy, optics, weight, balance, and air pressure. Skeletal models teach children the arrangement of bones in a human, a brontosaurus, and a stegosaurus. The toys are available in many stores, and the company asks that customers make an effort to find them at local shops before resorting to the mail.

Standard Hobby Supply
Telescopes, rock polishers, chemistry sets, and see-through models of internal-combustion engines. See entry under *Model Trains, Planes, Cars, and Boats*.

The Sycamore Tree
Science activity books, kits, and supplies, including books that present the biblical interpretation of fossils and dinosaurs. See entry under *Educational Supplies*.

Things of Science
1950 Landings Boulevard, Suite 202, Sarasota, FL 34231
Telephone: 813-923-1465
Free flier

Monthly science experiments mailed as part of a year-long subscription program. One recent kit gave materials and instructions for building a sundial; others have dealt with seed growth, magnetism, buoyancy, optics, and weather.

Three Rivers Amphibian, Inc.
688 Broadway, Suite 1, Massapequa, NY 11758
Telephone: 516-795-3794
Catalog price: free

Sells Grow-a-Frog kits and supplies. The basic kit includes a small aquarium, a decorative plastic plant, an instruction book, some frog food, and a "LIVE baby Grow-a-Frog" (or tadpole). The tadpoles are guaranteed to mature into frogs, and if one doesn't, the firm promptly sends out another. If yours grows to full froghood and seems lonely, the firm will also sell you a "frog friend." With two of the frogs and some luck, you might be growing little frogs all over again.

TOPS Learning Systems
10970 S. Mulino Road, Canby, OR 97103
Catalog price: free

Science and math activities designed to promote hands-on learning. The projects are aimed at classrooms, but there's no reason they can't be tried at home. Each kit comes with a set of task cards (but no equipment) for activities that demonstrate physical and mathematical principles. Energy transfer, for example, is explored in one card by swinging pendulums made from coins, thread, and tape. Study subjects include probability, electricity, seed growth, light, sound, pressure, metric measuring, depth perception, and graphing.

Treasured Toys
Science kits, a bird feeder, and other nature-study aids. See entry under *Toys*.

Troll
Dinosaur toys. See entry under *Toys*.

Unknown Products, Inc.
Ant farms and Sea Monkey kits. See entry under *Toys*.

Wild Bird Supplies
4815 Oak Street, Crystal Lake, IL 60012
Telephone: 815-455-4020
Catalog price: free

A one-stop source for bird-lovers, with feeders, feed, birdhouses, baths, and identification guides. Two feeders attach with suction to a window; 40 others dole out seed, suet, and syrup to meet the needs of orioles, woodpeckers, and hummingbirds. Several martin houses offer lodging for purple martins at various group rates. Smaller birdhouses provide nesting space for wrens, bluebirds, chickadees, and flickers.

World Wide Sea Shells
2007 S. Compton, St. Louis, MO 63104
Telephone: 314-865-0656
Catalog price: free

Specimen sea shells for collectors and nature enthusiasts. The list gives only the scientific names and countries of origin, so buyers will need some knowledge or a reference book to guide them through. The firm offers shells of more than 300 different species.

Sheets, Blankets, and Sheepskins

Artisans Cooperative
Children's quilts. See entry under *Gifts*.

Laura Ashley by Post
Quilts and sheets of Laura Ashley fabrics. See entry under *Children's Clothes*.

Baby & Company, Inc.
Lambskins. See entry under *Baby Needs*.

Baby Lamb Products, Inc.
31 Drakewood Lane, Novato, CA 94947
Telephone: 800-897-2578 or 415-897-2578
Catalog price: free

Whole New Zealand lambskins, fleece baby slippers, and lambskin car-seat liners are the specialties here. A sunshade for the car window, an extra rearview mirror for child-watching, lullaby tapes, a whale's-head cushion for the bathtub spout, and an infant bath seat make this an all-around source of baby needs. Adult-size fleece slippers and fleece mattress pads for full-size beds will work wonders for mom and dad too.

Between the Sheets Waterbed Products, Inc.
P.O. Box 1086, Addison, IL 60101-8086
Telephone: 800-238-9336 or 312-766-0080
Catalog price: $1

A color catalog of waterbed sheets, covers, and pads. Parents of young children may want to investigate the water-crib mattress.

Bighorn Sheepskin Company
11600 Manchaca Road, Austin, TX 78748
Telephone: 800-992-1650 or 512-280-1650
Catalog price: free

Sheepskin products for the whole family. For babies the firm sells sheepskin buntings, car-seat covers, booties (sizes 1-2 to 5-6), and lambskin rugs. A sheepskin-covered rocking sheep is the most comfortable-looking child's rocker we've seen. Cuddly sheepskin bears, rabbits, and lambs round out the children's end of the catalog. Adult-size slippers, jackets, mattress pads, and steering-wheel covers should help parents get through the winter.

Biobottoms
Flannel crib sheets. See entry under *Children's Clothes*.

Bright Future Futon Co.
3120 Central Avenue, S.E., Albuquerque, NM 87106
Telephone: 505-268-9738
Catalog price: free

Futons, from crib size to king size with seven stops in between. We've never used one as a regular bed, but we can speak from experience in recommending them as terrific guest beds. They're easy to store in a closet and give as comfortable a rest as any ordinary mattress. A futon might be just the thing for a child's room with limited space or as a resting and tumbling pad in a play area.

Cabin Creek Quilts
P.O. Box 383, Cabin Creek, WV 25035
Telephone: 304-595-3928
Catalog price: $2

Handmade quilts from a cooperative of West Virginia quilters, presented in a 16-page color catalog. Children's quilts are offered in a number of designs—mostly traditional patterns like Double Wedding Ring, Log Cabin, Flower Garden, and Dresden Plate—made with color-coordinated fabrics. Adult quilts and wall hangings come in an even greater choice of patterns. Prices are high when compared to mass-produced bedding, but

Album baby quilt from Cabin Creek Quilts

quite reasonable for handmade quilts. Most of the children's quilts are in the $100 to $150 range, and simple comforters can be had for about $60. Dolls, clutch balls, and bibs complement the quilts.

Clothcrafters, Inc.
Flannel sheets. See entry under *Cooking*.

Cloud Nine, Futons & Furnishings
142 Loma Alta, Oceanside, CA 92054
Telephone: 619-722-1676

Catalog price: free

Futons, nap mats, and comforters sold with coordinating sheets, comforter covers, crib bumpers, bolsters, and drapes. The futons come in six sizes, from crib to king size. Waterproof pads are sold for the smaller mattresses. The color catalog illustrates all the possibilities clearly, and a set of fabric swatches shows customers exactly what they'll get. Some play items are offered as well: an indoor tepee, a feltboard with felt shapes, and a reversible cape with Batman, Superman, or an initial-letter motif sewn on both sides.

The Company Store
500 Company Store Road, La Crosse, WI 54601
Telephone: 800-356-9367 or 608-785-1400

Catalog price: free

Down comforters, cotton sheets, fleece mattress pads, and other sleeptime comforts sold to fit beds large and small. Cribs can be outfitted with cotton and cotton flannel sheets in solid colors, cotton bumper pads and matching ruffles, down comforters, wool mattress pads, and boudoir pillows. Thick merino wool car-seat and stroller liners keep babies warm and comfortable outside the crib too. Bigger kids can be outfitted from an exceptional lineup of down parkas, coats, and vests—more than 30 styles to choose from, each of them in several colors. Down hats, mittens, muffs, and gaiters are sold to seal off any exposed areas. Prices are good for genuine down outerwear.

Cuddledown
42 N. Elm Street, P.O. Box 667, Yarmouth, ME 04096
Telephone: 800-323-6793 or 207-846-9781

Catalog price: free

Down comforters and pillows, cotton flannel sheets, and Acorn slipper socks for children and adults. The comforters and sheets come in crib size and in bed sizes from twin to king. The slippers start at a child's small (3½-5½) and go up to a men's 13.

Daisy Kingdom
134 N.W. 8th Avenue, Portland, OR 97209
Telephone: 800-234-6688 or 503-222-3817

Catalog price: $2

Nursery decor and children's clothing kits. The fabric has the cutting lines printed right on it, so no paper pattern is necessary (though adjustments must be made to some of the multi-sized clothing patterns). The nursery sets include sheets, comforters, bumpers, dust ruffles, mobiles, pillows, lamp shades, wall hangings, bibs, and high-chair covers, all with matching picture designs of rabbits, teddy bears, lambs, and other childhood motifs. Some of the same pictures reappear on ruffled bib collars, overall fronts, and jacket backs in the children's clothing line. All clothing and nursery items can also be ordered ready-made.

Direct-To-You Baby Products
Lambskins. See entry under *Baby Needs*.

The Disney Catalog
Mickey Mouse sheets. See entry under *Gifts*.

Domestications
Hanover, PA 17333-0040
Telephone: 717-633-3333

Catalog price: $2

Cartoon characters on crib sheets and children's linens. The Sesame Street gang drives a car around one crib set (sheets, comforter, bumpers, and pillow); the Disney Babies cavort on another. Twin- and full-size children's beds can be outfitted with Mickey Mouse, Alf, the California Raisins, and the Donald Duck family. Down comforters, pillows, and sheets without characters can be had in twin, full, queen, and king sizes.

Engineered Knits
143 W. 29th Street, #904, New York, NY 10001
Telephone: 212-629-0524

Free brochure

Cotton-lace baby blankets and pillows, with a subtle pattern of ducks, butterflies, teddy bears, or elephants.

Frostline Kits
Comforters in sew-them-yourself kits or ready-made. See entry under *Outdoor Gear*.

Garnet Hill
Cotton sheets, down comforters, and natural-fiber blankets. See entry under *Children's Clothes*.

Grey Owl Indian Craft Co.
Pendleton blankets in American Indian patterns. See entry under *Hobby Supplies*.

Hand in hand
Sheepskins. See entry under *Toys*.

Hearthside Quilts
P.O. Box 429, Route 7, Shelburne, VT 05482
Telephone: 800-451-3533

Catalog price: $2 (includes fabric swatches)

Quilt kits in traditional designs. The pieces are precut with steel dies to the precise shapes needed. All the

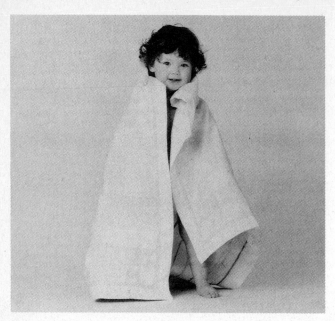

Child's quilt from Garnet Hill

buyer has to do is sew them together. Pattern pieces are 100 percent cotton. The backs can be either cotton or a cotton/poly blend. Pattern choices include Baby Wreath, Dresden Plate, Double Wedding Ring, Ohio Star, Bear Paw, Log Cabin, Lone Star, Country Star, Shoo Fly, and others. Most are available in crib to king sizes; a few are designed for dolls.

I Love You Drooly
109 W. Pippin Drive, Islamorada, FL 33036
Telephone: 305-852-8967

Catalog price: free

Musical sheets and bibs, activated by touching a "music spot." Choose from four tunes when you order: "It's a Small World," "Teddy Bears' Picnic," "Brahms Lullaby," and "Love Me Tender." Each musical device is said to be capable of playing its tune 10,000 times. The musical sheets come in sizes to fit cribs and playpens. The playpen sheets can also be ordered in silent editions.

Knit Wits
P.O. Box 1043, Westerly, RI 02891

Catalog price: free

Personalized baby afghans, with the name woven in white onto a pink or blue background. The blankets are made of acrylic yarn and come in either crib size or a smaller carriage size.

Lambskin, Inc.
1707 Third Avenue N., Dept. SP, Seattle, WA 98109
Telephone: 206-282-8552

Free brochure

New Zealand lambskins for soft and soothing baby bedding. The firm sells most of its fleeces shorn, with 1- to 2-inch fibers, but also offers them unclipped. A lambskin car-seat insert fits in both infant and toddler seats to provide a warm winter cover, as well as summer protection from hot metal and vinyl. The full fleeces measure roughly 28 × 33 inches and cost about $50.

Lamby
P.O. Box 5125, Petaluma, CA 94953
Telephone: 800-669-0527 or 707-763-4222

Free brochure

"Medical quality" lambskins imported from Australia. All of Lamby's fleeces are clipped to an even wool thickness of a little less than an inch.

Lancaster Towne Quilts
P.O. Box 277, Dept. SP, Strasburg, PA 17579
Telephone: 717-687-8600

Catalog price: $5

The $5 fee buys two exquisitely printed booklets, each showing about 40 hand-stitched Amish quilts. Customers can pick designs from the books and specify the colors they want, or order a custom quilt to be made from a photograph or drawing. The company sells the true Amish quilts pieced of solid colors, as well as the familiar calico quilts in traditional designs.

Landau
114 Nassau Street, P.O. Box 671, Princeton, NJ 08542
Telephone: 800-257-9445

Catalog price: $1

Wool sweaters and coats for mom and dad, washable wool blankets and long underwear for the children. Landau boasts that its Superwash woolens have a "soft, smooth, baby-pleasing texture" and backs up the claim with a fabric swatch in each catalog. Such comfort and warmth carries its price. A child's set of Superwash long underwear costs almost $60. The blankets, in solids and plaids with fringed or satin edges, are more affordable.

Maine Baby
A crib-size futon and a child's first bed made to fit a crib mattress. See entry under *Baby Needs*.

M. Matthews Handwoven Fabrics
21 Hillcrest Drive, Rochester, NH 03867
Telephone: 603-332-6779

Free brochure and fabric swatches

Handwoven baby blankets in a choice of 100 percent cotton or a cotton/angora blend. The spreads are generously sized and come in ivory as well as the obligatory pink and blue. The fabric swatches let shoppers see and feel exactly what they'll get. Ms. Matthews also weaves scarves, shawls, and tablewear.

Metrobaby
Cotton flannel crib sheets and bumpers, and down-filled crib quilts. See entry under *Baby Needs*.

Moncour's
4233 Spring Street, Suite 14, La Mesa, CA 92041
Telephone: 800-541-0900

Catalog price: $2

Crib sheets, comforters, bumpers, quilts, and baby-room accessories. Moncour's bumpers are extra thick. The sheets come in a range of pastels and bright solids, as well as some bold picture patterns. They're available in chintz, polyester/cotton broadcloth, and cotton flannel. Everything is well displayed in the 12-page color catalog.

David Morgan
11812 Northcreek Parkway N., Suite 103, Bothell, WA 98011
Telephone: 206-485-2132

Catalog price: free

Traditional clothing of the British Isles and Australia, such as waxed cotton rain gear and Welsh fishermen's smocks, along with work clothes and Indian jewelry from the Pacific Northwest. None of the clothes come in children's sizes, however. Keep thumbing through, and on the very last page you'll find New Zealand lambskins at some of the best prices around.

J. C. Penney Company Inc.
Mattresses, sheets, blankets, comforters, and accessories for cribs and full-size beds. See entry under *General Catalogs*.

Personal Statements
P.O. Box 576, Dept. MOS, Merrimack, NH 03054-0576
Telephone: 603-424-8650

Free brochure

Acrylic baby blankets embroidered with a child's name and birth date. Blankets come in pink, blue, maize, or white, finished with satin bindings and lace eyelet trim. Three-day rush delivery is available for an extra fee.

Pillow Pals, Inc.
P.O. Box 291, Harvey, LA 70059
Telephone: 504-367-9814

Catalog price: free

Pillow creatures with faces, arms, and legs. Head and body fit within the squat confines of a normal pillow, arms and legs burst out the sides. The basic Pillow Pals family is made up of a mother, father, and two pillow kids. A pillow panda, tiger, and puppy extend the group.

Quilts Unlimited
440A Duke of Gloucester, Williamsburg, VA 23815
Telephone: 804-253-8100

Catalog price: $5

Antique quilts and some new crib quilts and wall hangings. The crib quilts are available in a dozen traditional designs, including Double Wedding Ring, Lone Star, School House, and Carolina Lily. Amish crib quilts are made in dark solid colors, the others in brighter prints and pastels. More than 200 antique quilts fill the main part of the list, from early 19th-century relics to pieces from the 1930s and '40s. A packet of good-quality color photographs supplements the descriptive list to show each of the antiques.

Recreational Equipment, Inc.
Flannel sheets and down comforters. See entry under *Outdoor Gear*.

The Right Start Catalog
Sheepskins and bedding. See entry under *Baby Needs*.

Sears, Roebuck & Co.
Crib bedding, sheets, comforters, and blankets for children's beds can be found in the main "Home" catalog. See entry under *General Catalogs*.

The Sheepskin Factory
510 S. Colorado Boulevard, Denver, CO 80222
Telephone: 800-327-4337 or 303-329-8400

Free brochure

Unshorn New Zealand lambskins offered at competitive prices. The fleeces can be used as sleep pads in a crib, as portable nap mats, or as liners for car seats and strollers. Larger rugs and mattress pads are also sold, along with covers for steering wheels, seat belts, and drivers' seats.

Soft as a Cloud
Cotton sheets and afghans and an all-cotton crib mattress. See entry under *Children's Clothes*.

Spencer's
1579 Old Baxter Road, Chesterfield, MO 63017
Telephone: 314-537-3389

Long self-addressed stamped envelope for brochure

Unshorn New Zealand lambskins at good prices. The company claims the unshorn fleeces are softer than the clipped skins.

The Vermont Country Store
Cotton sheets, terry robes, and down pillows. See entry under *Baby Needs*.

Winganna Care Products
1612 Highway 93, Polson, MT 59860
Telephone: 406-883-2337

Free brochure

Baby blankets, sheared lambskins, and fleece car-seat covers imported from Wales. The baby blankets are woven of Shetland wool with a subtle check pattern, lightly brushed for softness, and finished with a satin edge. The lambskin car-seat covers are available in several color choices.

Shoes

After the Stork
Boots and shoes. See entry under *Children's Clothes*.

Alpine Ventures
P.O. Box 667, Granby, CO 80446
Telephone: 303-887-3882

Free flier

Makes Scooters booties for babies from birth to 18 months. The booties have quilted cotton/poly uppers with elastic at the ankle. All but the smallest sizes are made with non-slip soles.

Eddie Bauer
Insulated boots and fleece slippers. See entry under *Children's Clothes*.

Bear Feet
1911 Austin Avenue, Brownwood, TX 76801
Telephone: 915-646-0141

Catalog price: 50¢

Soft leather children's shoes and baby booties. The booties come in four styles, sizes 00 (newborn) to 3 (12 months). One has a leopard cuff; a ballerina bootie has a ribbon tie in pink, white, black, or polka dots. The shoes for older children come in three basic styles—a closed-toe sandal, a lace-tied shoe, and an elastic-cuff espadrille—with various decorations punched out of or sewn onto the toes.

Ballerina baby booties by Bear Feet

Biobottoms
Booties, sneakers, and shoes in smaller sizes. See entry under *Children's Clothes*.

Birth & Beginnings
Deerskin baby shoes. See entry under *Baby Needs*.

Country Spirit Crafts
P.O. Box 320, Talent, OR 97540
Telephone: 503-846-7391

Catalog price: long self-addressed stamped envelope

Deerskin moccasins for babies and young children. The soft leather is lined with a sheepskin innersole, and the tops come up above the ankle for a snug fit. Elk-hide sandals can be custom made for children from a foot tracing; they're shipped off the shelf in adult sizes. Sheepskin slippers are sold in sizes from child's 12 to men's 14.

Cuddledown
Acorn slipper socks. See entry under *Sheets, Blankets, and Sheepskins*.

Justin Discount Boots and Cowboy Outfitters
Cowboy boots. See entry under *Children's Clothes*.

Les Petits
Stylish shoes, sneakers, and boots imported from France. See entry under *Children's Clothes*.

Museum of the American Indian
Infant and toddler moccasins (sizes 1 to 6). See entry under *Baby Needs*.

Nature's Little Shoes
3089 Hidden Creek Lane, Escondido, CA 92026
Telephone: 619-747-9231

Long self-addressed stamped envelope for brochure

High-top moccasins made of soft leather, for infants and toddlers (sizes 1 to 7). The brochure comes with a handful of leather swatches to show both smooth and suede finishes in the nine colors offered. The shoes come in two styles, both with adjustable Velcro closures.

Olsen's Mill Direct
Shoes and sneakers, sizes 3 to 12. See entry under *Children's Clothes*.

Over the Moon Handpainted Clothing Co.
Swedish moccasins. See entry under *Children's Clothes*.

J. C. Penney Company Inc.
See entry under *General Catalogs*.

Price Is Rite Shoes
1840 Centre Street, West Roxbury, MA 02132
Telephone: 617-325-1250

Catalog price: free

A mail-order shoe store for women and children. Among the small-size offerings are white leather baby shoes, black wing tips, saddle shoes (pink and white or navy and white), New Balance sneakers, and patent leather dress shoes with satin bows.

Shoes & Socks, Inc.
1281 Andersen Drive, Suite D, San Rafael, CA 94901
Telephone: 800-228-1820

Catalog price: free

Children's shoes and socks. Tiny Capezio footwear is offered for infants, soft Babybotte shoes for babies learning to walk. Older children (up to child's size 12) can pick and choose from saddle shoes, T-straps, sparkling party shoes, classic Keds, leather high-tops, OshKosh trail shoes, sheepskin slippers, winter boots, and some black-and-red sport shoes with suave European styling.

Soft as a Cloud
Toddler footwear, including socks, Chinese silk shoes, flannel-lined cord booties, and several styles of soft leather slippers. See entry under *Children's Clothes*.

Soft Shoes
4295 Deming Road, Everson, WA 98247
Telephone: 206-592-5748

One first-class stamp for brochure

This company makes elk-hide slippers, moccasins, and walking shoes, and knit slippers of cotton and handspun wool. The elk-hide footwear can be ordered with sheepskin insoles for greater warmth and comfort, and with crepe outersoles for outdoor wear. Sizes run from infant to adult.

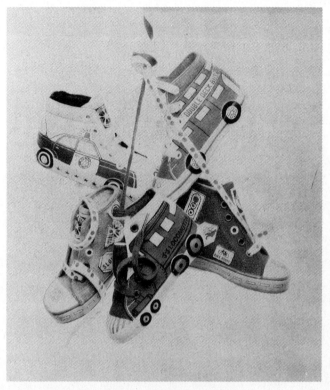

Action-Jacks sneakers from Shoes & Socks, Inc.

Soft Star Shoes
P.O. Box 1629, Wimberley, TX 78676
Telephone: 512-847-3460

Long self-addressed stamped envelope for brochure

Shoes for babies and young children, sizes 1 to 12. Moccasins, sandals, and boots are made of soft cowhide in bright colors; the boots are lined with shearling lamb fleece.

Soft Steps
P.O. Box 827-S, Corvallis, OR 97339
Telephone: 503-753-1455

Free brochure

Children's shoes made of supple leather in three styles: a basic walking shoe in white or tan, a flexible saddle shoe, and a closed-toe sandal. At a small extra charge, crepe soles can be added for greater durability on rough surfaces. Sizes range from 4 to 8, in regular and wide. A sizing diagram helps parents get just the right fit.

Sports Equipment

ABC School Supply
Balls and other sports equipment for young children. See entry under *Educational Supplies*.

Austad's
4500 E. 10th Street, P.O. Box 1428, Sioux Falls, SD 57196-1428
Telephone: 800-759-4653 or 605-336-3135
Catalog price: free

Fifty color pages of golfing equipment, including junior-size clubs and a child-size indoor putting green.

Back to Basics Toys
Roller skates and sleds. See entry under *Toys*.

Chaselle, Inc.
9645 Gerwig Lane, Columbia, MD 21046
Telephone: 800-CHASELLE (800-492-7840 in MD) or 301-381-9611
Catalog price: free (specify "Pre-School & Elementary School Materials" or "General School & Office Products" catalog)

Balls and equipment for all kinds of sports. The catalog for younger children has freestanding basketball hoops of less than regulation height, Nerf balls, and the red utility balls that are such a fixture of school playgrounds. See main entry under *Educational Supplies* for a description of Chaselle's other offerings.

English Garden Toys
Trampolines, basketball hoops, soccer goals, pogo sticks, and stilts. See entry under *Swings and Climbing Sets*.

J. L. Hammett Co.
Balls, rackets, sticks, and goals; a good source of sporting goods. See entry under *Educational Supplies*.

Hoover Brothers, Inc.
See entry under *Educational Supplies*.

Just for Kids!
Sports equipment for young children, including a scaled-down soccer net, golf clubs, and a baseball batting set. See entry under *Toys*.

Kaplan School Supply Corp.
See entry under *Educational Supplies*.

Racing Strollers, Inc.
516 N. 20th Avenue, Yakima, WA 98902
Telephone: 800-548-7230 or 509-457-0925
Free brochure

Makes the Baby Jogger, an aluminum-frame stroller with three big bicycle wheels. It can be used for running or for walking over uneven ground. We've been pushing various models for more than three years, and can report that they're wonderful running accessories and great strollers for beach walks. They're almost effortless to push, and most children love the ride. A two-seat model carries twins or siblings; The Walkabout is made with smaller wheels for slipping into car trunks that won't hold the original model.

The Baby Jogger, made by Racing Strollers, Inc.

Sports Bookshelf
P.O. Box 392, Ridgefield, CT 06877
Telephone: 203-438-3055
Catalog price: $1

Sports books, photographs of baseball and football players, and audiocassettes of great baseball games. Each new catalog features special deals on such things as media guides, team flags, and calendars.

Taffy's-by-Mail
Dance, parade, and exercise wear. See entry under *Costumes*.

Teco
1122 Industrial Drive, Matthews, NC 28105
Telephone: 704-847-4455

Free brochure

Kits for building in-ground trampolines. The customer must dig a 6 × 12-foot hole to a depth of 3 feet and buy some lumber and bolts at a local lumberyard. Teco supplies the plans, the mat, and the springs.

Trampoline World
P.O. Box 808, Highway 85 N., Fayetteville, GA 30214
Telephone: 404-461-9941

Catalog price: free

Freestanding trampoline kits. A tiny model designed for jogging in place makes a good bouncer for younger children. Larger rectangular and octagonal trampolines have 14-foot mats for serious leaping and aerial gymnastics.

Stickers and Rubber Stamps

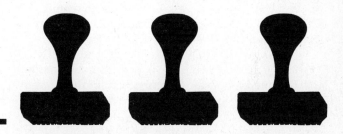

Aladdin Stamps & Celebrations
P.O. Box 354, Elk Grove, CA 95759
Telephone: 916-686-5970

Catalog price: two first-class stamps

Fanciful stamps, mostly small designs, of flowers, teddy bears, hearts, stars, shells, dinosaurs, butterflies, and seasonal motifs. Most of the stamps were priced at $3 to $4 in 1989; alphabet sets at $16 and number sets at $5.50 (check current rates before sending an order).

All Night Media
P.O. Box 2666, San Anselmo, CA 94960
Telephone: 415-459-3013

Catalog price: $2, deductible from first order

From the ridiculous to the sublime, All Night Media's thick catalog seems to have a stamp for every taste. We're particularly taken with the Stamp-A-Zoo, Stamp-A-Farm, Stamp-A-Face, and Creature Features sets. Each gives the stamper 18 pieces of animals or faces to put together in different combinations. With the Stamp-A-Zoo set, impressionists can make elegators, allipards, and pandaffes; the Stamp-A-Farm set scrambles the barnyard to make sheepigs, cowhorses, or chickgoats. The catalog also sells realistic animal and plant stamps, more than a dozen Winnie-the-Pooh stamps, plenty of wacky cartoons, holiday stamps, name stamps, alphabet sets, ink pads, colored pens (for fancy multicolor impressions), embossing powder for metallic images, and glitter glue. Now get stamping!

American Arts & Graphics, Inc.
Stickers and rubber stamps. See entry under *Posters and Decorations*.

Badge A Minit
348 N. 30th Road, Box 800, LaSalle, IL 61301
Telephone: 800-223-4103 or 815-224-2090

Catalog price: check for current price

The premier supplier of button-making tools and supplies. A starter set for about $30 can get you going making buttons from photographs, children's artwork, or anything else you care to pin to yourself. Badge A Minit

Creatures made with All Night Media's Stamp-A-Farm set

© All Night Media, Inc.

Stamp Out Your Children's Clothes

Rubber stamps make for terrific correspondence, but they can also be used to decorate clothes. With some fabric paint and a little practice, they can turn plain children's T-shirts into multicolor extravaganzas. A chain of elephants always looks nice marching around a child's waist; add some palm trees, monkeys, and alligators and you have a jungle shirt. A more cosmopolitan look can be had by combining skyscrapers, cars, and airplanes.

Any rubber stamp will work, and any permanent fabric paint. Simply brush the paint onto the stamp—not too heavily or it will fill in the stamp's holes and grooves, and not too lightly or you won't get a strong impression. Some people prefer to spread the paint on a piece of fine-grain foam and use it as they would a conventional stamp pad. Try brushing and stamping on a piece of scrap fabric until you master the technique. Then stamp out any design you want. Repeat images to make stripes, collar ornaments, and borders, or combine a number of stamps in a larger picture or an overall pattern.

To fix the design permanently, let the ink dry; then iron with a hot iron (laying an old sheet or pillowcase between the iron and your art) or dry in a very hot clothes dryer. Our home dryer doesn't get hot enough to fix the paint, but the one at the local laundromat does—you'll have to experiment a little to find what works. Wash the rubber stamps with water while they're still wet. An old toothbrush helps to get them completely clean.

Rubber stamps can be ordered from any of the suppliers listed here. Fabric paints are sold by Dick Blick Art Materials, Nasco, and Sax Arts & Crafts (see entries under *Art Supplies*), or at local art-supply stores.

---

backs up its basic supplies with transfer lettering, special background papers and foils, accessories for making key chains and pendants, power button presses, and hundreds of preprinted button designs.

Bizzaro
P.O. Box 16160, Rumford, RI 02916
Telephone: 401-728-9560

Catalog price: $1, refunded with order

A funky blend of old and new images made into rubber stamps, with slight favoritism shown toward graphics from the 1920s and '30s. A flip through the pages of the catalog turns up a veritable zoo of animals, both realistic and caricatured, along with planets, spaceships, cowboys, mummies, robots, flappers, and flowers. A stamp of a photo corner for snapshots looks like a useful one, as do the many relating to mail and postage. Bizzaro sells stamp pads, stamp-carving kits, watercolor markers, and metallic embossing powders. Postcards with pressure-sensitive backs let you turn any photograph or rubber stamp masterpiece into a piece of mail; blank jigsaw puzzles can be used to turn letters into major aggravations.

California Pacific Designs
P.O. Box 2660, Alameda, CA 94501
Telephone: 415-521-7914

Catalog price: free

Puffy full-color nature stickers, sold in packets with educational information on the back. A set of dinosaur stickers, for example, comes six to a card with an explanatory text about the reptiles' rise and extinction, and their discovery through fossil remains. Among the other sets are butterflies, tropical fish, whales, sharks, birds, shells, cats, dogs, and horses. All are beautifully colored and realistic depictions.

Circustamps
P.O. Box 250, Bolinas, CA 94924

Catalog price: $1

A complete 19th-century circus in rubber stamps. Order by the piece, or splurge and buy a complete set with performing elephants, lions, tigers, and horses, circus wagons, clowns, jugglers, tightrope artists, ringleader, and various props. These are big stamps and look like a great deal of fun, but we must warn you that they aren't cheap. A performing pig will cost about $5, a clown $6, and an elephant $8.50. The boxed sets start at $45.

The tiger act from Circustamps

Current
P.O. Box 2559, Colorado Springs, CO 80901
Telephone: 800-525-7170 or 719-594-4100

Catalog price: free

Stickers, gift wrap, party supplies, and greeting cards for almost every occasion. More than 100 different birthday cards are sold in the color catalog, along with baby congratulations cards, anniversary cards, and scores of Valentine and Christmas cards (but no birth announcements). Stickers range from dinosaurs and horses to stars and balloons. One fun-looking set provides eight cards with blank faces and 100 stickers of eyes, ears, noses, glasses, and other features for finishing the pictures.

I-Z Industries
P.O. Box 735-CM, Acton, MA 01720

Catalog price: free

Animal and heart stickers printed with amusing messages or custom-printed with your name and address.

Kidstamps
P.O. Box 18699, Cleveland Heights, OH 44118
Telephone: 216-932-9237

Catalog price: free

Rubber stamps by great children's book illustrators and of great children's book characters. On the roster of artists are Tomie dePaola, Nicole Rubel, Ray Cruz, Edward Gorey, Beatrix Potter, James Marshall, Sandra Boynton, and Syd Hoff. The characters include Curious George, the Tin Man, Tweedledee and Tweedledum, George and Martha, and Squirrel Nutkin. As if that weren't enough, the catalog goes on to offer several ornamental alphabets, a rubber-stamp village, a number of bookplate stamps, and plenty of funny picture messages for letters and envelopes. The company will custom-make stamps of your child's artwork, a club logo, a signature, or anything else you may want to impress in ink.

Personal Stamp Exchange
345 S. McDowell Boulevard, Suite 324, Petaluma, CA 94952
Telephone: 707-763-8058

Catalog price: $2.50

Rubber stamps with a strong folk and country flavor. Ornamental borders, flowers, and hearts rub shoulders with cats, birds, teddy bears, dolls, and ducks. A dozen Raggedy Ann and Andy stamps are featured, and several wonderful bookplate designs. Musical instruments, airplanes, dragons, dinosaurs, alphabets, and holiday motifs offer something for almost every taste. Two pages of teachers' stamps mete out praise and criticism with humorous pictures: "Terrific" is spelled in the teeth of a smiling tyrannosaurus; "That's great" on the television belly of a robot.

Posh Impressions
30100 Town Center Drive, Suite V, Laguna Niguel, CA 92677-2048
Telephone: 714-495-8242 (800-531-7766 to order)

Catalog price: $3, refunded with order

From a cartoon armadillo to an old-fashioned Santa Claus, Posh Impressions covers most of the rubber stamp bases. Realistic jungle beasts share the page with goofy rabbits and teddy bears. Planes, trains, trees, and flowers come in a similar jumble of styles. Stars, moons, and rainbows get a page all to themselves. Party graphics rate two pages, and teacher and business comments take up four. (A "Final Notice" stamp features a hooded executioner with a double-bladed axe.) Shoppers will find dinosaurs, dragons, and plenty of art for stamping out the major holidays.

Rubber Stamps of America
P.O. Box 567-SP, Saxtons River, VT 05154
Telephone: 802-869-2622

Catalog price: $1

This catalog starts off gently and beautifully, with 15 pages of serenely realistic images from nature: falling leaves, raindrops, butterflies, flowers, rabbits, birds, whales, snails, bears, and dinosaurs. From there things start to get a little crazy. Left and right footprints come as two stamps, the sign-language alphabet as one. An ice-cream cone comes with a separate stamp-on scoop

A sampling of Rubber Stamps of America's wares

for multicolor ice-cream dream towers. Coffee makers, blenders, sea monsters, dancing elephants, and surfing frogs combine in a somewhat quirky inventory of images. Message stamps range from "Ooohh" and "Aaahh" to a "Vote for me" button and a simulated postal cancellation from the Starship Enterprise. Other necessities include holiday stamps, stamp pads in unusual colors, alphabet and number sets, and blank calendars ready for stamping.

Stamp Magic
P.O. Box 60874, Longmeadow, MA 01116
Telephone: 413-567-1919 or 0284

Catalog price: $2, refunded with order; $1 for specialized catalogs (see below)

A mix of wacky and realistic rubber stamps, with the balance tilted gently toward silly cartoon images. An old-fashioned Santa or a Boynton-style reindeer can be ordered for Christmas correspondence. Creatures ferocious and farcical mingle on the nonseasonal pages. Cute kittens make their stand next to beautiful porpoises and crazy-looking sea monsters. Flowers, leaves, butterflies, footprints, musical instruments, and bookplate stamps all leave their impressions here. If your interests are focused, try ordering one of the three specialized catalogs. The separate books of cats, teddy bears, and teachers' stamps cost $1 each, refunded with an order.

Stickers Only
P.O. Box 09285, Columbus, OH 43209

$1 for brochure and samples

Stickers from Sandylion Sticker Designs, Mrs. Grossman's Paper Company, Cardesign, Lisa Frank, and other popular sticker producers. Inquirers receive a packet of stickers showing some of the lines and an illustrated brochure that pictures samples from them all. The price list lays out all 200 choices.

Swings and Climbing Sets

Big Toys
7717 New Market, Olympia, WA 98501
Telephone: 800-423-0082 (800-752-7511 in WA)

Catalog price: free

Log-construction play structures that won't make an eyesore of your backyard. Start with a safe 4-foot deck for little ones, and build on as they grow. Add towers, swings, slides, tunnel chutes, climbing nets, tire swings, rings, trapezes, and parallel bars. A plastic spiral slide looks like an especially entertaining touch. If you have the room and the money, Big Toys sells some amazing sprawling sets for schools and city parks that can accommodate dozens of kids without crowding. All sets require assembly, and uprights must be sunk 2 feet into the ground. Some will need to have the legs set in concrete.

CedarWorks
Rt. 1, Box 640, Rockport, ME 04856
Telephone: 207-236-3183

Catalog price: free

Swings and climbing sets made of cedar and ash. Cedar (4 × 4) is used for the frames because it weathers well without requiring paint or chemical treatment. Ash (1 5/16) is used for ladder rungs because of its strength. All parts are sanded smooth to eliminate splinters, and the corners have been rounded. The smallest set, a climbing gym without swings, measures 52 × 52 inches and is 8 feet tall. The largest takes up 8 × 18 feet of yard space. If you have even more room to spare, most of the swing sets and climbing gyms can be combined to make larger systems. Rope ladders, rings, sliding poles, trapezes, canvas swing seats, and slides are offered separately so that customers can design the unit exactly to their needs. The firm also sells a 10-foot cedar seesaw.

Chaselle, Inc.
9645 Gerwig Lane, Columbia, MD 21046
Telephone: 800-CHASELLE (800-492-7840 in MD) or 301-381-9611

Catalog price: free (specify "Pre-School & Elementary School Materials" or "General School & Office Products" catalog)

Steel-frame swing sets, aluminum climbing gyms, and colorful indoor-outdoor Rotoplay play sets made of molded polyethylene. See main entry under *Educational Supplies* for a description of Chaselle's other offerings.

Child Life Play Specialties, Inc.
55 Whitney Street, P.O. Box 6159, Holliston, MA 01746
Telephone: 800-462-4445 or 508-429-4639

Catalog price: $1, refunded with first order

Swings and climbing sets made of pressure-treated Douglas fir and Alaskan yellow cedar, painted with a green enamel. Ladder rungs are made of unspecified hardwood. One small climbing set measures 5 × 4 × 8 feet tall. The roomiest swing set stretches to 28 feet long. Optional accessories include a canvas-covered "tree house" (4 × 5 feet) to perch on top of the set, an 8-foot slide to stick on the side, and the usual assortment of swing seats, rope ladders, rings, and trapezes. The firm offers a toddler gym with steps, a slide, and a 4-foot climbing tower; a playhouse for older kids, with a railed roof deck; sandboxes in several sizes; and wooden blocks.

Children's Playgrounds Inc.
P.O. Box 1547, Cambridge, MA 02238-9990 or
P.O. Box 370, Unionville, ON L3R 2Z7, Canada
Telephone: 617-497-1588 (U.S.) or 416-475-7648 (Canada)

Catalog price: free

Indoor and outdoor play structures. For indoor use the firm offers Quadro play sets, which come as kits and consist of heavy red plastic tubing and sturdy black connectors. By combining different lengths of tubing with different angled connectors and accessories, the Quadro system allows for an almost infinite number of construction possibilities. Optional pieces include slides, floor platforms, rope ladders, and roof and wall panels. Order kits to make prescibed play sets, or by the piece to design your own creations. The outdoor play sets are made of pressure-treated jack pine or red pine and come with some amazing features: suspension bridges, balance beams, tube slides, tire swings, playhouses, chin-up bars, and steering wheels. With climbing walls covered with rubber tires, wire cables strung loosely for balancing and climbing, and wobbly "totter bridges," these sets provide endless fun (as we can attest from visiting two local playgrounds built by the firm). The biggest units, at 50 × 100 feet, are likely to be of more interest to schools and city park departments than parents, but some of the smaller models would fit in a yard. We should note that the outdoor sets are built for heavy use, with heavy-duty hardware and thick cuts of wood, and they're priced accordingly ($1,500 and up). The Quadro sets fall closer to the range of most family budgets.

Community Playthings
Indoor and outdoor climbing sets for young children. See entry under *Educational Supplies*.

Constructive Playthings
See entry under *Educational Supplies*.

Creative Playgrounds Ltd.
P.O. Box 10, McFarland, WI 53558
Telephone: 608-838-3326

Catalog price: free

TimberGym play structures, made of pressure-treated southern yellow pine. Uprights are of 4 × 4 lumber, horizontal supports of 2 × 6. Designs range from small

A Quadro play set from Children's Playgrounds Inc.

climbing sets to a 12 × 20-foot swing-and-tower combination topped with a vinyl-covered playhouse. Slides, rope ladders, turn bars, parallel bars, rings, and trapezes can be added for extra appeal. The company also sells a 5 × 5-foot sandbox, a seesaw, and a tire swing. The sets are sold through a number of regional distributors, which will reduce the freight cost for those living far from Wisconsin. When you write, the closest dealer will respond. Buyers will have to assemble the sets themselves, and the work involves digging post holes to set the uprights firmly in place.

English Garden Toys
P.O. Box 786, Indianola, PA 15051
Telephone: 800-445-5675 or 412-767-5332

Catalog price: free

Climbing and swing sets made of tubular steel, in a range of sizes for indoor and outdoor use. Tents, platforms, slides, rope ladders, even a log cabin, can be ordered separately or built into elaborate constructions with the basic metal frames. Among the more unusual extras here are a plastic water slide, a simple roller coaster with track that sits on the ground, and trampolines (from a tiny nursery model to an impressive outdoor leaper). For young children the firm sells seesaws, play tunnels, plastic sandboxes, water trays and small merry-go-rounds. Older children can be equipped with basketball hoops, soccer goals, pogo sticks, stilts, and a 15-foot rope ladder to be tied to a tree.

Florida Playground & Steel Co.
4701 S. 50th Street, Tampa, FL 33619
Telephone: 800-444-2655 or 813-247-2812

Catalog price: free

Galvanized steel swing sets, merry go-rounds, slides, and climbers. This is a reasonably priced alternative to the wood sets, and from the testimonials in the brochure it seems the structures last for decades. Legs are made extra-long to be sunk into concrete footers, so be prepared for a little digging if you order. The ladder rungs on the climbers are welded to the uprights, and the swing sets are basically four legs and a top bar; so after the legs are planted, assembly is just a matter of tightening a few bolts.

Grey Owl Indian Craft Co.
Full-size Indian tepees. See entry under *Hobby Supplies*.

GYM*N*I Playgrounds, Inc.
P.O. Box 96, Laurel Bend, New Braunfels, TX 78130
Telephone: 800-232-3398 or 512-629-6000

Catalog price: free

Wooden play sets made of pressure-treated lumber. The firm promises only that it uses hickory for the ladder rungs and elsewhere "the right wood in the right place." Options include a two-story play fort with a sandbox on the ground floor, a simple three-seat swing set, and a deluxe fort, swing, and slide combination that comes with tire swing, glider, and two slides, and that takes up a 26 × 25-foot chunk of the yard. A sandbox and a child's picnic table can complete the amusement park.

J. L. Hammett Co.
Heavy-duty steel and wood climbing sets, slides, merry-go-rounds, and swings. The familiar steel schoolyard equipment is offered, along with some more inviting wood and steel sets from Playworld (made with 6 × 6 wooden beams) and colorful playhouses and climbing towers from Rabo and The Children's Factory. A number of smaller indoor sets are also offered. See entry under *Educational Supplies*.

TimberGym Fun Center from Creative Playgrounds Ltd.

Hand in hand
Toddler swing seats. See entry under *Toys*.

HearthSong
Sells Tree Play climbing and swing sets made of 4 × 4 redwood timbers. One small model measures 8 × 8 × 9-feet tall and includes two swings, a pair of rings, a rope ladder, and a railed top walkway. Platforms, canvas tents, and slides are some of the other options. See entry under *Toys*.

Hoover Brothers, Inc.
See entry under *Educational Supplies*.

Isis Innovations
177 Thornton Drive, Hyannis, MA 02601
Telephone: 508-790-5992 (800-245-5224 to order)
Catalog price: free

Climbing gyms, swing sets, and aerial playhouses made of 4 × 4 and 2 × 6 redwood timbers with 1⅝-inch maple ladder rungs. Some of the smaller climbing sets are freestanding; the larger sets must have the legs set in the ground for stability. These are handsome, sturdy play sets, with extra attention paid to safety. No treated lumber is used—the redwood is naturally resistant to decay. Post ends have been sanded to eliminate sharp edges and corners, and bolt holes are recessed. Optional equipment includes a climbing net, a rope ladder, a horse swing, a slide, movable platforms, an infant swing, a sliding pole, and a trapeze.

Just for Kids!
A log cabin playhouse. See entry under *Toys*.

Kaplan School Supply Corp.
See entry under *Educational Supplies*.

KiddyKube, Inc.
P.O. Box 488, Cypress, TX 77429
Catalog price: free

Makes a combination playhouse/climbing gym for preschoolers. The 32-inch wooden cube is brightly painted, fits indoors without major disruption or remodeling, and is split into two floors of toddler fun. The ground floor holds a winding passage for crawling and hiding. The top floor is half enclosed with wooden bars for imaginative play. It can all be yours for less than $200 postpaid.

Nasco
Metal climbing sets. See entry under *Educational Supplies*.

The Natural Baby Co.
Wooden swing and climbing sets from The Original Wooden Gym at very good prices. See entry under *Toys*.

Play-Well Equipment Co.
655 S. Raymond Avenue, Pasadena, CA 91105
Telephone: 818-793-0603 or 213-681-1908
Catalog price: free

Swing and climbing sets made of Douglas fir, finished with a nontoxic varnish. Dozens of different sizes and combinations are available, from a two-swinger with simple ladders at each end to a four-swing set with climbing tower, slide, playhouse, and firemen's pole. The extras can be added in almost any combination, so customers can build exactly the set they want from the list of components. The structures have wide stabilizing braces on the bottom, and the uprights do not have to be sunk into the ground. To fill in the rest of the yard, the company sells sandboxes (with or without sun canopies), picnic tables, tetherball poles, balance beams, and seesaws.

Rainbow Play Systems
5980 Rainbow Parkway, Prior Lake, MN 55372
Telephone: 800-447-2553 or 612-447-2553
Catalog price: free

Redwood climbers, swing sets, and playhouses. The frames look sturdily made of 4 × 4 lumber, sanded to eliminate splinters and sharp corners, and the sets are attractive. A deluxe swing and climbing set with slide, two swings, rope ladder, rings, and a canvas-covered play area above can be had for about $2,500, including delivery and installation. The firm's playhouse is especially cute. Two small windows on the first floor open over real window boxes (flowers not included). Shoppers can choose from such extras as sandbox, balance beam, tire swing, firemen's pole, trapeze, and steering wheel to come up with the ideal set for their needs.

Sears, Roebuck & Co.
Choose from more than a dozen wood and steel play sets, or take the middle path: a steel swing set with a brown enameled "wood-look" frame. A couple of the smaller steel swing sets can be had for under $100. The wood sets, framed with pressure-treated 2 × 4 lumber, are somewhat more expensive but still very reasonably priced. A cunning playhouse is sided with milled lumber to look like a log cabin. Smaller slides and play structures are offered for preschoolers. All are listed in the specialty "Toys" catalog. See entry under *General Catalogs*.

Shelter Systems
P.O. Box 67, Aptos, CA 95001
Telephone: 406-662-2821
$1 for brochure

An inexpensive climbing gym for preschoolers, made of plastic pipe. The base of the structure forms a triangle 6½ feet on each side.

Tryon Toymakers

A swing horse with a hand-painted horse's head. See entry under *Toys*.

U-Bild

Playhouses. See entry under *Furniture*.

Wood Built of Wisconsin, Inc.

P.O. Box 92-SP, Janesville, WI 53547
Telephone: 608-754-5050

Catalog price: free

This firm fills a needed niche in the playground market: plans and hardware for home-built play sets. Do-it-yourselfers can save hundreds of dollars by buying standard lumber at a local yard and building their own swing and climbing sets with Wood Built's hardware and accessories. The firm sells framing brackets, precut ash dowels for ladder rungs, swing seats and hangers, steel slides, climbing nets, rings, trapeze bars, cables and mounts for suspension bridges, and hardware for tire swings (but not the tires).

Woodplay Incorporated

P.O. Box 27904, Raleigh, NC 27611-7904
Telephone: 919-832-2970

Catalog price: free

This outfit began as a custom builder of wooden play equipment for schools and playgrounds and has since shifted into making backyard play sets. Frames are made of 4 × 4 redwood timbers, with steel top bars across the swing sets. The simplest model has room for just two swings. Larger sets have slides, rope ladders, lookout towers, climbing ropes, firemen's poles, and room for up to five swings. For any leftover yard space, choose from sandboxes, seesaws, and balance beams.

Woodset, Inc.

P.O. Box 2127, Waldorf, MD 20601
Telephone: 800-638-9663 or 301-843-7767

Catalog price: free

Wooden play structures made of pressure-treated 4 × 4 and 4 × 6 southern yellow pine. These look to be as sturdy and well made as any on the market, and the company's catalog is sprinkled with testimonials claiming just that. (Parents concerned about the safety of pressure-treated lumber are encouraged in the catalog to call the treatment's manufacturer, the Osmose Wood Preserving Company of America, at 800-522-9663.) Buyers can put together the set of their dreams from the company's many modular components: climbers, open platforms, enclosed platforms, swing frames, ladders, towers, slides, firemen's poles, climbing ropes, rope ladders, and balance beams. For the swing sets the firm offers infant seats, tire mounts, toddler buckets, trapezes, rings, and a swinging horse in two sizes to fit one or two riders. Legs of the structures will have to be sunk into the ground. (The firm believes that horizontal ground supports present a tripping hazard and rarely find a sufficiently level spot to sit on.)

Dennis Wyatt

Tepeze, 700 Hoover Street, Johnson City, TN 37604
Telephone: 615-928-3333

Free flier

Makes a fabric play tepee printed in red, brown, and white with American Indian motifs.

Yards of Fun

P.O. Box 119, N. Manchester, IN 46962
Telephone: 800-228-0471

Catalog price: free

Wooden play sets distributed by a network of local dealers. When you write you'll receive a color catalog showing all the possibilities, and with it the name and address of a local distributor. The swing and climbing sets are made of pressure-treated 2 × 4 yellow pine with hardwood ladder rungs. The structures are designed with cross bars at the base (supported by 2 × 2 braces) so the legs don't have to be planted in holes. Other suppliers use heavier materials, but these sets present an affordable alternative. Swings sets can be outfitted with rings, trapezes, sling seats, toddler seats, or gliders. Extras include decks, covered "penthouses," slides, firemen's poles, rope ladders, and sandboxes. The firm also makes an attractive playhouse with shutters, window boxes, and a chimney.

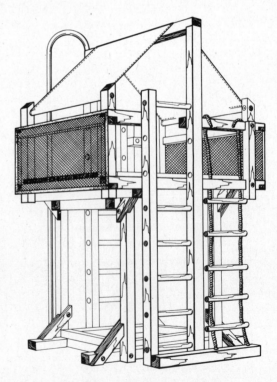

Woodplay's Clubhouse climbing set

Teddy Bears

AG Industries
3832 148th Avenue N.E., Redmond, WA 98052
Telephone: 206-885-4599

Catalog price: free (request retail price list)

Precut packaged paper airplanes, a radio-controlled sailboat, and an innovative soft toy called Snapazoo that folds and snaps to make 25 different creatures. The Snapazoo looks like a wonderful toy for the creative child—a soft, cuddly version of Japanese origami.

Artisans Cooperative
Teddy bears and sock monkeys. See entry under *Gifts*.

Bear-in-Mind
53 Bradford Street, West Concord, MA 01742
Telephone: 508-369-1167

Catalog price: $1

More than 150 different teddy bears from Gund, Steiff, Merrythought, North American Bear, Bearly There, and many other makers. Paddington Bear is here in four different sizes—from a 3-inch miniature to a 40-inch giant—and so is Smokey the Bear. Children's T-shirts, bearfoot slippers (they make your feet look like giant paws), and books for collectors fill the rest of the booklet.

Bears to Go
900 North Point, San Francisco, CA 94109
Telephone: 415-775-9828

Catalog price: $1

Bears by Gund, Mac Pohlen, Mill Creek Creations, and other artists, many of them with price tags that will put them out of reach of most children. The Gund bears are among the cheapest of the collection. About 30 different bears are pictured in the color catalog.

Bighorn Sheepskin Company
Stuffed animals made of sheepskin. See entry under *Sheets, Blankets, and Sheepskins*.

Claire's Bears & Collectibles
56 E. Main Street, Newark, DE 19711
Telephone: 302-731-0340

Catalog price: $2 (specify Steiff bears or Breyer horses catalog)

The $2 fee buys you a copy of the manufacturer's color catalog and a price list from Claire's. For someone who is looking for a specific or hard-to-find Steiff bear, or for collectors of Steiff or Breyer, Claire's looks like a good place to shop. The firm also sells painted metal soldiers, Brio trains, Ravensburger puzzles, Kiddicraft baby toys, and Baby Björn baby and toddler products. For additional catalog information, see entry under *Toys*.

Cracker Barrel Old Country Store
Teddy bears. See entry under *General Catalogs*.

Cynthia's Country Store, Inc.
11924 W. Forest Hill Boulevard, Suite 1, West Palm Beach, FL 33414
Telephone: 407-793-0554

Catalog price: $2

Steiff bears. The $2 catalog fee buys a pared-down edition of Steiff's catalog: 12 color pages of fuzzy bears and other animals, featuring both new designs and reproductions of older toys. If you want to see more, send $5 for the larger "Great Bear" catalog.

The Disney Catalog
Mickey Mouse, Donald Duck, Pinocchio and other Disney soft dolls. See entry under *Gifts*.

Walt Disney World
Mickey Mouse, Pluto, Jiminy Cricket, Lady, Bambi, Dumbo, the Seven Dwarfs, and a host of other Disney characters as soft dolls. See entry under *Gifts*.

The Doll Cottage
Steiff bears. See entry under *Dolls*.

Dollsville Dolls & Bearsville Bears
Steiff bears. See entry under *Dolls*.

Gail Wilson Duggan Designs, Inc.
Box D, Summer Street, Charlestown, NH 03603
Telephone: 603-826-5630

Catalog price: free

Charming little bear families, rabbits, and folk dolls made from bits of felt and calico. Buy the ingeniously assembled kits and make them yourself, or order them ready-made. Furniture, a house, and additional clothing can be ordered in kit form only. We've put some of these bears together and are impressed with the cleverness of the designs and the clarity of the instructions.

Gail Wilson Duggan's bears in the complete furnished house

The Enchanted Doll House
See entry under *Dolls*.

The Fantasy Den
25 Morehouse Avenue, Stratford, CT 06497
Telephone: 203-377-2968

Catalog price: $1.75

Hundreds of teddy bears and other furry animals from makers all over the world. Gund is prominently represented in the black-and-white catalog, as are Alresford, Binkley, and North American Bear (with its V.I.B.'s: Bearb Ruth, Albeart Einstein, Clara Bearton, etc.). A series of Whatchamacallits from Down Under includes a fuzzy kangaroo, a koala, a wombat, a platypus, and a Tasmanian devil. Stickers, posters, and bear-making supplies should meet the needs of the beary obsessed.

Lamkin, Inc.
P.O. Box 25607-D, Sarasota, FL 34277
Telephone: 813-923-6856

Free brochure

Sells a soft stuffed lamb, named Li'l Lambkin, that plays the song "Guidance" when his rear hoof is pressed. He's joined by several bears, including Choo-Choo Bear, who's dressed in an engineer's costume, and Bearasota Bear, in beach garb. All are made of "the softest materials," according to the brochure.

Leonard Bear Learning, Ltd.
Teddy bears sold as part of an early learning program. See entry under *Educational Supplies*.

Munchkin Outfitters
Stuffed animals. See entry under *Children's Clothes*.

Pillow Pals, Inc.
Pillow creatures. See entry under *Sheets, Blankets, and Sheepskins*.

The Soft Menagerie
240 Driftwood Lane, Guilford, CT 06437
Telephone: 203-453-0656

Catalog price: free

A collection of soft animals that ventures beyond bears to rope in beavers, pigs, rabbits, lambs, squirrels, dogs, ferrets, seals, otters, and ducks. Bear-lovers will of course find a handful of their favorite furry beasts, from cute koalas to Humphrey Beargart. The creatures come from Gund, Dakin, Douglas, Steiff, North American Bear, and other makers. Prices start at about $15.

Stuf'd 'n Stuff
10001 Westheimer, Dept. S, Houston, TX 77042
Telephone: 713-266-4352

Catalog price: $1, refunded with order

Teddy bears share the stage here with such stuffed beasts as rattlesnakes, armadillos, vultures, puppies, and tigers. A line of plush hand puppets includes a turtle, an alligator, a frog, a skunk, and an eagle. Though they enjoy the company, the bears manage to dominate the scene. Scores of the furry ursines are offered from well-known and sought-after makers. Steiff and Gund bears trample through the pages, joined by such rarities as a bear with genuine mink fur, close-outs from defunct makers, and costly limited-edition creatures. The catalog is also liberally sprinkled with inexpensive animals, some as cheap as $10. Regular mailings fill customers in on the latest offerings and spread news of happenings in the stuffed-toy world.

Toys

ABC School Supply
Wooden blocks, construction sets, trucks, and toddler toys. See entry under *Educational Supplies*.

Abilities International
Old Forge Road, Elizabethtown, NY 12932-0398
Telephone: 518-873-6456

Catalog price: $1

An exceptional collection of developmental toys, practical clothes, and baby products. Each item in this 32-page color catalog has obviously been chosen with care. Offerings range from a large crib mirror and a freestanding play bar for infants to a wooden trike for toddlers, a roller-coaster bead maze, a plastic schoolhouse in a shoe, and a beautiful wooden train with lumber car, circus car, passenger car, and caboose. For older kids the company sells an easel (with a colorful paint suit to keep the mess under control) and a puppet theater with wipe-clean surfaces on the front that can be decorated and redecorated for each performance. Block sets, Quadro climbing sets, games, puzzles, and books add to the appealing array.

Recognizing that life is not all play, the firm also sells a changing table, a porta-crib, a baby bouncing seat, a breast pump, a selection of attractive bibs, a baby backpack, a stroller bag, and the best-looking baby travel bag we've seen. It has insulated compartments for bottles, a waterproof padded changing mat, and zippered compartments for keys, tickets, wallet, and other adult necessities.

Afterschool
1401 John Street, Manhattan Beach, CA 90266
Telephone: 213-545-1073

Catalog price: free

Toys and project supplies for school-age child-care centers that can also be ordered by parents. Large nylon-covered foam blocks make building projects fun for younger kids; the covers come off for easy washing, and Velcro corners hold the blocks together when stacked. Foam balls, flying disks, "boppers," and "pushers" help keep indoor roughhousing under control. For a more contemplative atmosphere, parents might want to order an ant farm, a solar graphics kit (with sun-sensitive paper), a grow-your-own butterfly kit, or a model of an erupting volcano (it spews safe "sparks" a few inches above the crater).

All But Grown-Ups
Maple blocks and roller-coaster bead mazes. See entry under *Furniture*.

Kinderworks' Kindercolor Express!!! bead maze, sold by All But Grown-Ups

The Ark Catalog
4245 Crestline Avenue, Fair Oaks, CA 95628
Telephone: 916-967-2607

Catalog price: $1

Toys and books chosen to promote imaginative and creative play. The oversize color catalog displays some exquisite wooden toys, including a paddle-wheel boat, a sailboat, a train, a castle set, and carved wooden figures of such fairy tale characters as the Bremen Town musicians, Little Red Riding Hood and the wolf, Hansel and Gretel, and the Pied Piper and his rats. Shoppers will also find modeling beeswax, pewter birthday-candle holders, a puppet theater, a pattern for making a tepee, a Goldilocks cookie-cutter set, a child-size loom, and several soft dolls. The wide-ranging nature of the catalog continues in the book section, with titles on myths and folk tales, art for children, the Waldorf philosophy of education, and such storybooks as *Old Mother West Wind*, *The Borrowers*, and *Children of Green Knowe*.

Artisans Cooperative
Hand-crafted push toys. See entry under *Gifts*.

Back to Basics Toys
3715 Thornapple Street, Chevy Chase, MD 20815
Telephone: 800-356-5360

Catalog price: $2

A dream catalog of toys for the babies of baby boomers. Flipping through these pages is like stepping back into childhood in the 1950s. Raggedy Ann and Andy dolls, Lincoln Logs, Tinkertoys, Radio Flyer wagons (both the standard red model and the wood-body Flying Eagle), Lionel trains, Mickey Mouse and Donald Duck puppets, ant farms, American Flyer sleds, and the Ertl pedal-driven tractor all assault us with waves of fond memories. And to give our heartstrings an extra yank, the catalog is designed to look like an old *Life* magazine, the pictures printed in a nostalgic sepia color. The catalog does acknowledge the passage of time and gives credit to some technological advances of the past three decades. Along with the older toys, shoppers will find an ultra-modern set of roller skates with polyurethane wheels and self-lubricating ball bearings, a roller-coaster bead maze, and a radio-controlled model sailboat. Timeless entries include an excellent maple block set, an oak rocking horse with spring mounts, a puppet theater, a magic set, a dollhouse open on all sides, and tabletop soccer, baseball, and shuffleboard games.

Benny's Express
495 Tiogue Avenue, P.O. Box 38, Coventry, RI 02816
Telephone: 800-456-1700

Catalog price: free

A catalog of tools, toys, and useful items for grown-ups that carries several pages of products for children. Lots of riding toys are offered (wagons, trikes, training bikes, and pedal carts), a good collection of Fisher-Price toys, a dozen board games, and such items as a microscope, a Tonka dump truck, a working child's typewriter, and the Etch-A-Sketch drawing toy.

Chaselle, Inc.
9645 Gerwig Lane, Columbia, MD 21046
Telephone: 800-CHASELLE (800-492-7840 in MD) or 301-381-9611

Catalog price: free (specify "Pre-School & Elementary School Materials" catalog)

Hardwood blocks, Lego, Duplo, and Waffle block sets, dress-up costumes, puppet theaters, indoor and outdoor playhouses, toy trucks and trains, marble tracks, plastic play food, and lots of educational toys in a big school-supply catalog. See main entry under *Educational Supplies* for a description of Chaselle's other offerings.

Cherry Tree Toys
Parts for wooden toys. See entry under *Woodworking*.

Childcraft
20 Kilmer Road, Edison, NJ 08818
Telephone: 800-367-3255 (800-631-5657 to order)

Catalog price: free

"Toys that teach" is the motto here, and the collection blends playthings that are pure fun with those that could be labeled educational. On the fun end of the spectrum are squirting sunglasses, dinosaur piñatas, dress-up costumes, miniature racing car sets, walkie-talkie telephone receivers, and a rocking car. More educational are alphabet and map puzzles, color-in flash cards, and magnet sets. In between we discover telescopes, microscopes, chemistry sets, art supplies, magnetic picture blocks, hand-held electronic games, hardwood building blocks, child-size tables, chairs, and fold-out sofas, plastic dinosaurs, horses, and jungle animals, playhouses, trikes, scooters, sleds, even a battery-powered sports car that holds children up to 75 pounds.

Claire's Bears & Collectibles
56 E. Main Street, Newark, DE 19711
Telephone: 302-731-0340

Catalog price: $2 (specify imported toys or Brio wooden trains catalog)

The $2 fee buys you a packet of manufacturers' catalogs and a price list from Claire's. The Brio catalogs include an amazing array of specialty track, landscaping, and accessories, as well as Brio's other childhood toys. If you write for the imported toys catalog, you'll get color booklets with complete inventories of Kiddicraft toys, Ravensburger games and puzzles, Metal-Models hand-painted soldiers, and Baby Björn baby supplies (crib mobiles, potties, changing pads, booster seats, and portable cribs). Claire's also sells Steiff bears and Breyer horses. See entry under *Teddy Bears* for additional catalog information.

Cloud Nine, Futons & Furnishings
An indoor tepee, a feltboard with felt shapes, and a reversible cape with Batman, Superman, or an initial-letter motif sewn on both sides. See entry under *Sheets, Blankets, and Sheepskins*.

Community Playthings
Wooden blocks, hardwood riding trucks, foam blocks, and other nursery- and elementary-school toys. See entry under *Educational Supplies*.

Constantine's Woodworker's Catalog
Plans and parts for making wooden toys. See entry under *Woodworking*.

Constructive Playthings

Maple blocks, hollow cardboard construction blocks, Brio train sets, bath toys, and other playthings for young children. See entry under *Educational Supplies*.

Crate & Barrel

P.O. Box 3057, Northbrook, IL 60065-3057
Telephone: 312-272-3112

Catalog price: $2

Crate & Barrel has earned a reputation as a supplier of finely designed housewares at reasonable prices. Its catalogs also offer children's furniture and toys. The inventory in past editions has included wooden puzzles, easels, and Radio Flyer's row cart and wagon. See additional entry under *Furniture*.

Creatively Speaking

Wooden trucks. See entry under *Dolls*.

Cumberland General Store

Rt. 3, Box 81, Crossville, TN 38555
Telephone: 800-334-4640 or 615-484-8481

Catalog price: $3

A hefty 250-page catalog that looks for all the world like an 1890 Sears or Montgomery Ward catalog, but with current prices and merchandise that's really in stock. The average suburban parent may not have much use for blacksmith's anvils, goat harnesses, cuspidors, or plow points, but it certainly is fun to browse and imagine. Cookie cutters, *McGuffey's Reader*s, sleds, farm machinery for the sandbox, Radio Flyer wagons, and a high chair that folds down to a baby rocker come a little closer to modern family needs. The music department is stocked with dulcimers, mandolins, banjos, fiddles, and harmonicas.

John Deere Catalog

1400 Third Avenue, Moline, IL 61265
Telephone: 800-544-2122 or 309-765-7878

Catalog price: free

A home and garden catalog that is big on items with early John Deere trademarks. Two old-fashioned sandbox tractors in John Deere green and yellow should appeal to budding farmers. A pedal tractor with trailer looks like the ultimate riding machine for a five-year-old. Gardening parents will find much to tempt them in the rest of the catalog.

Design Group

Wooden-toy patterns. See entry under *Woodworking*.

Before Buying a Toy, Try Borrowing It from a Library

Fifty years ago, the first toy library was founded in Los Angeles. Today the Los Angeles County Toy Loan Program maintains a circulating inventory of some 10,000 toys, distributed through a system of local libraries. Similar toy libraries, on a smaller scale, lend toys to families throughout the United States. The YWCA of Metropolitan Chicago, for example, sponsors a toy-mobile program, as does the Broward County Library in Fort Lauderdale, Florida. Public libraries, children's museums, schools, childcare resource centers, and hospitals operate toy lending programs in more than 300 cities and towns across the country.

While many parents borrow books from local libraries, far fewer are aware of the benefits of toy libraries. Not only do the loan programs provide access to quality toys that parents might not otherwise be able to afford, they're generally staffed by people with some background in child development who can give advice on appropriate toys and beneficial play.

Check with your local library to see if they operate a toy lending program or know of any toy libraries in your area. If they don't, you can contact the U.S.A. Toy Library Association (2719 Broadway Avenue, Evanston, IL 60201; telephone 312-864-8240) for further information. The association publishes an annual directory of toy libraries, which it sells to nonmembers for $23.

Duncan Toy Co.
P.O. Box 165, Baraboo, WI 53913
Telephone: 800-356-8396

$1 for *Yo-Yo & Spin-Top Trick Book*

Duncan won't sell its yo-yos to individual customers by mail, but it will mail its 48-page trick book, which is full of diagrams and instructions for making the yo-yos perform. The first few pages go over the familiar sleeper and walk-the-dog stunts. By the end of the book you'll be splitting the atom and doing whirlybirds with both hands.

Early Learning Centre
P.O. Box 821, Lewiston, ME 04243-0821
Telephone: 800-255-2661 or 800-752-3433

Catalog price: free

The Early Learning Centre temporarily suspended its mail-order catalog in the spring of 1989, but may revive it in the fall or in 1990. When available, the catalog offers toys, puzzles, musical instruments, books, and cassettes for young children (to about age 8), all chosen with the idea that play should "promote sound intellectual and physical development." The color catalog is filled with fun in the form of Brio trains, wire bead mazes, Duplo sets, magnetic marbles, Lincoln Logs, costumes, infant toys, crystal-making kits, and a periscope for young spies. The company is well stocked with art and craft supplies, including paints, easels, finger crayons, dinosaur templates, and a child's loom. It sells a dozen board games, a good selection of simple jigsaw puzzles, and a small shelf of quality books. Among the musical instruments are a wooden guitar, a mouth organ with color-coded keys, a plastic kazoo, and a set of chimes. At the company's stores, we've noticed that a trampoline with a padded grab-on bar is a huge hit with three- and four-year-olds. It's in the catalog for about $60.

Educational Teaching Aids
See entry under *Educational Supplies*.

Emily's Toy Box
163 Main Street, P.O. Box 48, Altamont, NY 12009
Telephone: 518-861-6719

Catalog price: $1

A home business that has assembled a choice collection of toys, games, puzzles, and art supplies. The main catalog is produced with a typewriter and a copy machine, but though it lacks color and slick graphics, it commands careful study. Papa Don's hardwood baby rattles are offered for the very young, wooden shape sorters and stacking blocks for toddlers. Young children can be outfitted with water toys, tea sets, Brio-compatible trains (at good prices), wooden village and farm sets, pattern blocks, marble mazes, and magic sets. Games and puzzles form a special strength. With the photocopied catalog comes the complete color catalog from Lauri, which is full of puzzles, peg sets, stringing toys, and building kits, all made of colorful crepe-rubber sheets. Ravensburger puzzles and games are sold for starting puzzlers—the backs have finger holes to help push out the pieces. Among the games are picture dominoes, Baby Animal Lotto, a Chinese checkers set with an oak board and wooden pegs, several cooperative games from Family Pastimes, and a few of Aristoplay's board games. (See separate entries for Family Pastimes and Aristoplay under *Games and Puzzles*.) The art-supply offerings include short fat crayons from Scola, Brunzeel colored markers and pencils (in beautiful metal boxes), watercolor crayons, face paints, foam paint, and Dover coloring books.

The Enchanted Doll House
Sells a miniature castle, Brio trains, and other toys. See entry under *Dolls*.

Environments, Inc.
Maple blocks, shape sorters, and other toys for younger children. See entry under *Educational Supplies*.

Bill Ding stacking clowns from Environments, Inc.

Exploratorium Store, Mail Order Dept.
Toys relating to science and mathematics. See entry under *Science and Nature*.

Fisher-Price
Consumer Affairs, 636 Girard Avenue, E. Aurora, NY 14052-1880

Catalog price: free (request "Bits and Pieces" catalog)

Hundreds of spare parts for Fisher-Price toys, listed (without pictures) by the name and model number of the toys that they accompany. If you lose or break any of the loose pieces from a Fisher-Price toy set, this is the source for replacements. With wheels, little people, barnyard animals, shapes for shape sorters, battery covers, and dishes for the kitchen set, Fisher-Price is ready and willing to make your toys whole again. If you want to see everything Fisher-Price makes, send $2 for the complete toy catalog. Be forewarned that while the "Bits and Pieces" catalog *is* a mail-order source, the toy catalog *is not*. It's just a reference guide to help you find what you want in stores.

Free Play
996 Walnut Street, Newton, MA 02161
Telephone: 617-332-8234

Catalog price: $1

Handcrafted toys that encourage learning through imaginative play. The six-page catalog lays out the choices: wooden puzzles (some with peg handles and some without), wooden block sets with storage wagons, a wooden lawn-mower push toy, a ring stacker, a puppet theater with wipe-clean coloring panels, puppets, ethnic dolls, and a tiny tooth-fairy sack for that magical exchange of teeth for cash.

The Gifted Children's Catalog
2922 N. 35th Avenue, Suite 4, Phoenix, AZ 85061-1408
Telephone: 602-272-1853 (800-528-6050 to order)

Catalog price: $2

Toys and educational equipment that any child, gifted or otherwise, can enjoy playing with and learning from. The approach here is far from heavy-handed. The pedal-powered Kettcar isn't likely to put competitive pressure on a child, nor is an ant farm, a Colorforms set, or a periscope. A microscope, a chemistry set, star charts, and other tools for learning are offered for children who are ready and curious. Pop-up books, a chime instrument, a screen-printing kit, a leaf press, and a beginner's chess set (pieces are marked with name and moving restrictions) all present opportunities for learning fun.

Gil & Karen's Toy Box
3975 Kim Court, Sebastopol, CA 95472
Telephone: 707-823-8128

Catalog price: free

A wooden-toy catalog with some special treats. Other companies sell colored wooden blocks, but here they come in a personalized wooden storage box. And the blocks themselves are a cut above the competition. Made by T. C. Timbers in New York, they're stained with a nontoxic dye and glossed with a wax finish. That means no paint to chip or chew. Personalized wooden tote boxes and toolboxes are made to look like classic carpenter's boxes. A rhythm band set comes in a similar carrying case. Alphabet, number, and name puzzles are sold, along with geometric pattern blocks, a painted wooden train set, a simple dollhouse and play barn, and a set of castle blocks. A wooden train whistle, a pull-string top, a Jacob's ladder toy, rubber stamp dinosaurs and alphabets, and build-them-yourself fire-engine and sailboat models offer even more fun.

Personalized block set from Gil & Karen's Toy Box

Growing Child
P.O. Box 620, Lafayette, IN 47902
Telephone: 317-423-2624

Catalog price: free

An exceptional collection of playthings, books, and recordings for children from birth to about age ten. The catalog is not massive, but the selections have been made with care. For babies the company offers teethers, rattles, nesting cubes, shape sorters, a basic construction set, pull toys, a starter set of blocks, and simple puzzles. For preschoolers it sells pattern blocks, peg boards, hand puppets, a puppet theater with wipe-clean surfaces, big cardboard brick blocks, games, more puzzles, and a make-believe kit stuffed with such things as play money, tickets, price tags, award ribbons, and order pads. School-age children will find maple block sets, a wooden easel, paints with spill-proof cups, a block-and-marble building set, Cuisenaire rods, magnets, more complex puzzles, and lots of little cars, people, and building sets for pretending. The book collection ranges from old classics like *Pat the Bunny*, *Millions of Cats*, and *Winnie the Pooh* to newer titles like *The Jolly Postman* and *Pigs in Hiding*, with some good activity books and handbooks for parents as well. The music section includes recordings by Tom Glazer, Hap Palmer, and Mister Rogers, and some simple instruments for beginning musicians.

Hagstrom's Sales Ltd.
2 Dunwoody Park, Atlanta, GA 30338
Telephone: 404-393-0363

$3 for brochure

For the child who has everything, or the parent who's never grown up, real gasoline-powered cars in reduced scale. The color brochure lays out the incredible choices: a Tin Lizzie, a 1920s-era fire engine, a 1932 roadster, an Indy racing car, a Corvette, a Jeep, a Mercedes 500 SL, and a few others. Most have electric starters, disc brakes, and working head- and tail-lights. All have 3- to 11-horsepower Briggs and Stratton engines. Prices start at about $1,300.

Hammacher Schlemmer
147 E. 57th Street, New York, NY 10022
Telephone: 800-543-3366 or 212-421-9000

Catalog price: $2

Toys, housewares, gadgets, and gifts for those who can afford some luxuries. Recent catalogs have held such treats as a genuine arcade pinball machine ($7,500), a commercial flight simulator which lets the home pilot take off and land at 80 different U.S. airports ($150), and a covered sled for pulling children over snow while hiking or skiing ($500). Down a bit closer to ordinary checkbook limits were a roller-coaster bead maze (dubbed "The Child's Elemental Skills Trainer"), a musical humidifier, the Gerry baby gate, and a couple of unusual construction sets. Order the catalog and see what they've come up with this time.

J. L. Hammett Co.
Toys for babies and preschoolers. See entry under *Educational Supplies*.

Hancock Toy Shop
97 Prospect Street, Jaffrey, NH 03452
Telephone: 603-532-7504

Catalog price: $1

Handcrafted wooden toys and children's furniture made of rock maple and other hardwoods. Blocks, trucks, doll furniture, and play kitchen equipment make up the toy list. Chairs, toy chests, tables, and storage shelves are the furniture options.

Hand in hand
First Step, Ltd., Catalog Center, 9180 LeSaint Drive, Fairfield, OH 45014
Telephone: 800-543-4343

Catalog price: free

A color catalog of necessities and delights for young children and their parents. A bath page holds no-slip bath and shower mats, a shampoo visor, and a floating bathtub carousel puzzle. Parents who really want to indulge their children with water fun can order a free-standing water table for about $300. Sheepskins, a soft play area for toddlers (called The Nest), a clock for children learning to tell time, toddler and preschool swing seats, a toy car-phone, a cushioned shopping-cart seat, and an assortment of videos and tapes combine in a nice blend of fun, learning, and practicality.

HearthSong
P.O. Box B, Sebastapol, CA 95473
Telephone: 800-325-2502 or 707-829-1550

Catalog price: $1

A catalog of special things for young children that crosses over all our neat and tidy categories. A browse through these beautiful color pages turns up a finger puppet theater, a tiddlywinks set packaged in a wooden mushroom, a set of easy games for children too young to count, rustic wooden blocks made from branches, a fireplace popcorn popper, Stockmar watercolor paints and modeling beeswax, Weleda baby soaps and creams, silk scarves for dress-up play, a children's lyre, and a German dollhouse that can be outfitted with little wood-headed dolls and furniture made of branches. Everything in the catalog seems to be aimed at bringing children (and parents) together in creative, noncompetitive play. A separate catalog of play structures is available on request.

Tops, felt balls, and a millipede pull toy from HearthSong

Heartwood Arts
Rt. 1, Box 126, Modena, NY 12548
Telephone: 914-883-5145

Catalog price: free

Wooden animals, figures, villages, and castles made with rounded contours and bold-grained wood for a look that fits our image of J. R. R. Tolkien's Middle Earth. The workshop sells much of its output to Waldorf schools, but all parents are welcomed customers. An outdoor playhouse with a twisting, pointed roof and irregular rounded shingles would certainly be the talk of the neighborhood, though its price will likely keep it out of reach of most parents.

High Q Products
9272 E. Pioneer Drive, Dept. SP, Parker, CO 80134
Telephone: 303-841-0932

Catalog price: free

Learning toys and games, mostly for young children. The company sells shape sorters and alphabet blocks for toddlers, wooden arithmetic blocks, plastic dinosaurs, games that teach reading skills and ancient history, and a puzzle map of the United States complete with rivers and mountain ranges. Counting books are offered for the youngest children, books of math puzzles for older minds, and books for parents to help guide children in their learning activities.

Hoover Brothers, Inc.
Construction toys, wooden blocks, and other educational toys. See entry under *Educational Supplies*.

The Horchow Collection
P.O. Box 620048, Dallas, TX 75262-0048
Telephone: 214-484-3558 (800-527-0303 to order)

Catalog price: $5, refunded with purchase

Extravagant gifts, including some for the younger set. Don't expect a whole toy store, however. A recent catalog set aside just 3 of its 64 pages for children's merchandise. The selection included a set of child-size golf clubs, a Mickey Mouse gumball machine, a red plastic easel, and a couple of dress outfits.

Ice Wood Designs
P.O. Box 901, Kalispell, MT 59901
Telephone: 406-755-1818

Free brochure

Makes an affordable wooden rocking horse (under $50, some assembly required), a wooden logging truck with a load of small-scale logs on its bed, and a block truck packed with 32 colored wooden blocks.

Imagination Toys
302A Main Street, P.O. Box 230, Spring, TX 77383
Telephone: 713-350-2926

Catalog price: $1, refundable on first order

A small collection of quality toys, mostly for preschoolers. The ten-page catalog includes maple blocks with an optional storage cart; a roller-coaster bead maze; wooden name, map, and alphabet puzzles; a train cap and whistle; a wooden train set; a chime instrument; a wooden rocking horse; and more. The company sells some things that aren't in the catalog, including many toys from Playmobil, Marklin, Movit, T. C. Timber, and Brio. It will send brochures from any of these manufacturers on request, and will do its best to fill special orders.

Johnson & Johnson Child Development Toys
6 Commercial Street, Hicksville, NY 11801-9955
Telephone: 800-645-7470

Free information

Some of the best toys available for babies and preschoolers, sold through a "toy-of-the-month club" program. Johnson & Johnson's Red Rings, Spinner Rattle, Star Teether, and Balls-in-a-Bowl toys are almost required playing for today's babies. While many toys appeal to parents with cute pictures and babyish designs, these are designed to meet a child's learning and development needs. Babies newly able to grab love the Red Rings toy, made of concentric plastic rings joined by a flexible rubber strip. It's attractive to young eyes, easy to grab, satisfying to chew, and eventually a challenge to pass from hand to hand. The Star Teether and the Spinner Rattle both fit the needs of sitting babies who like to hold and shake objects. Each has a flexible ring for chewing, a spinning part that's fascinating to watch and entertaining to manipulate, and a not-too-loud rattle that sounds with a good shake. The Balls-in-a-Bowl toy is perfect for babies and toddlers who like to play with balls, put things away, or empty things out. The clear plastic balls have bright spinners inside that flash and turn when they're moved. Babies put them in and take them out of a round bowl that looks like a fishbowl tipped slightly to the side—sized and angled precisely for young investigators.

Members receive two new toys every few weeks, each mailing geared to the age of the child. Roughly 20 toys are offered in all, for children from birth to 18 months.

Johnson & Johnson's Balls-in-a-Bowl toy

A Playcling set from Julia & Brandon

Julia & Brandon
791 Roscommon Drive, Vacaville, CA 95688
Telephone: 707-446-8838

Catalog price: free

An excellent collection of toys, necessities, safety devices, and learning puzzles for toddlers and preschoolers. Among the household needs are the Sassy hook-on portable high chair, two sets of dinnerware for beginning eaters and those a bit more advanced, a cushioned shopping-cart seat, and a light-switch extender that lets little ones flip wall switches themselves. The color catalog offers such safety devices as doorknob covers that turn for large hands but not for small ones, screw-on outlet covers, and padded bathtub spout covers. Among the toys are nesting beakers, a ring rattle, an egg-carton shape sorter, see-through spin tops, bathtub and beach toys, and Brio trains. The puzzles include crepe-rubber counting puzzles, simple cardboard jigsaw puzzles from Ravensburger, and starter wooden puzzles with knobs and finger holes. A special feature of the catalog, which we haven't seen elsewhere, is Playcling play sets. Each set comes with a packet of vinyl figures that stick to specially treated picture boards. One shows a fire station with firemen and fire trucks, another a natural scene with stick-on wildlife, and another a map of the world with detachable countries. All look like aids to imaginative and educational play.

Julia Toys
Plans and parts for making wooden toys. See entry under *Woodworking*.

Just for Kids!
75 Paterson Street, P.O. Box 15006, New Brunswick, NJ 08906-5006
Telephone: 800-443-5827

Catalog price: free

A delightful catalog overflowing with toys, games, costumes, dolls, party supplies, and other amusements for young children. Three seasonal editions each run to almost 100 color pages. The spring-summer booklet carries wading pools, sand molds, bunny costumes, and a child-size soccer net. The fall-winter edition includes holiday tights, cookie cutters, and dinosaur sweatshirts. The holiday catalog highlights attractive gift ideas. All three editions hold such playthings as Claibois walkers and riding toys, wooden blocks, a log cabin playhouse, plastic dinosaurs and jungle animals, ballerina and gladiator costumes, feather boas, Sesame Street and Disney toys, a tabletop soccer game, child-size golf clubs, and beginners' electronic keyboards. Among the more sensational offerings are a pedal-powered Ferrari Testarossa and a foam chair in the shape of a dinosaur that folds out to make a nap mattress. For birthdays the catalog offers party hats and horns, paper plates, balloon decorating kits, cake molds, animal candle holders, and windup party favors.

Klockit
Parts, plans, and kits for making wooden toys. See entry under *Hobby Supplies*.

Lauri, Inc.
Replacement pieces for Lauri play sets. See entry under *Games and Puzzles*.

Learning Materials Workshop
58 Henry Street, Burlington, VT 05401
Telephone: 802-862-8399

Free brochure

Beautiful and innovative wooden construction sets that seem to delight parents almost as much as they absorb the attention of children. Karen Hewitt started her novel

Arcobaleno blocks from Learning Materials Workshop

Six Billion Lego Bricks a Year

When Lego, a Danish company, introduced the first version of its tiny blocks in 1949, it found the public resistant to plastic toys. But today Lego bricks, with their clever stud-and-tube coupling system, are among the world's most popular toys. The company produces 6 billion bricks each year (enough to circle the globe 14 times) to a standard of precision that is virtually unmatched in the toy industry. The allowable variation in the diameters of studs and tubes is a mere .02 millimeter, less than a thousandth of an inch.

One woman who knows the capacities of the Lego system well is Francie Berger, senior model designer of the company's American division. She, along with several other designers and builders, snaps together elaborate creations of Lego bricks for store displays, advertisements, and promotional events. Her crew has built a 13-foot replica of the U.S. Capitol, a 3-foot-tall skateboarding pig, and an outsize jack-in-the-box that, while it didn't exactly jump, swayed impressively over its open lid.

Berger stepped naturally into this dream job. She played with Lego bricks as a child and somehow never stopped. For her architecture thesis at Virginia Tech she designed a town and farm of Lego blocks. When she applied for a job after college, the then manager of Lego's design department asked her to build a ball, saying "To build something round using square elements is the ultimate challenge." Several hours later she rolled a perfect yellow ball across the floor to his desk.

Families who want to see a permanent display of Lego mastery need only travel to the Legoland park in Billund, Denmark. Here model engineers have put brick to brick to create replicas of Mount Rushmore, a Space Shuttle launch, a Norwegian fishing village, and the residence of the Danish royal family. A safari ride travels past life-size African elephants, giraffes, and zebras (made entirely of Lego bricks, of course); a boat cruise takes in Egyptian monuments. People who've gone report being awestruck and inspired.

Lego sets are sold by most of the major educational-supply companies (see *Educational Supplies*), and by Chaselle, Inc. and Treasured Toys, both described in this section. The Lego Shop-at-Home Service offers a good selection of both basic and advanced building sets.

Legoland Black Monarch's Castle set

line of toys in the late 1970s with a set called Thingamabobbin, made in part of wooden spools salvaged from nearby mills. She's since built the line up to include Arcobaleno, an elegant, brightly colored set of nesting arches that can be formed into spirals, domes, and bridges, and Prismatics, a block-and-prism set for creative visual play. Several other sets combine stacking and interlocking mechanics with lovely colors and patterns. Parents may have trouble sharing these blocks with their children.

LEGO Shop-at-Home Service
555 Taylor Road, P.O. Box 640, Enfield,
CT 06082-3298
Telephone: 203-749-0706 or 2291

Catalog price: free

Parts, accessories, and selected building sets for adding to your home Lego, Duplo, and Fabuland collections. The catalog is a good source for new sets and unusual pieces not easily found in stores. A book section offers storybooks featuring the Fabuland characters and idea books for building with Lego.

Archie McPhee & Company

Beautiful metal windup toys from Hungary, potato guns that shoot "harmless potato pellets," rubber lizards, and other curious playthings. See entry under *Magic Tricks and Novelties*.

Maine Baby

A toddler block set in a bright red carpenter's box. See entry under *Baby Needs*.

Marvelous Toy Works
2111 Eastern Avenue, Baltimore, MD 21231
Telephone: 301-276-5130

Catalog price: free

A full line of wooden toys, including some of the best (and best priced) maple block sets we've seen. One small set comes in a cloth bag, a slightly larger one stores in a low wagon, and an ample beginner set fits into a deep wagon with a high push handle just right for toddlers. A castle block set comes with corner towers, entrance gates, and notched rampart blocks to top the walls. Even larger sets are offered for young contractors and for classrooms, and 40 different block shapes are offered by the piece. The catalog goes on to picture wooden cars and trucks, rocking horses, board games, wagons, dollhouses, and doll cradles.

The Metropolitan Museum of Art

See entry under *Gifts*.

Mill Pond Farms
P.O. Box 203, Rochester, MA 02770
$1 for brochure

Wooden toys and children's furniture imported from New Zealand, all made of pine with a natural finish. A rocking horse actually swings on its own stand. A wood-framed blackboard is mounted like an easel. Many of the toys are for riding and pushing preschoolers: a push cart, a wagon, a wheelbarrow, and a trike, all four made with big wooden wheels. A simple dollhouse is open on one side for easy access. A crib, high chair, stroller, and cradle are made in reduced size for dolls. A table with two chairs looks about the right size for four- to six-year-olds.

Modern Homesteader

Miniature metal tractors. See entry under *Children's Clothes*.

The Montgomery Schoolhouse, Inc.
Montgomery, VT 05470
Telephone: 802-326-4272

Catalog price: free

Brightly painted wooden cars, trains, and trucks, along with wooden baby rattles and two nifty circus toys. Some of the little vehicles have moving pieces and removable parts: a crane's grapling hook goes up and down; special train cars can be loaded with logs, automobiles, or colored blocks; and a dump truck has a hinged back for dumping. A circus elephant stands on its hind legs or rolls on all fours with a basket of clowns on its back. Take the clowns down and pull a string, and the elephant rolls a ball with its trunk.

Bowling Elephant toy from The Montgomery Schoolhouse, Inc.

Mountain Craft Shop
American Ridge Road, Rt. 1, New Martinsville,
WV 26155
Telephone: 304-455-3570

Catalog price: free

Folk toys and games, handmade dolls and puppets, and doll furniture. The catalog is really a price list with tiny drawings of the scores of items sold. Some of the more complicated action toys, like the climbing bear ("pull down, bear climbs up") and the ball trick ("educated ball on a string") aren't really done justice by the tiny sketches. A phone call might clear up just what they do. Mechanical toys like the whimmydiddle, the do-nothing machine, and the limber jack find a place on the list alongside noisemakers like the bullroarer and rattle gun and traditional puzzles like the buttonholer and the penny pincher. Pop guns and slingshots can be had, as well as marble chutes, a toy parachute, and a hand-carved Noah's ark toy. The doll list sticks to traditional country designs like buckeye, clothespin, and corncob dolls, and such stuffed toys as sock monkeys and Raggedy Ann and Andy. The doll furniture ranges from a rocking chair in ¼-inch scale to a cradle in 6-inch scale. A doll log cabin built of tiny individual logs would make a fun alternative to a Victorian home.

Mountain Toy Makers
Box 51-S, Long Lake, NY 12847
Telephone: 518-624-6175

Catalog price: $1

Wooden toys made of pine, maple, and birch, presented in a beautifully photographed catalog. These are mostly stylized vehicles—cars, trucks, trains, and a helicopter—in a natural wood finish. One car has oval wheels for a crazy tip-and-bump ride. A car-carrier truck holds five little cars; a logging tractor comes with a load of logs. A family of ducks works as a pull toy: the bill of one hooks to the tail of the next in line to form a chain, the front duck equipped with a pull string. Oval wheels give them an appealing waddle. Most of the smaller vehicles are priced in the $5 to $10 range.

Mueller-Wood Kraft, Inc.
23425 W. Wall, Lake Villa, IL 60046
Telephone: 312-395-0005

Catalog price: free

Mueller-Wood Kraft makes a wonderful wooden circus train with separate cars for an elephant, a lion, a giraffe, and a gorilla, the wooden animals included in the set. Another work train pulls a crane, a hopper, a block-filled gondola, and a tool car stocked with repair tools. On a much larger scale is a rubber-tired riding train—with two cars and a caboose it stretches 7 feet. Several trucks carry cars and wooden blocks; a camper pulls a boat trailer; and an assortment of brightly painted boats really float and sail. The shop makes a crib, a cradle, and a high chair for dolls, and sells parts for making wooden trains.

Museum Books Mail-Order
Art-related toys. See entry under *Gifts*.

Museum of Fine Arts, Boston
Marbles, magnetic blocks, and other toys. See entry under *Gifts*.

The Museum of Modern Art
Bauhaus blocks, Colorforms, and other toys relating to modern art and design. See entry under *Gifts*.

The Natural Baby Co.
Rt. 1, Box 160, Titusville, NJ 08560
Telephone: 609-737-2895

Catalog price: $20 deposit for "Toypack"

A one-stop source for all kinds of childhood pleasures, from wooden toys to bicycles to books. Send a check for $20 and the company will mail you a fat packet of manufacturers' catalogs. If you send the catalogs back within five weeks, with or without an order, the company will tear up your deposit check. The prices are very good and the selection is impressive. To get all these choices at such savings, customers have to come up with a $60 minimum order, and they must be willing to wait a bit longer than usual for orders to arrive—the Natural Baby Co. does not keep the merchandise in stock, but forwards the orders to the manufacturers.

What's in the packet? The complete Radio Flyer catalog of wagons, riding toys, and bicycles; a fleet of exquisite wooden trucks and trains from Circlewood Products; Family Pastimes' collection of noncompetitive games; an assortment of wooden puzzles from Nashco

The pile of catalogs that comes in The Natural Baby Co.'s "Toypack"

Products; and Alcazar Records' extensive catalog of children's music. And that's just for starters. The 40-page Galt Toys catalog holds some wonderful art supplies for younger children, beautiful board games like Catch-a-Mouse and the Royal Game of Goose, Octons construction sets, play kitchen sets, plastic dinosaurs, and science toys. Barron's brochure of lovely wooden books is tucked into the pile, with its board books made of real boards. Taylor Made Toys manufactures wooden blocks, a duck push toy with flappy feet, wooden boats and trucks, several animal pull toys, and a menagerie of smaller creatures carved from wood. Its set of toy trees shows the shape of a dozen different species, each piece carved from the correct wood. Wooden swing and climbing sets from The Original Wooden Gym come in an array of different models, all at very good prices. The Nantucket Rocking Dory looks much like the Gloucester Rocker but sells for $100 less. And the Beka Wood Products catalog presents a maple block set in a handsome wagon, several wooden easels, an innovative child's desk, and a choice of play theaters with wipe-clean panels. In addition to all that, the Natural Baby Co. sends a list of more than 100 favorite children's books which it sells at a discount. If you've made it through this description, you're probably ready to tackle the "Toypack."

North Star Toys
617 North Star Route, Questa, NM 87556
Telephone: 505-586-0112

Free brochure

Wooden trucks and trains, little animals on wheels, and colorful wooden puzzles. A magic wand might make a nice present for a make-believer, rattles and stacking toys will keep the babies happy, and a gear toy made of four different woods is beautiful as well as fun. North Star Toys uses a nice mix of light and dark woods in its toys. All but the puzzles are given a natural oil finish.

Orbix Corporation
6329 Mori Street, McLean, VA 22101
Telephone: 703-759-4645

Free brochure

Makes a set of toys called Odd Balls. These are colored plastic wedges of three types that can be clipped onto a central disk to form hexagons, triangles, squares, stars, diamonds, and all sorts of other curious shapes.

Papa Don's Toys
Walker Creek Road, Walton, OR 97490
Telephone: 503-935-7604

Catalog price: free

"Quality hardwood toys, made fresh daily," proclaims the cover of Papa Don's color catalog, and that humor and good cheer come through in the toys. For babies the outfit makes six different rattles and two hanging crib gyms. All combine natural-finished wood with brightly painted balls and beads to hold the interest of little explorers without offending parents' eyes. A wooden lawn-mower toy tumbles colored balls around a wooden cage when it's rolled across the floor, and several stackers mix pretty wooden bases with rainbow-hued rings. Letters on wheels can be ordered to spell a child's name. Wobble-rolling pull toys (with off-center wheels) tip and bounce as they're dragged along. A wooden train set comes at a good price, as do some simple little cars and trucks. Papa Don's wooden block set includes number blocks wood-burned on four sides and some odd-shaped pieces for topping towers with a flourish.

PlayFair Toys
1690 28th Street, Boulder, CO 80301
Telephone: 303-440-7229

Catalog price: $2

"Nonsexist, nonviolent, educational toys for all ages," boasts the PlayFair catalog, and indeed toy guns and Barbie dolls would not fit into this mix. Separate spring and fall catalogs spell out the progressive choices: wooden puzzles, hobbyhorses, a wooden easel, Brio train sets, a few infant amusements, a tabletop loom, some furniture for young children, some cassettes and simple musical instruments, a small climbing set, and toys for pool and beach, to give a few examples. The "all ages" part of the company's sales line stretches the facts a bit. A puppet theater, an electric rock tumbler, or an insect-collecting kit might hold the attention of an older child, but for the most part PlayFair's collection is geared to preschoolers and primary-school children.

The Playmill
Wooden pull toys and crayon holders in animal shapes. See entry under *Posters and Decorations*.

Renew America Catalog
1400 16th Street N.W., Suite 710, Washington, DC 20036
Telephone: 202-466-6880

Catalog price: $2

Solar-powered toys, cards and gift wrap made from recycled paper, and nontoxic flea and tick controls for pets. A hot-air balloon slowly rises when exposed to the sun; a construction kit has a motor driven by a small photovoltaic cell; and an experiment set offers dozens of investigations into solar power.

The Ridge Company, Inc.
P.O. Box 2859, 1535 S. Main Street, South Bend IN 46680
Telephone: 219-234-3143

Free brochure

The Ridge Company makes a giant construction set for building playhouses, forts, puppet theaters, and other toys that children can play in or on. The set takes the concept of Lincoln Logs and other interlocking construc-

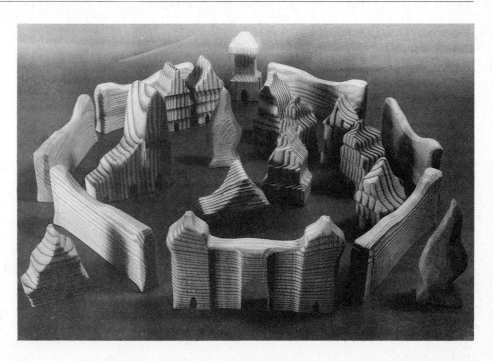

Hans Schumm's medieval village set

tion sets and enlarges it to children's full size. Called Building Boards, the set comes with notched wooden boards up to 4 feet long. Shorter boards and connectors allow children to build windows and doorways in their structures, or to add airplane propellors and smokestacks to other play creations.

Hans Schumm Woodworks
Rt. 2, Box 233, Ghent, NY 12075
Telephone: 518-672-4685

Free brochure

Wooden farm animals, a lumber truck, and a medieval village made from pine and fir with striking grain patterns. Only the truck is given a traditional smooth, sanded finish. The other pieces are wire-brushed to accentuate the grain, then lightly sanded to eliminate rough spots. The treatment gives Mr. Schumm's work an attractive earthy look.

F. A. O. Schwarz
P.O. Box 182225, Chattanooga, TN 37422-7225
Telephone: 800-426-TOYS

Catalog price: $5

A dazzling collection of toys in a slick oversize catalog that matches the image of F.A.O. Schwarz's glittery stores. If you're searching for the biggest stuffed animals and the most remarkable mechanical toys, this is the place to shop. F.A.O. Schwarz goes in for the sensational and the dramatic. In 1988, for example, the catalog sold no individual children's books, but offered a collection of 500 books and 50 videos. The price: $6,000. Extravagant gas- or battery-powered cars for children are a regular feature, as are gorgeous rocking toys (Mickey Mouse and Babar were featured in 1988), lovely dolls, and Marklin electric trains. The company's costume sets are among the best (and the most expensive) around, and the selection of plush animals is always first-rate. Mixed in among the high-priced items are lots of good basic toys, such as Lincoln Logs, wooden blocks (colored and natural), pull toys, board games, and art supplies.

Sears, Roebuck & Co.
A specialized "Toys" catalog presents just the playthings from Sears' huge inventory. See entry under *General Catalogs*.

Sensational Beginnings
P.O. Box 2009, 430 N. Monroe, Monroe, MI 48161
Telephone: 800-444-2147 or 313-242-2147

Catalog price: free

Toys and books for babies and young children. Sensational Beginnings has one of the best collections of toys for babies and toddlers, starting with black-and-white visual amusements and moving through hanging crib toys, teethers, rattles, stacking toys, shape sorters, hammering toys, and knobbed puzzles. Babies in the shaking and biting stage, for example, are offered a choice of seven wooden rattles in various arrangements of links, rinks, loops, beads, and bells; ten washable cloth rattles, some of which strap to car seats and cribs; and six teething and finger-manipulation toys. Toddlers can be furnished with big soft blocks, Radio Flyer wagons and trikes, small indoor climbing sets, and several styles of tables, chairs, and benches. The book selection is strong on board books and early read-alouds, and includes some titles for parents on baby care and games to play with young children.

Shibumi Trading, Ltd.
Japanese folk toys. See entry under *Gifts*.

Small World Toys
Science toys. See entry under *Science and Nature*.

Spinning Fool Top
P.O. Box 1987, Fairfield Glade, TN 38555
Telephone: 615-484-8941
Free brochure

Makes a wooden top that is set into action with a pull string and a handle that releases when fully unwound. A good yank can set the top spinning for up to 12 minutes.

Stocking Fillas
Party favors and stocking gifts. See entry under *Gifts*.

Sun Sparks
P.O. Box 325, Ganges, BC V0S 1E0, Canada
Free brochure

Sells a set of 12 colored tiles that combine to make rainbow-hued geometric patterns, symmetrical designs, letters of the alphabet, or pictures of animals and people. The diamond-shaped tiles are designed to fit together in many different ways, using the angles found in such natural forms as honeycombs, crystals, and flowers. The tiles can be arranged in free-form designs on a tabletop, or following the guidance of pattern cards that come with each set.

The Timberdoodle
Fischertechnik construction kits. See entry under *Educational Supplies*.

Timbers Woodworking
Patterns and parts for making wooden toys. See entry under *Woodworking*.

Toy Designs
Wooden-toy kits, parts, and plans. See entry under *Woodworking*.

The Toy Factory
88878 Highway 101, Florence, OR 97439
Telephone: 503-997-8604
Catalog price: free

In 1988 The Toy Factory changed its mail-order catalog into a smaller newsletter with discussion of many of the favorite toys in the company's two retail stores. Prices and shipping costs are no longer listed; mail-order customers must call or write for the cost of specific items. Special features of the list include a name puzzle cut from a single pine board, puppets and marionettes, alder blocks, and Brio train sets. Books and parts are sold for making wooden toys.

Toys and Joys
Wooden cars and trucks sold as kits or in blueprint form. See entry under *Woodworking*.

Toys to Grow On
P.O. Box 17, Long Beach, CA 90801
Telephone: 213-603-8895 (800-542-8338 to order)
Catalog price: free

A catalog that culls learning toys from the school-supply world and presents them with playthings that are pure

Kids' Business set from Toys to Grow On

fun. On the educational side are maple block sets, fossil and sea-shell kits, wooden puzzles, and an attractive collection of art supplies and easels. On the lighter side are windup monster toys, sticker collections, a game called Space that floats a ball on a jet of air, a battery-powered recording studio, and a box of gags that includes a squirting flower and a whoopee cushion. There's plenty of fun in between, too, with costume sets, a rubber-stamp alphabet, a starter woodworking kit, a periscope, and a set of wooden fruit and vegetables that can be "sliced" in half at Velcro joints with a toy knife.

Treasured Toys
418 Main Street, Amesbury, MA 01913
Telephone: 508-388-1907
$1 for brochure

A carefully chosen collection of toys, books, puzzles, and other playthings, organized by age group on four efficient catalog pages. A single color photograph on each page displays the choices in a group shot; surrounding text explains what's what. The crawling and toddling set can play with covered foam bouncers, stacking beakers, wooden rattles, a foam play nest, a maple block set in a wooden wagon, and a roller-coaster bead maze. Preschoolers can pass the time with simple matching games, alphabet and number puzzles, a wooden easel, Italian modeling dough, giant animal-shape soft blocks, a Kinderworks table-and-chair set, plush puppets, a wooden Noah's ark set, and a traditional painted rocking horse. School-age children can tinker with science kits, mechanical Lego sets, oversize puppets, wooden tops, and such nature-study aids as a bird feeder, a hand-held microscope, and a bug box. Books are scattered among the toys on each page, among them *Goodnight Moon*, *The Jolly Postman*, and several educational activity books.

Trend-Lines, Inc., Woodworking Supplies
Parts and plans for toymakers. See entry under *Woodworking*.

Tryon Toymakers' Dream Box

Troll
100 Corporate Drive, Mahwah, NJ 07430
Telephone: 201-529-8000 (800-247-6106 to order)
Catalog price: free (specify "Learn & Play" or "Family Gift" catalog)

Troll publishes several different catalogs, each mailed in three seasonal editions. "Learn & Play" is stocked with toys, books, costumes, and educational playthings. Past editions have held shape sorters, wooden castle blocks, wood and crepe-rubber puzzles, pogo sticks, an inflatable play tunnel, toy cookware, a canvas tepee, a crystal-growing kit, and a small-scale sewing machine for making doll clothes. The "Learn & Play" catalog is also a hot spot for dinosaur-lovers. It carries plastic dinosaur models, wooden skeleton kits, a brontosaurus play tent, dinosaur hand puppets, pterodactyl and stegosaurus costumes, and a giant tyrannosaurus mask that fits over a child's entire head. The "Family Gift" catalog carries soft baby rattles, dinosaur cookie cutters, crayon-shaped drinking cups, books, cassettes, and a much smaller selection of toys.

Tryon Toymakers
1851 Redland Road, Campobello, SC 29322
Telephone: 803-457-2017
Catalog price: $1

Hand-painted wooden toys in a mix of original and traditional designs. A "Dream Box," painted with stars, holds a crowd of miniature figures: a flying pig, a cowboy, a princess, a magic wand, a wolf, an ice-cream cone, and more. A stacking sandwich adds a tasty twist to a familiar toy. Instead of rings, it piles up whole-wheat bread, lettuce, tomato, Swiss cheese, salami, and a pickle—all crafted of wood. A rocking horse follows a more traditional path. Shaped from an old-fashioned pattern, it's decorated with red, white, and blue flowers. A hobbyhorse and swing horse are both topped with the same friendly horse head, a pair of stilts is built with woodpeckers for footrests, and a child's wheelbarrow is painted with gardening rabbits on the sides. For make-believe play, Tryon Toymakers offers a house and barn to fit its miniatures, three different magic wands, an ironing board with a wooden iron, and a doll bed. A garage is scaled for Matchbox and Tonka cars. A paper-making kit, a toolbox, and a two-step stairway are genuinely useful. Wooden puzzles start out simple and advance to a complete map of the United States.

U-Bild
Patterns for building pull toys, boats, and riding toys. See entry under *Furniture*.

Unknown Products, Inc.
P.O. Box 225, Midwood Station, Brooklyn, NY 11230
Telephone: 718-377-7808
Catalog price: free

Unknown Products has unearthed a nostalgic collection of playthings from the 1960s, which parents may have trouble sharing with their children. In the doll department are Mr. Potato Head kits, bendable Gumbi and Pokey figures, and Troll dolls. Under pets, look for Amazing Live Sea Monkeys, ant farms, and pink plastic lawn flamingos. Play-Doh, Slinkies, and Uncle Fester's magic light bulb all find a home here, along with buttons of popular cartoon characters. T-shirts are sold in adult sizes.

Woodcraft

Parts for making wooden toys. See entry under *Woodworking*.

The Woodworkers' Store

Plans, parts, tools, and lumber for making wooden toys. See entry under *Woodworking*.

Woodworks

4500 Anderson Boulevard, Fort Worth, TX 76117
Telephone: 817-281-4447

Catalog price: $1

Turned hardwood pieces, some of which can be used as toys and others made into toys. Wooden fruit can go straight from the mail into the toy box; the company sells life-size apples, pears, mushrooms, acorns, and cherries. Wooden eggs in various sizes can be used plain or decorated with paint. Blocks, balls, checkers, bracelets, and wooden hearts stand ready for imaginative play. Wheels, axles, beads, smokestacks, barrels, and little people (in three styles) can be turned into toy trains and trucks with passengers. Owners of Victorian houses may want to examine the inventory of full-size spindles, finials, and spandrel balls.

Travel

American Youth Hostels, Inc.

See entry under *Associations, Travel*.

Bicycle USA

League of American Wheelmen, 6707 Whitestone Road, Suite 209, Baltimore, MD 21207
Telephone: 301-944-3399

$4 for TourFinder edition

The March/April issue of *Bicycle USA* provides a directory of organized bicycle tours in the coming months. Many are open to teenagers, a few to younger children.

Bike Vermont Tours

P.O. Box 207, Woodstock, VT 05091
Telephone: 802-457-3553

Catalog price: free

Inn-to-inn bicycle tours of Vermont that last a weekend or five weekdays. The weekend tours offer alternative routes for those who want to travel different distances. The weekday tours cover about 30 miles each day. Children under 12 must be accompanied by a parent. Some of the inns do not allow children under five.

Country Cycling Tours

140 W. 83rd Street, New York, NY 10024
Telephone: 212-874-5151

Catalog price: free

Bicycle and walking tours for adults and families. From April to October, join day trips in New York, New Jersey, and Connecticut, weekend trips throughout New England, or longer excursions like an island-hopping adventure off the coasts of Massachusetts and Rhode Island or a bike tour of northern Virginia. Winter bicycle trips to the Caribbean are arranged, as are walking and bike tours in Europe and China. Children must be old enough to cycle 10 to 15 miles in a day, walk 7 to 10 miles, or be small enough to carry.

Families Welcome!

1416 Second Avenue, New York, NY 10021
Telephone: 800-472-8999 or 212-861-2500

Free brochure

A travel service specifically for families with children. Packages include accommodations, baby-sitting or "kids only" activities, and an information packet with such things as the locations of all-night pharmacies, lists of restaurants that welcome children, events and entertainments for children, suggested sightseeing routes, and special shops to visit. Regular destinations include London, Paris, New York, the English countryside, and the Carribbean. Cruises, dude-ranch vacations, treks to the Himalayas, and adventures almost anywhere can be specially arranged.

Family Travel Times
See entry under *Parents' Magazines, Travel*.

Rascals in Paradise
185 Berry Street, Suite 5503, China Basin, San Francisco, CA 94107
Telephone: 800-443-0799 or 415-442-0799

Free flier; $2 for complete brochure

Family travel packages. In 1989 the company sent groups to Mexico, the Caribbean, Hawaii, Tahiti, Thailand, Nepal, Bali, the Galapagos Islands, and other exotic spots. One tour took families on barges through the canals of France, another in camper vans through New Zealand. Each group is made up of three to six families. Baby-sitting services and teachers are made available to occupy the younger travelers for part of each day.

Videos

The ChildLife Video Collection
QualityLife Video Publishing, P.O. Box 800, Dept. CL, Boulder, CO 80306-0800
Telephone: 303-440-9109

Catalog price: $2

Almost 200 videos for children, from *Baby Songs* to the Beatles' *Yellow Submarine*. Lots of story classics make the list, including *The Elephant's Child*, narrated by Jack Nicholson, Captain Kangaroo's *Favorite Adventure Stories*, and the Children's Circle productions of *The Foolish Frog, Corduroy Bear*, and *Make Way for Ducklings*. Other videos teach children how to raise a puppy and take care of a horse, how to use money wisely, and how to juggle. The list goes on to Martin Luther King's "I have a dream" speech, the documentary *Say Amen, Somebody*, and a film of Tchaikovsky's *Nutcracker Suite*.

Children's Book & Music Center
2500 Santa Monica Boulevard, Santa Monica, CA 90404
Telephone: 800-443-1856 or 213-829-0215

Catalog price: free

A delightful and comprehensive 80-page catalog of children's books, records, tapes, videos, and musical instruments. Around 50 videos are offered, about evenly divided among story, music, and educational tapes. The business carries a dozen Sesame Street videos, a puppet production of *Carnival of the Animals*, Wee Sing and Raffi videos, and such story videos as *The Steadfast Tin Soldier*, narrated by Jeremy Irons, *The Tale of Peter Rabbit*, narrated by Meryl Streep, and a number of Children's Circle productions. For a description of the musical and book offerings, see separate entries under *Music* and *Books*.

Critic's Choice Video, Inc.
800 Morse Avenue, Elk Grove Village, IL 60007
Telephone: 800-367-7765 or 800-544-9852

Catalog price: $2

Almost 2,000 videos for sale, all in VHS. Choose from hundreds of recent and classic Hollywood films; instructional tapes that teach children juggling, cooking, sign language, and other skills; educational tapes that deal with science, reading, and such problems as divorce and drug abuse; and classic children's features from Disney and other producers. Other tidbits include a *Candid Camera* tape, Martin Luther King's "I have a dream" speech, and a number of *National Geographic* and *Nova* videos.

Direct-To-You Baby Products
Instructional videos for new mothers; music and activity videos for babies and young children. See entry under *Baby Needs*.

The Disney Catalog
Disney films and cartoon collections. See entry under *Gifts*.

Early Advantage Programs for Children
47 Richards Avenue, Norwalk, CT 06857
Telephone: 800-243-5160

Catalog price: free

Runs The Children's Circle Video Reading Program, a video/book club that encourages children to read by pairing animation videos with popular books. Subscribers to the program receive 12 videos a year, animated versions of such children's stories as *Curious George Rides a Bike*, Maurice Sendak's *Really Rosie*, William Steig's *Dr. De Soto*, and Tomi Ungerer's *Moon Man*. Along with the videos come paperback copies of the books. According to the literature, children are inspired by the television stories to read the books. The videos have been praised in *Parents* magazine, the *New York Times*, the *Los Angeles Times*, and other publications.

Family Home Entertainment, IVE
P.O. Box 2279, Los Angeles, CA 90051
Telephone: 800-PLAY-FHE or 805-499-5827

Catalog price: free

Videos for children and parents, sold mainly through direct-response ads in magazines. The firm will send a complete list to those who write. It includes some 200 tapes, from Rambo, G.I. Joe, and Thundercats cartoons to *A Cricket in Times Square* and *On Being a Father*.

Feeling Fine Programs
3575 Cahuenga Boulevard W., Suite 440, Los Angeles, CA 90068
Telephone: 213-851-1027

Catalog price: free

Books and videos on fitness and health, including a few on pregnancy, childbirth, and breastfeeding. *What to Expect When You're Expecting* is on the book list; *The American College of Obstetricians and Gynecologists Postnatal Exercise Program* is one of the videos. Dr. Art Ulene has had a hand in many of the offerings, as author, producer, or narrator.

KIDVIDZ
618 Centre Street, Newton, MA 02158
Telephone: 617-965-3345 or 277-8703

Free brochure

This firm has produced three appealing videos for children: *Hey What About Me!*, a guide for brothers and sisters of new babies, *Kids Get Cooking*, an entertaining kitchen lesson, and *Squiggles, Dots, and Lines*, a video drawing class.

Lorimar Home Video
5959 Triumph Street, Commerce, CA 90040
Telephone: 800-323-5275 or 714-474-0355

Catalog price: free

Lorimar has produced a set of parenting videos in co-operation with *Parents* magazine that discuss the rearing of infants, toddlers, and preschoolers. For children, Lorimar sells videos of the Care Bears, Alvin and the Chipmunks, and Silverhawks. One intriguing series, made in conjunction with *Scholastic* magazine, combines videos with Colorforms play sets.

Movies Unlimited
6736 Castor Avenue, Philadelphia, PA 19149
Telephone: 800-523-0823 or 215-722-8298

Catalog price: $9.95, $5 discount coupon with first order

Six hundred pages of videos should be enough for even the most demanding viewers. The 60-page family section holds Disney features, Lassie films, and collections of shorts featuring Betty Boop, the Little Rascals, Shirley Temple, and Gumbi. Modern characters like Strawberry Shortcake, the Care Bears, Thundercats, and She-Ra are well represented as well. Mister Rogers and Sesame Street anchor a good-size educational section. Another 40 pages of documentaries includes hundreds of informative productions on nature, history, art, and other subjects. And for pure entertainment, sort through the thousands of feature films.

National Gallery of Art, Extension Programs
Videos, films, and slide programs on fine art loaned at no charge. See entry under *Posters and Decorations*.

ParentCare Ltd.
2515 E. 43rd Street, P.O. Box 22817, Chattanooga, TN 37422
Telephone: 800-334-3889

Catalog price: free

Videos "for parents who care." Two 32-page color catalogs are mailed during the year. Each holds a nice selection of Disney classics, educational tapes, cartoon collections, and feature films, including some meant for mom and dad. Learning tapes like Bill Cosby's *Picture Pages* series and Sesame Street's *Learning About Numbers* have an important place here, but they're outnumbered by productions of pure and harmless entertainment. Eight Shirley Temple films are listed, two Babar films, and a video of stunt-driving monster trucks. A final section of the catalog is stocked with great films that everyone should see, like *Kind Hearts and Coronets*, *A Night at the Opera*, and *Gone With the Wind*.

Publishers Central Bureau
VHS- and Beta-format videos. See entry under *Books*.

Signals Catalog
274 Fillmore Avenue E., St. Paul, MN 55107
Telephone: 800-669-9696

Catalog price: $2

"A catalog for fans and friends of public television." The Signals catalog sells videos, books, T-shirts, toys, and other merchandise related to popular programs on public television. *Sesame Street* generally rates a few videos in the catalog, as do *Mr. Rogers' Neighborhood, Ramona,* and *Nova.* Program-related items have included Beatrix Potter books and china; a sidewalk play set with jump rope, jacks, and chalk; the two-potato clock; a glow-in-the-dark star chart T-shirt; and a handmade wooden Noah's ark.

Special Effects Merchandise
475 Oberlin Avenue S., P.O. Box 823, Lakewood, NJ 08701
Telephone: 800-245-1007

Catalog price: free

The Paramount movie company's consumer mail-order catalog—filled, naturally, with videos of Paramount films and with Paramount memorabilia. Among the videos are all of the *Star Trek* movies and television shows, the Monty Python films, some of John Wayne's best works, *Raiders of the Lost Ark,* and *Big Top Pee-wee. Star Trek* and Pee-wee Herman sidelines take up a fair chunk of the catalog, running the gamut from a U.S.S. *Enterprise* crew uniform and Pee-wee pajamas to a Chairry doll chair and a *Star Trek* sleeping bag. A child-size director's chair emblazoned with the Paramount logo might make a nice gift for a budding filmmaker; a chalk clapboard would look great in the home videos. Posters, records, dolls, and clothes celebrate some of Paramount's most popular productions.

Video F.I.N.D.S.
P.O. Box 4990, Stamford, CT 06907-0990
Telephone: 800-443-7359

Catalog price: free

Hollywood films get top billing in this catalog, from old favorites like *Top Hat* and *A Day at the Races* to the most recent video releases. Shirley Temple's films are here, as are the greats from Disney and such children's movies as *Willy Wonka and the Chocolate Factory, Charlotte's Web,* and Leslie Ann Warren's *Cinderella.*

Waldenvideo
P.O. Box 9497, New Haven, CT 06534-0497
Telephone: 800-443-7359

Catalog price: free

Disney, *Sesame Street, Peanuts,* and *Looney Tunes* share space here with *National Geographic* specials, *Star Trek* episodes, Beatles movies, and classic Hollywood films. Seasonal editions bring shoppers fresh selections and the latest price reductions.

Woodworking

Armor Products
P.O. Box 445, East Northport, NY 11731
Telephone: 516-462-6228

Catalog price: $1

Wooden-toy plans and kits, with hardware and supplies for the home craftsman. Five different rocking horse plans are offered; several trains, cars, and trucks; and more than a dozen different dollhouses in kit form. Wooden wheels, axle pegs, smokestacks, barrels, and little people can get a woodworker going even without plans.

Barap Specialties
835 Bellows Avenue, Frankfort, MI 49635
Telephone: 616-352-9863

Catalog price: $1

Supplies, kits, and plans for home woodworkers. Toy makers will find little wooden wheels, axle pegs, smokestacks, headlights, and people. Young craftspeople will find clock and lamp kits and hardware for dozens of other projects.

Chaselle, Inc.
9645 Gerwig Lane, Columbia, MD 21046
Telephone: 800-CHASELLE (800-492-7840 in MD) or 301-381-9611

Catalog price: free (specify "Pre-School & Elementary School Materials" or "General School & Office Products" catalog)

The "Pre-School & Elementary School Materials" catalog offers a child-size set of woodworking tools, including saws, a 7-ounce hammer, screwdrivers, pliers, drills, a level, a clamp, and a file. Wood pieces for projects are also sold, and, for those who want to go all the way, a 24-inch-high workbench. The larger "General School &

Office Products" catalog has tools for older children. See main entry under *Educational Supplies* for a description of Chaselle's other offerings.

Cherry Tree Toys
408 S. Jefferson Street, P.O. Box 369, Belmont, OH 43718
Telephone: 614-484-4363

Catalog price: $1

Parts, kits, plans, tools, books, and supplies for toy and miniature makers. The 60-page color catalog is a treasure trove of dollhouse and dollhouse furniture kits, toy truck and train kits, children's clocks, and turned wooden parts. Cherry Tree sells toy wheels in more than 20 different styles and sizes, toy people in six different designs, and dozens of choices in smokestacks, axles, pegs, and spindles.

Completed truck kit from Cherry Tree Toys

Constantine's Woodworker's Catalog
2050 Eastchester Road, Bronx, NY 10461
Telephone: 212-792-1600

Catalog price: $1

More than 100 pages of tools, finishes, kits, and books for the home woodworker. Toy makers will find a dozen idea books, several books of plans, and a little lumberyard of wheels, smokestacks, pegs, and other wooden pieces. Books and plans are also offered for making rocking horses, cradles, dollhouses, and children's furniture.

Design Group
P.O. Box 514, Miller Place, NY 11764
Telephone: 516-928-2644

Catalog price: $1

Blueprint patterns for making wooden toys, games, and puzzles. Animal pull toys are offered in a dozen fun designs, from a grasshopper to a stegosaurus. The toy truck fleet includes a road grader, a dump truck, and a wrecking crane; the bathtub armada takes in a paddleboat, a sailboat, a tugboat, a towing barge, and a ferryboat with little cars. Jigsaw puzzles, name trains, animal-shape crayon holders, and a toy gas station are among the other ideas for sale. A library at the back of the catalog offers books for making dolls, dollhouses, needlepoint dollhouse rugs, and wooden toys.

Environments, Inc.
A workbench and tools for younger children. See entry under *Educational Supplies*.

The Fine Tool Shops
P.O. Box 7091, Portsmouth, NH 03801
Telephone: 800-533-5305 or 603-433-0409

Catalog price: $1, earns $5 discount on first order

Hand tools, workbenches, and Makita power tools for the home woodworker. The selection is heavy on chisels, planes, files, carving knives, wood-turning tools, and devices for measuring and marking. Furniture makers and sign carvers should find what they're looking for. Parents planning to make simple toys may be outclassed.

Industrial Arts Supply Co.
Plans and parts for making wooden toys. See entry under *Hobby Supplies*.

Julia Toys
1283 Avery Court, St. Louis, MO 63122

Catalog price: $1, deductible from first order

Plans and parts for making wooden toys. Among the designs are a cute elephant pull toy, an old-fashioned touring car, a World War I biplane, a fire truck, a bulldozer, and a foot-flopper push duck. Several other cars are shown in the catalog, all detailed enough to make interesting playthings but simple enough that they're possible to build. All necessary wheels, axles, passengers, smokestacks, plugs, and ornaments are offered in a list at the back of the booklet.

Nasco
Woodworking tools and kits. See entry under *Art Supplies*.

Sears, Roebuck & Co.
Tools. See entry under *General Catalogs*.

Timbers Woodworking
Timbers Building, Carnelian Bay, CA 95711-0850
Telephone: 916-581-4141

Catalog price: $1

Patterns and plans for wooden toys and furniture. The catalog lists several cradle and rocking-animal designs, a few storage chests, and a captain's bed. For toy makers, the firm sells truck and train patterns and a good selection of wooden wheels, pegs, smokestacks, oil drums, headlights, and little people.

Children's workbench from Environments, Inc.

Toy Designs
P.O. Box 441, Newton, IA 50208
Telephone: 515-792-6551

Catalog price: $1.50

Patterns, kits, and parts for crafting wooden toys. Cars and trucks create the most traffic in the catalog, with a wide choice of steamrollers, log trailers, fire trucks, dump trucks, sports cars, Model T's, and moon rovers. Trains (including an alphabet train), boats, and buildings add a little variety to the mix. A collection of Sig aircraft kits really fly, powered by twisted rubber bands.

Toys and Joys
P.O. Box 628, Lynden, WA 98264
Telephone: 206-354-3448

Catalog price: $1

Wooden cars and trucks sold either as kits or in blueprint form. Some are stylized vehicles with circular holes for windows and simplified details. Others are beautifully realistic scale models. A 40-inch tractor-trailor rig looks like a wood-grained miniature, as does a 1930 Ford "Woodie" station wagon, a 1929 Coca-Cola truck, and a modern four-wheel-drive Bronco. Wheels, axles, headlights, air horns, barrels, and smokestacks come in a wide range of styles and sizes.

Trend-Lines, Inc., Woodworking Supplies
375 Beacham Street, P.O. Box 6447, Chelsea, MA 02150-0999
Telephone: 617-884-8882 (800-343-3248 to order)

Catalog price: free

A woodworker's catalog that goes heavy on the power tools. The wood parts list includes Shaker pegs and turned pieces for making wooden toys: wheels, axles, smokestacks, headlights, barrels, and people. Books on toy making, a few toy plans, and several Victorian dollhouse kits round out the parents' pages.

Woodcraft
41 Atlantic Avenue, P.O. Box 4000, Woburn, MA 01888
Telephone: 800-225-1153 or 617-935-5860

Catalog price: $3; free shorter catalog

An arsenal of precision tools for serious woodworkers. Specialized planes and spokeshaves, sculptor's rifflers, Swiss carving sets, and Makita power tools will set some craftspeople drooling. Toy makers can pick and choose from four styles of wooden wheels in dozens of sizes; wooden people with dresses, jackets, or hats; round balls, beads, and cam wheels; and a wagon load of barrels, milk cans, and drums.

The Woodworkers' Store
21801 Industrial Boulevard, Rogers, MN 55374-9514
Telephone: 612-428-2199 (orders only)

Catalog price: $1

A woodworking catalog aimed more at the home handyman than the fine craftsman. Hand and power tools can serve novice or expert; hardwood lumber, turned pieces, veneers, inlay bandings, finishes, and adhesives provide the raw materials. What distiguishes this catalog from others is the big selection of hardware: hinges, catches, locks, shelf supports, drawer slides, casters, cabinet knobs, and much more. The usual wooden toy parts are sold—wheels, axles, chimneys, barrels, and balls—along with plans for rocking horses, pull toys, and children's furniture. Several plans from Strom Toys provide the expert with a chance to make such exceptional playthings as a child's sleigh, a wagon with a curved wooden handle, an old-fashioned stove miniaturized in wood, and a beautiful stroller.

Woodworks
Parts for making wooden toys. See entry under *Toys*.

Associations

For Parents and Families

American Family Society
P.O. Box 70095, Washington, DC 20088
Membership: $25 annually

A group that strives to promote traditional family values through awards, a monthly *Great American Family Update*, and media publicity. No local chapters or meetings.

American Mothers, Inc.
The Waldorf-Astoria, 301 Park Avenue, New York, NY 10022
Telephone: 212-755-2539
Membership: $10 annually

Founded in 1935, American Mothers, Inc., works to preserve American family values. The group selects a National Mother of the Year, promotes the observance of Mother's Day, and sponsors local chapters which hold meetings and maintain support networks.

Christian Family Movement
P.O. Box 272, Ames, IA 50010
Telephone: 515-232-7432
Membership: $25 annually for families

A network of small groups of parents who seek to promote a Christian way of life for families. Members receive a program book, which is used as a guide at meetings, and a subscription to the newsletter *Act*.

The Compassionate Friends, Inc.
P.O. Box 3696, Oak Brook, IL 60522-3696
Telephone: 708-990-0010
Membership: no membership dues or fees

An international self-help organization offering friendship, understanding, and support to parents who have experienced the death of a child. More than 600 local chapters meet regularly throughout the U.S. and Canada. A quarterly newsletter is available for $10; a catalog of books and audiocassettes is free.

Depression After Delivery
P.O. Box 1282, Morrisville, PA 19067
Telephone: 215-295-3994
Membership: by contribution (optional)

"A self-help mutual aid support group" for women who are experiencing postpartum depression or psychosis. A letter or telephone call will bring a list of names of women in your area who have experienced postpartum illness and who are willing to listen and help. A periodic newsletter is also mailed to those who contact the group.

Family Communications
Marketing Dept., 4802 Fifth Avenue, Pittsburgh, PA 15213
Telephone: 412-687-2990
Membership: no members

Family Communications, producers of *Mister Rogers' Neighborhood*, publishes an array of specialized pamphlets written by Fred Rogers, as well as some helpful books and videos. Send $2 for a copy of the catalog and ten of the pamphlets (including "Talking With Families About Discipline," "When Your Child Goes to School," and "When You Have a Child in Day Care").

Family Resource Coalition
230 N. Michigan Avenue, Suite 1625, Chicago, IL 60601
Telephone: 312-726-4750
Membership: $30 annually

An organization that works for preventive, community-based family support and parent education services throughout the U.S. and Canada. Anyone can contact the coalition for information about parent organizations in their area. Parents interested in organizing, lobbying, or forming support groups may want to join. Membership brings a newsletter and a journal, access to a clearinghouse of information on family support programs, expert advice on program development, and discounts on publications.

Fatherhood Project

c/o Bank Street College of Education, 610 W. 112th Street, New York, NY 10025

Telephone: 212-663-7200

Membership: no membership dues or fees

"A national research, demonstration, and dissemination project designed to encourage wider options for male involvement in childrearing." The project offers information on fatherhood-related topics and referrals to resources for and about fathers. A directory of programs, services, and resources, titled *Fatherhood U.S.A.*, is available for sale.

Interracial Family Alliance

P.O. Box 16248, Houston, TX 77222

Telephone: 713-454-5018

Membership: $25 annually ($15 for newsletter only)

A group that seeks to "affirm, through education, the dignity and equality of every racial group; to support and serve the interracial family and multiracial individual; and to strengthen the bonds among groups in our multi-ethnic society." Counseling services, educational programs, and social activities are available to members in the Houston area. The newsletter *Communique* is mailed nationally.

Kindred Spirits

c/o 92nd Street Y, 1395 Lexington Avenue, New York, NY 10128

Telephone: 212-996-1105

Membership: $50 annually

A group for young divorced and widowed parents and their children. Activities include sporting events, walking tours, camping trips, travel packages, parties, and brunches. A newsletter is sent out three times a year.

Mothers at Home

See entry for *Welcome Home* under *Parents' Magazines, General Parenting and Family*.

Mothers Without Custody

P.O. Box 56762, Houston, TX 77256-6762

Telephone: 713-840-1622

Membership: $18 annually

A nationwide support organization for women living apart from their children, whether because of court rulings, voluntary exchange of custody, state intervention, a child's abduction by an ex-spouse, or any other reason. Local groups meet in some areas. All members receive the bimonthly newsletter *Mother-to-Mother* and an opportunity to attend the annual meeting.

National Coalition Against Domestic Violence

P.O. Box 15127, Washington, DC 20003-0127

Telephone: 202-293-8860 (hot line: 800-333-7233)

Membership: $15 minimum suggested, based on ability to pay

A membership organization working to end violence in the lives of women and children. A 24-hour hot line is open for those needing information about shelters or programs. The coalition distributes literature on domestic violence and sends members a quarterly newsletter, *The Voice*.

National Committee for the Prevention of Child Abuse

332 S. Michigan Avenue, Suite 950, Chicago, IL 60604

Telephone: 312-663-3520

Membership: by contribution

An organization dedicated to ending child abuse. Through volunteer efforts in local chapters, the committee supports community-based child abuse prevention programs. As part of its public-awareness drive, the national office offers a free catalog of publications aimed at parents, children, and teachers. Special Spider Man comic books teach children how to deal with emotional and sexual abuse; materials for parents offer techniques for avoiding abuse and help in coping with isolation and stress.

National Council on Child Abuse and Family Violence

1155 Connecticut Avenue N.W., Suite 300, Washington, DC 20036, and 6033 W. Century Boulevard, Suite 400, Los Angeles, CA 90045

Telephone: 800-222-2000, 202-429-6695, or 818-914-2814

Membership: no members

A group formed to help strengthen community agencies that provide service to victims of family violence. The council provides emergency referral information to victims (at 800-222-2000), recruits and trains volunteers for local programs, and offers publications on the subject of family violence.

Parents Anonymous

6733 S. Sepulveda, Suite 270, Los Angeles, CA 90045

Telephone: 800-421-0353 or 213-410-9732

Membership: no membership dues or fees

A network of self-help support groups for parents who feel isolated, overwhelmed, or afraid of the ways they find themselves dealing with their children. Local groups—and there are hundreds throughout North America—meet weekly for about two hours. Call or write for the meeting times of groups in your area.

Parents Without Partners, Inc.
8807 Colesville Road, Silver Spring, MD 20910
Telephone: 800-638-8078 or 301-588-9354

Membership: $23 annually

An organization of more than 180,000 single parents in approximately 900 local chapters throughout the U.S. and Canada. Members receive *The Single Parent* magazine (also available separately, see entry under *Parents' Magazines, General Parenting and Family*) and the opportunity to participate in local chapter activities, shop from the group's list of publications, and let their children take part in the annual PWP Youth Program.

The Salvation Army
Attn: Captain Margaret E. Dougherty, Territorial Headquarters, 120 W. 14th Street, New York, NY 10011
Telephone: 212-337-7200

Membership: no members

The Salvation Army offers a free "Catalog of Resources for Strengthening the Family." Included are publications on child safety, child abuse prevention, alcohol and drug abuse (written for young people), and child growth and development.

Single Mothers by Choice
P.O. Box 1642, Gracie Square Station, New York, NY 10028
Telephone: 212-988-0993

Membership: $40 for first year, $30 for annual renewal

An association providing support and information to single mothers and single women considering motherhood. The group defines a single mother by choice as "a woman who *decided* to have or adopt a child, knowing she would be her child's sole parent, at least at the outset." Women in the New York City area have the option of paying a higher membership fee ($60) and attending monthly meetings in Manhattan. Those who live outside the area receive the quarterly newsletter; an information packet with articles about single motherhood, artificial insemination, and adoption; and a list of local members who have expressed an interest in meeting others.

Stepfamily Association of America
602 E. Joppa Road, Baltimore, MD 21204
Telephone: 301-823-7570

Membership: $35 annually

An advocacy, information, and support organization that focuses on the needs of stepfamilies. Local chapters provide self-help programs, education, and friendship to members. The national office mails a quarterly newsletter, hosts an annual conference, and offers a catalog of books for parents and children.

U.S.A. Toy Library Association
2719 Broadway Avenue, Evanston, IL 60201
Telephone: 312-864-8240

Membership: $40 annually

An association of toy libraries that can refer parents to lending centers in their area. See description on page 193.

Working Mothers Network
1529 Spruce Street, Philadelphia, PA 19102
Telephone: 800-648-8455 or 215-875-1178

Membership: $15 annually

A national organization of working mothers that offers seminars and workshops; a gift-buying service; a quarterly newsletter, *Exchange*, with advice on work, home, child rearing, and money matters; and an annual resource guide.

For Children and Youth

Big Brothers/Big Sisters of America
230 N. 13th Street, Philadelphia, PA 19107
Telephone: 215-567-7000

Membership: by contribution

A volunteer organization that matches adult volunteers with children—primarily from single-parent households—who can benefit from the adult friendship and guidance. The pairings are made by the group's 483 affiliated agencies throughout the U.S.

Boy Scouts of America
1325 Walnut Hill Lane, P.O. Box 152079, Irving, TX 75015-2079
Telephone: 214-580-2000

Membership: contact a local council for information

Parents and children interested in Cub or Boy Scouting should contact a local council for information on the nearest unit and how to join. If one can't be found in your local telephone directory, the national office can help you find it. Children unable to join a Cub Scout pack or Boy Scout troop because of distance, time conflicts, or physical handicaps may want to participate in the Lone Scout Program. Contact a local council or the national office for details.

Camp Fire, Inc.
4601 Madison Avenue, Kansas City, MO 64112
Telephone: 816-756-1950

Membership: $10 annually

A coeducational youth program, once known as Campfire Girls. Members now progress through four different levels as they grow older, starting with Starflight for children aged five to eight, and moving through Adventure, Discovery, and Horizon. Those interested in joining should contact a local council or ask the national office for help in finding one.

Girl Scouts of the U.S.A.
830 Third Avenue, New York, NY 10022
Telephone: 212-940-7500

Membership: $4 annually

A volunteer organization open to girls aged 5 through 17. Parents with children interested in joining should contact a local council by looking up the number in a telephone directory. The national office can help if no council is listed in your area.

Orphan Foundation of America
1500 Massachusetts Avenue N.W., Suite 448, 14261
Ben Franklin Station, Washington, DC 20044-4261
Telephone: 202-861-0762

Membership: $45 annually

An advocacy and support organization for orphans and foster children. The foundation's various projects provide residential facilities, scholarships, job training, emergency cash grants, and professional counseling services.

ADOPTION

Adoptive Families of America, Inc.
3333 Highway 100 N., Minneapolis, MN 55422
Telephone: 612-535-4829

Membership: $24 annually, US$34 in Canada

A national support organization for adoptive and prospective adoptive families that can direct members to local support groups and adoption agencies. Members receive the bimonthly magazine *OURS*, which includes parenting articles, firsthand experiences of member parents, a pen-pal directory for young readers, pictures and descriptions of waiting children, pictures of adopted children sent in by members, a regular list of adoptive parent support groups, and an annual list of adoption agencies. The group also sells books and other parenting resources by mail, including children's and adult books about adoption, ethnic dolls, language tools, and activities for family sharing.

Americans for International Aid and Adoption
877 S. Adams, Birmingham, MI 48009
Telephone: 313-645-2211

Membership: see below

An organization that places children from Korea, India, Chile, Taiwan, and Honduras in adoptive homes. The group's efforts concentrate on "waiting children," those whose age, racial background, or medical condition has left them waiting for homes. Interested parents receive a free information packet with a questionnaire. Those who want to proceed further send a $25 filing fee. Formal application, transportation, and home study/supervision fees bring the total cost of adoption to more than $5,000.

Committee for Single Adoptive Parents
P.O. Box 15084, Chevy Chase, MD 20815
Telephone: 202-966-6367

Membership: $15 for two years

A source of information for prospective and actual single adoptive parents. Members receive a list of agencies and direct sources of adoptable children in the U.S. and abroad who will accept single applicants. Periodic updates put members in touch with single adoptive parent groups. A *Handbook for Single Adoptive Parents* is sold for an additional fee.

Families Adopting Children Everywhere
P.O. Box 28058, Northwood Station, Baltimore, MD 21239
Telephone: 301-239-4252

Membership: by contribution (optional)

A support group for people seeking to adopt children. Information about domestic and foreign adoptions is sent free of charge to anyone who writes. Members can order a bimonthly newsletter for $15 and can borrow adoption exchange books with pictures and descriptions of waiting children.

Information for Private Adoption
P.O. Box 6375, Washington, DC 20014
Telephone: 202-722-0338

Membership: $40 annually

A volunteer support group for parents who are looking to adopt or who have adopted privately. Meetings and workshops are held in the Washington, D.C., area. Parents elsewhere can subscribe to the quarterly newsletter for $10 without becoming active members.

International Concerns Committee for Children
911 Cypress Drive, Boulder, CO 80303
Telephone: 303-494-8333

Membership: by contribution (optional)

A group that helps families interested in adopting children from other countries. The committee puts most of its efforts into finding homes for "waiting" children who have not been placed quickly because they are older, have medical conditions, or have siblings and want to stay together. The committee is not a child-placing agency. It provides information and counseling and does not charge parents for its services.

Latin America Parents Association
P.O. Box 72, Seaford, NY 11783
Telephone: 718-236-8689

Membership: $40 for first year, $30 thereafter

A parent support group for those considering Latin American adoption and for families that have already adopted Latin American children. New members receive a 150-page workshop packet. All members receive the quarterly newsletter *Que Tal?* and invitations to monthly meetings and activities.

National Adoption Center
1218 Chestnut Street, 2nd Floor, Philadelphia, PA 19107
Telephone: 800-TO ADOPT or 215-925-0200

Membership: by contribution (optional)

An association that works for adoption of difficult-to-place children—older children, those with physical or emotional disabilities, siblings who want to be placed together, and children of minority culture groups. A bibliography of children's books on adoption is available for $5.

National Committee for Adoption
1930 17th Street N.W., Washington, DC 20009-6207
Telephone: 202-328-1400

Membership: by contribution (optional)

An advocacy and informational agency that works on behalf of adoptive parents and children. The committee sells a number of books and pamphlets, including Jaqueline Plumez's *Successful Adoption: A Guide to Finding a Child and Raising a Family* and the committee's own *Adoption Factbook*. The publication list is free. For a self-addressed envelope and two first-class stamps the organization will mail a list of member agencies.

North American Council on Adoptable Children
1821 University Avenue, Suite S-275, St. Paul, MN 55104
Telephone: 612-644-3036

Membership: $25 annually

A group that works on behalf of children waiting for adoption, among them older children, children with mental, emotional, or physical handicaps, and minority children waiting for same-race parents. The council is not a placement agency, but works with agencies to help match children with qualified parents. Membership brings the quarterly newsletter *Adoptalk*. The group also sells a number of books, including *Understanding My Child's Korean Origins*, *Adopting the Older Child*, and such children's books as *Why Was I Adopted?*

Resolve, Inc.
5 Water Street, Arlington, MA 02174
Telephone: 617-643-2424

Membership: $30 annually

A national support group for couples with infertility problems. The organization publishes fact sheets on medical issues and maintains a directory of infertility specialists for couples who are trying to have a child. Local groups across the country provide additional information, as well as support for those coming to terms with the prospect of remaining childless.

CHILDBIRTH

ASPO/Lamaze
P.O. Box 952, McLean, VA 22101
Telephone: 800-368-4404 or 703-524-7802

Membership: $15 annually for families

An organization of certified childbirth educators. The national office refers expectant parents to local teachers and works to promote an optimal childbirth and early parenting experience for families. Members receive a discount on books and films about childbirth preparation. The bimonthly magazine *Genesis*, a benefit of membership, is written for childbirth instructors and professionals.

American College of Obstetricians and Gynecologists
c/o ACOG Resource Center, 409 12th Street S.W., Washington, DC 20024-2188
Telephone: 202-638-5577

Membership: professional only

The American College of Obstetricians and Gynecologists publishes dozens of informational patient education pamphlets. The Resource Center can tell you which ones have the information you need and will send out sample copies on request. The staff there will also provide lists of ACOG members and subspecialists for women who need an obstetrician or gynecologist.

Association for Childbirth at Home, International
P.O. Box 430, Glendale, CA 91209
Telephone: 213-667-0839 or 663-4996

Membership: $25 annually

An "international organization supporting home birth and dedicated to parents' right to decide where and with whom they will give birth." The group sponsors childbirth classes; refers parents to doctors and midwives in their area; and offers members a number of books, pamphlets, and research papers on the subject of home birth. Members also receive the quarterly newsletter *Birth Notes*.

Cesarean Prevention Movement
P.O. Box 152, Syracuse, NY 13210
Telephone: 315-424-1942

Membership: $25 annually

An organization that works to prevent unnecessary cesarean births by disseminating information about the benefits of vaginal birth. The national office puts out a quarterly newspaper, *The Clarion*, and offers a big list of books on pregnancy and childbirth. (Members get a discount on the books, but anyone can order from the catalog.) Local chapters provide additional information and support.

International Childbirth Education Association
P.O. Box 20048, Minneapolis, MN 55420-0048
Telephone: 612-854-8660

Membership: open to childbirth educators only

Sends out a free catalog of publications, called "Bookmarks," which offers books on pregnancy, childbirth, and breastfeeding.

Maternity Center Association
48 E. 92nd Street, New York, NY 10128-1397
Telephone: 212-369-7300

Membership: no members

"A national non-profit health agency founded in 1918 to demonstrate the value of prenatal care in reducing maternal and infant mortality." The association's birth center is open to any family in the New York City area. Its publications list is sent free on request. It includes books and pamphlets on childbirth preparation, childbearing choices, and infant care.

NAPSAC International
Rt. 1, Box 646, Marble Hill, MO 63764
Telephone: 314-238-2010

Membership: $15 annually, $17 outside U.S.

An association dedicated to promoting safe family-centered birth programs. Members receive the quarterly journal *NAPSAC News* and can participate in the activities of local chapters throughout the U.S. and Canada. Anyone can order from the group's extensive list of books and pamphlets, sent free on request.

National Association of Childbearing Centers
Rt. 1, Box 1, Perkiomenville, PA 18075
Telephone: 215-234-8068

Membership: no members

A national organization of birth centers that offer delivery in a homelike setting, along with pregnancy and childbirth education programs. Send a long self-addressed stamped envelope for a list of birth centers in your area.

DIVORCE

National Council for Children's Rights
2001 O Street N.W., Washington, DC 20036
Telephone: 202-223-6227

Membership: $25 annually

A membership organization that works to strengthen families and reduce the trauma of separation and divorce for children. The council publishes reports and educational materials (many advocating joint custody after divorce), holds an annual conference, and files court briefs on behalf of a child's right to two parents. The publication list, mailed free, includes a couple of books for children (*The Divorce Workbook: A Guide for Kids and Families* and *The Written Connection*, a system of parent-child communication), audio- and videocassettes, and works for parents and professionals on joint custody and family relations.

Parents Sharing Custody
435 N. Bedford Drive, Suite 310, Beverly Hills, CA 90210
Telephone: 213-273-9042

Membership: no members

A group that offers information and education to parents by telephone on the issues of divorce, family relocation, and parenting in joint-custody homes.

EDUCATION AND CHILD CARE

Association for Childhood Education International
11141 Georgia Avenue, Suite 200, Wheaton, MD 20902
Telephone: 800-423-3563 or 301-942-2443

Membership: $36 annually

A nonprofit association for parents, educators, and child-care professionals, dedicated to "advancing a comprehensive view of childhood education from birth through early adolescence—in the classroom and beyond." Members receive the journal *Childhood Education*, published five times a year with a dozen thoughtful articles in each issue, as well as access to local chapter meetings throughout the U.S. and Canada, and to the association's books and pamphlets.

Canadian Alliance of Home Schoolers
195 Markville Road, Unionville, ON L3R 4V8, Canada

Membership: $15 annually

A group founded in 1979 to provide support and information to Canadian home-schoolers. Members receive *Childs Play*, a newsletter; prospective home-schoolers are mailed an information package. The group also maintains a resource center and publicizes the benefits of deschooling in the media.

Child Care Action Campaign
99 Hudson Street, Suite 1233, New York, NY 10013
Telephone: 212-334-9595

Membership: $20 annually

An advocacy group whose goal is "to set in place a national system of quality, affordable child care, using all existing resources, both public and private." The membership fee helps the campaign with its lobbying and public-education efforts, and entitles members to free copies of the organization's informational papers. Non-members can order papers as well (on such topics as "Family Day Care: How to Find and Evaluate It" and "Temporary Care for the Mildly Sick Child"), but only three at a time.

Child Welfare League of America, Inc.
440 First Street N.W., Suite 310, Washington, DC 20001
Telephone: 202-638-2952

Membership: by contribution

An advocacy organization that works to improve services for deprived, dependent, or neglected children. The League's catalog of books and publications is available to anyone. It includes works on day care, adoption, developmental disabilities, foster care, and child abuse and neglect.

Children's Defense Fund
122 C Street N.W., Washington, DC 20001
Telephone: 202-628-8787

Membership: by contribution

An advocacy group that strives "to provide a strong and effective voice for the children of America who cannot vote, lobby, or speak for themselves." The fund pays particular attention to the needs of poor, minority, and handicapped children. A list of publications is available for the asking. Included are such studies as "Your Child's School Records," "How Congress Stands on Children," and "The Child Care Handbook: Needs, Programs, and Possibilities," as well as a number of reports on the problem of teen pregnancy.

Committee for Children
172 20th Avenue, Seattle, WA 98122
Telephone: 206-322-5050

Membership: by contribution

A nonprofit organization that provides classroom material, training, videos, and original research in the prevention of child abuse and youth violence. A catalog of classroom kits and videos is free for the asking.

Institute for Responsive Education
605 Commonwealth Avenue, Boston, MA 02215
Telephone: 617-353-3309

Membership: no members

An organization that works to "encourage citizen participation as an essential ingredient in school improvement." The institute conducts research studies, works as a consultant to school districts and community organizations, and offers parents a number of books and pamphlets on community involvement in education. The book list, sent free on request, includes titles like *A Citizen's Notebook for Effective Schools*, *Education Through Partnership*, and *The Home-School Connection*.

National Association for Family Day Care
815 15th Street N.W., Suite 928, Washington, DC 20005
Telephone: 202-347-3356

Membership: $12 annually

A professional organization of day-care providers that lobbies for legislation affecting children, sponsors group liability insurance plans, and offers information to members and the public. Parents are encouraged to join. Membership brings a subscription to the quarterly newsletter and a discount on the association's books (which include *Child Care in a Family Setting: A Comprehensive Guide to Family Day Care* and *Explore and Create: Activities for Young Children*). The book list is free to anyone who writes.

National Association for the Education of Young Children
1834 Connecticut Avenue N.W., Washington, DC 20009
Telephone: 800-424-2460 or 202-232-8777

Membership: $50 annually, $70 outside U.S.

A professional organization dedicated to improving the quality of services provided young children and their families. Members receive the bimonthly journal *Young Children* (see entry under *Parents' Magazines, Education and Child Care*) and a chance to participate in the activities of local affiliate groups. Anyone can write for and order from the association's "Early Childhood Resources Catalog" of books, posters, and brochures.

National Coalition of Alternative Community Schools
58 School House Road, Summertown, TN 38483
Telephone: 615-964-3670

Membership: $20 annually

For parents, educators, and others interested in alternative schooling. Membership includes a subscription to the quarterly *National Coalition News*, with news of conferences and coalition activities, articles about successful schools and programs, and a regular section on home schooling. Members also get a discount on the *National Directory of Alternative Schools*—a descriptive list of 600 schools—and the biennial journal *Skole*.

The National PTA
700 N. Rush Street, Chicago, IL 60611-2571
Telephone: 312-787-0977

Membership: $7 for subscription to *PTA Today*

A support organization for local parent-teacher associations. The magazine *PTA Today*, published monthly from October to May, focuses on a different theme in each issue. In the past these have included educational excellence, communicating with your child, and strengthening the family. A list of publications is sent free on request. In it parents will find fliers on subjects like discipline, tests, and latchkey children, along with aids to organizing and running a local PTA.

Southern Association of Children Under Six
P.O. Box 5403, Brady Station, Little Rock, AR 72215
Telephone: 501-227-6404

Membership: dues vary, call or write for address of state affiliate

An organization of early childhood educators, child development specialists, and parents concerned about the well-being of young children and their families. Most members join through affiliate groups in the 13 southern states, and can participate in state and local activities. Those outside the region can pay a reduced membership fee and receive the quarterly journal *Dimensions*, which features articles on the practical application of child development theories.

Family Planning

Choice
125 S. 9th Street, Suite 603, Philadelphia, PA 19107
Telephone: 215-592-7644

Membership: by contribution

An organization that provides information on reproductive health care and child care, with a goal of ensuring equal access to quality care regardless of income. Referrals are offered to those living in the Delaware Valley. The group's publications are mailed to anyone who asks. They include "Changes," a handbook for teenagers on puberty, and "Oh No, What Do I Say Now," a series of fliers to help parents answer a child's questions about sex.

The Couple to Couple League
3621 Glenmore Avenue, P.O. Box 111184, Cincinnati, OH 45211-1184
Telephone: 513-661-7612

Membership: by contribution

An international organization "devoted to helping engaged and married couples develop the practical art of natural family planning." The league promotes and teaches the Sympto-Thermal Method, which we are assured is *not* the calendar rhythm method. Several free and inexpensive pamphlets are offered, as well as the league's manual, *The Art of Natural Family Planning*.

Planned Parenthood Federation of America, Inc.
810 Seventh Avenue, New York, NY 10019
Telephone: 212-541-7800

Membership: by contribution

A family planning organization that, through local affiliates, provides medical services to millions of men and women in the U.S. and abroad. A free publication list is available from the Marketing Department, with pamphlets on various birth control methods, sexuality and pregnancy (directed at teenagers), and reproductive health care.

Resolve, Inc.
A national support group for couples with infertility problems. See entry under *Associations, Adoption*.

Health

Al-Anon/Alateen
Family Group Headquarters, P.O. Box 862, Midtown Station, New York, NY 10018
Telephone: 800-356-9996, 800-245-4656 in NY, or 800-443-4525 in Canada

Membership: no membership dues or fees

An international organization of support groups for those whose lives have been affected by the compulsive drinking of a family member or friend. Alateen is a fellowship of young Al-Anon members, usually teens. Those who call or write the national office receive a directory of local information services which can give a schedule of meetings in the area. Al-Anon publishes a number of books and pamphlets, including a booklet for younger children, and three bimonthly newsletters.

American Academy of Pediatrics
Publications Dept., 141 Northwest Point Boulevard, P.O. Box 927, Elk Grove Village, IL 60009-0927
Telephone: 312-228-5005

Membership: professional only

The American Academy of Pediatrics publishes a number of informational books and pamphlets about children's safety and health, but most are available only in bulk quantities to physicians. Parents can order the *Child Health Record* ($2, #HE0004) for tracking a child's medical and developmental history; a poster-size *First Aid Chart* ($2.50, #HE0008) with advice for common childhood emergencies on one side and for choking and CPR on the other; the booklet *Breastfeeding: A Gift of Love* ($1.25, #HE0018); and a kit called *TIPP: The Injury Prevention Program* ($3.50), stocked with tips for parents of children from birth to age 12.

American Optometric Association
Communications Center, 243 N. Lindbergh Boulevard,
St. Louis, MO 63141
Telephone: 314-991-4100

Membership: professional only

A professional association that publishes a number of informational pamphlets, sent free to anyone who mails a request with a long self-addressed stamped envelope. Of interest to parents are "Your Baby's Eyes," "Your Preschool Child's Eyes," "Your School-Age Child's Eyes," and "Toys, Games and Your Child's Vision."

American Red Cross
National Headquarters, Washington, DC 20006

Membership: by contribution

Check your telephone directory for the nearest local chapter, or write to the national headquarters. The Red Cross sponsors CPR classes and programs like Adopted Grandparents, Hobby Pals, Gamemates, Wheelchair Escorts, and Nature Walk Companions (which pairs senior and junior citizens).

Children in Hospitals
31 Wilshire Park, Needham, MA 02192
Telephone: 617-482-2915

Membership: $10 annually

An organization of parents and health-care professionals that seeks to loosen hospital regulations regarding children and visitors, with the goals of allowing parents to live in with hospitalized children and allowing children greater access to hospitalized parents. Members receive a quarterly newsletter.

Juvenile Diabetes Foundation International
432 Park Avenue S., New York, NY 10016
Telephone: 800-223-1138 or 212-889-7575

Membership: by contribution

A health agency that raises funds for research into the cause, cure, treatment, and prevention of diabetes. The foundation offers 18 free brochures on different aspects of the disease, including "Pregnancy and Diabetes," "Your Child Has Diabetes," "Your Baby Has Diabetes," and "What You Should Know About Diabetes." Canadians can write the foundation at 4632 Yonge Street, Suite 201, Willowdale, ON M2N 5M1, or call 416-224-2633.

National Clearinghouse for Alcohol and Drug Information
P.O. Box 2345, Rockville, MD 20852
Telephone: 301-468-2600

Membership: no members

Under the wing of the U.S. Department of Health and Human Services, NCADI prepares and distributes publications on alcohol and drug abuse. The catalog and most of the material in it are free. Families may want to take a look at *Parents: What You Can Do About Alcohol and Drug Abuse, When Cocaine Affects Someone You Love,* or a series of kits designed for the children of alcoholics.

National Institutes of Health
Division of Public Information, Office of Communications, OD, Editorial Operations Branch,
Bethesda, MD 20892

Membership: no members

The National Institutes of Health puts out a free "Publications List" that combines works issued by a number of Institutes and Divisions. Under the National Institute of Child Health and Human Development, for example, the list offers "Facts about Childhood Hyperactivity," "Infantile Apnea and Home Monitoring," and dozens of other papers and fact sheets for parents and health professionals. From the National Institute of Dental Research come "Baby Bottle Tooth Decay" and "Dental Tips for Diabetics." The National Cancer Institute puts out scores of educational booklets, as does the National Institute of Allergy and Infectious Diseases.

The National Mental Health Association
1021 Prince Street, Alexandria, VA 22314-2971
Telephone: 703-684-7722

Membership: $15 for newsletter

A voluntary organization that works to improve diagnosis, care, and treatment for children and adults with mental illness, and to increase public support for the funding of effective services. The association's Information Center provides referrals and answers questions by telephone or mail. The Publications Division puts out a free "Publications and Merchandise Catalog," stocked with informational booklets, program directories, and advocacy guides.

National Society to Prevent Blindness
500 E. Remington Road, Schaumburg, IL 60173
Telephone: 312-843-2020

Membership: by contribution

A group that started more than 80 years ago as a movement to prevent needless blindness from a disease called babies' sore eyes, now routinely treated at birth with drops of silver nitrate. Since then the organization has worked on many other aspects of public-health eye care and blindness prevention. Parents can write for a free kit, the "Family Home Eye Test," which is an excellent screening tool, and for a complete catalog of publications.

National Stuttering Project
4601 Irving Street, San Francisco, CA 94122-1020
Telephone: 415-566-5324

Membership: $25 annually

An organization that provides support and information to people who stutter, as well as referrals for those seeking professional help. Members receive the monthly newsletter *Letting Go*, which can be supplemented by *Letting Go Jr.* for young people who stutter (ages 6 to 13). The project's other publications include brochures for parents and teachers of children who stutter, a guide to finding a speech pathologist, and a series of helpful tapes.

The Will Rogers Institute
785 Mamaroneck Avenue, White Plains, NY 10605
Telephone: 914-761-5550

Membership: by contribution

A nonprofit health organization specializing in medical research and public health and safety education. The institute distributes a number of free booklets, among them "Shots for Tots," "Buckle Up Your Kids for Safety," "Tips for Working Parents," "What Everyone Should Know About Child Abuse," and "What Every Parent Should Know About Drugs and Drug Abuse."

The Sudden Infant Death Syndrome Alliance
330 N. Charles Street, Suite 203, Baltimore, MD 21201
Telephone: 301-837-8300

Membership: by contribution

An alliance of several national and local organizations that fund research into the cause of sudden infant death syndrome. The SIDS Alliance mails out a free brochure of facts about sudden infant death syndrome, and can put parents in touch with a nationwide network of support groups or with local organizations that offer information by telephone.

Hobby

American Numismatic Association
818 N. Cascade Avenue, Colorado Springs, CO 80903-3279
Telephone: 719-632-2646

Membership: $32 annually, $11 for children under 18

A national association of coin collectors. Adult members receive a monthly magazine, *The Numismatist*, and have access to a mail-order lending library of over 30,000 reference works. Young members receive the magazine *First Strike* and can participate in home-study courses and apply for scholarships.

The Far and Wide Recording Club
c/o Mrs. Elaine Walsh, 52 Oakley Court, Ancaster, ON L9G 1T5, Canada

Membership: $8 annually

An association that seeks to promote "friendship worldwide through taping." Members are put on a rotation to send and receive tapes according to their interests. Club sections include Chit Chat Round Robins, Old-Time Radio Shows, Big Band Sounds, Scratchy Old Records Show, Musical Memories, Musical Souvenirs, and Country & Western Highways & Byways.

League of American Wheelmen
Suite 209, 6707 Whitestone Road, Baltimore, MD 21207
Telephone: 301-944-3399

Membership: $22 annually

A century-old membership organization run by and for bicyclists. Members receive the magazine *Bicycle USA*, published nine times a year, with articles about all aspects of bicycling. A special TourFinder edition in March lists tours and tour operators.

National Association of Rocketry
2140 Colburn Drive, Shakopee, MN 55379

Membership: $19 annually, $15 for members aged 16 to 21, $12 for members under 16

An organization for model-rocketry enthusiasts (or "spacemodelers"). Members receive the monthly magazine *American Spacemodeling*, a directory of model manufacturers, a rule book for rocketry contests, and an invitation to the annual meet. Manuals, booklets, plans, and patches are sold to members at a discount.

National Button Society
Miss Lois Pool, Secretary, 2733 Juno Place, Akron, OH 44313
Telephone: 216-864-3296

Membership: $15 annually, $2 for Junior members (18 and under)

An association of button collectors. Members receive five issues of the society bulletin each year, with collector information, a calendar of button events, a section devoted to Junior members, a roster of members and local clubs, and dealer advertisements. The society can help put collectors in touch with state societies and local clubs.

Universal Autograph Collectors Club
P.O. Box 6181, Washington, DC 20044-6181

Membership: $18 annually, $15 for students in grades K–12

An association of autograph collectors. Members receive the bimonthly journal *The Pen and Quill*, which gives addresses of the famous and near-famous, explains how to spot forgeries and machined signatures, and keeps readers posted on news of the autograph world.

Special Needs

(The National Information Center for Children and Youth With Handicaps, described on page 225, is a good source of general information. The center can refer parents to other national, state, and local groups.)

American Council of the Blind
1010 Vermont Avenue N.W., Suite 1100, Washington, DC 20005
Telephone: 800-424-8666 or 202-393-3666
Membership: dues vary, check with state or local affiliate

A national membership organization for the blind and visually impaired, with local affiliates in each state. Call or write the national office for the address of the nearest group. The council offers information and referrals to members, directing them to services, counseling, educational programs, and sources of equipment. Staff lawyers offer advice on legal issues, particularly problems of discrimination. ACB Parents, an affiliated group, is made up both of parents of visually impaired children and visually impaired parents of sighted children.

American Foundation for the Blind
15 W. 16th Street, New York, NY 10011
Telephone: 800-323-5463 or 212-620-2000
Membership: professional only

A professional organization that provides information and referrals to the blind and visually impaired. The foundation can direct parents to programs, services, and sources of equipment.

American Speech-Language-Hearing Association
10801 Rockville Pike, Rockville, MD 20852
Telephone: 800-638-8255 or 301-897-8682
Membership: professional only

A professional association for speech-language pathologists and audiologists. Parents of children with speech, language, or hearing problems can write or call for a directory of programs and professional services available in their area, and for pamphlets on such subjects as otitis media and language development, stuttering, articulation problems, and child language. One titled "How Does Your Child Hear and Talk?" helps parents decide whether to seek professional help.

Association for Children and Adults with Learning Disabilities
4156 Library Road, Pittsburgh, PA 15234
Telephone: 412-341-1515
Membership: $20 annually, $25 outside U.S.

An organization for parents of children with learning disabilities, with more than 800 local chapters. The national headquarters offers more than 500 books, pamphlets, and papers for sale; provides a film rental service; and publishes a members' newsletter.

Association for Retarded Citizens
2501 Avenue J, P.O. Box 6109, Arlington, TX 76005
Telephone: 817-640-0204
Membership: $15 annually

A volunteer organization devoted to improving the welfare of children and adults with mental retardation and their families. Interested parents can find a local chapter in the telephone book or can contact the national office. The association sponsors parent support groups, recreational activities, citizen advocacy programs, public education efforts, and employment programs. Members receive *the arc*, a bimonthly newspaper, and can order from a list of books and fact sheets.

Alexander Graham Bell Association for the Deaf
3417 Volta Place N.W., Washington, DC 20007
Telephone: 202-337-5220
Membership: $35 annually

A century-old organization dedicated to improving the "educational, vocational, and personal opportunities for all hearing-impaired people." Members receive *Our Kids Magazine*, designed for parents, *The Volta Review*, an educational journal, and the newsletter *Newsounds*. They can also order from the association's extensive catalog of publications, take advantage of a lending library, and participate in the activities of local affiliates.

Canadian Rehabilitation Council for the Disabled
1 Yonge Street, Suite 2110, Toronto, ON M5E 1E5, Canada
Telephone: 416-862-0340
Membership: $30 annually

A coordinating body for more than 80 member groups across Canada that provide a network of services and programs for people with physical disabilities. Members receive the quarterly *Rehabilitation Digest Journal* and the newsletter *Access*. The national office also maintains an Information Resource Centre and publishes a number of books and pamphlets.

The Center on Human Policy
724 Comstock Avenue, Syracuse University, Syracuse, NY 13244-4230
Telephone: 315-443-3851
Membership: no members

The Center on Human Policy offers dozens of informational reports on the integration of people with severe disabilities into community life. The list of publications is sent free to anyone who writes; prices for the papers range from 50¢ to $5. The center also answers letters and telephone calls from parents who want information on and help in finding appropriate services for their children with disabilities.

Epilepsy Foundation of America
4351 Garden City Drive, Landover, MD 20785
Telephone: 301-459-3700 (800-EFA-1000 for answers to questions about seizure disorders)
Membership: by contribution

A nonprofit agency dedicated to the welfare of people with epilepsy. Its activities include advocacy, education, support of research, and delivery of services to those with epilepsy. Members receive a monthly newspaper and the opportunity to buy antiepileptic medication at a discount. The foundation's books, pamphlets, films, and videos can be ordered by anyone; a list is mailed on request.

The House Ear Institute
256 S. Lake Street, Los Angeles, CA 90057
Telephone: 213-483-4431
Membership: by contribution

The House Ear Institute funds research on hearing impairment and trains specialists in diagnosis, treatment, and rehabilitative techniques. It maintains a hot line—800-352-8888, or 800-346-8888 in California—for parents who suspect their child might have a hearing problem or who want information about raising and educating a hearing-impaired child. Two nontechnical booklets are also available: *Practical Suggestions for Persons With a Hearing Impairment* and *Communicative Assistive Devices* (which reviews devices and lists sources). An affiliated group of parents of deaf children, Parent-to-Parent, shares information about hearing aids and implants, and offers members support and guidance. Parent-to-Parent also sponsors family camps and a young-adult work program.

Human Resources Center
I.U. Willets Road, Albertson, NY 11507
Telephone: 516-747-5400
Membership: by contribution

A nonprofit organization that provides education, training, and employment programs to people with disabilities, from infants to adults.

Mobility International U.S.A.
P.O. Box 3551, Eugene, OR 97403
Telephone: 503-343-1284
Membership: $20 annually

A group that arranges for people with disabilities to travel abroad and participate in international educational exchange programs. The organization publishes the quarterly newsletter *Over The Rainbow*, provides travel information and referral services, and helps members choose and apply to international exchange programs.

National Association for Parents of the Visually Impaired
2180 Linway Drive, Beloit, WI 53511
Telephone: 800-562-6265 or 608-362-4945
Membership: $10 annually

A membership organization that provides support and information to parents of visually impaired children. Members receive a quarterly newsletter with information on service agencies, updates on legislative activity, and reviews of new books and equipment. The association publishes a number of books, including *Mainstreaming the Visually Impaired Child* and several resource guides.

National Association of the Deaf
814 Thayer Avenue, Silver Spring, MD 20910
Telephone: 301-587-1788
Membership: $25 annually

An advocacy group for deaf and hearing-impaired people. The association, through affiliated Junior NAD chapters, supports student organizations, camps for teens and younger children, workshops, seminars, and training for young members. It maintains a complete file of service providers throughout the country and can make referrals, though it provides no direct client services such as placement or counseling. The association has a thick catalog of publications, free for the asking. In it are children's books, greeting cards with messages in finger spelling, posters, and books on parenting, rehabilitation, and sign language. Members receive subscriptions to the monthly tabloid *NAD Broadcaster* and to the quarterly *Deaf American* magazine.

National Center for Learning Disabilities
99 Park Avenue, New York, NY 10016
Telephone: 212-687-7211
Membership: $25 annually

Members receive the annual magazine *Their World*, which in past years has included articles with such titles as "What to Do If Your Child Is Misdiagnosed," "Learning Disabilities and Preschoolers," and "Dyslexia Is My Gift!"—an architect's story of how his learning disability has helped him in his work.

National Easter Seal Society
70 E. Lake Street, Chicago, IL 60601
Telephone: 312-726-6200 or 4258
Membership: by contribution

The National Easter Seal Society and its local affiliates provide direct services for the rehabilitation of people with disabilities. Each year the group helps more than a million children and adults. A free catalog of publications includes dozens of books and leaflets with such titles as "An Approach to Motherhood for Disabled Women," "A Handbook on Stuttering," "Dos and Don'ts for Parents of Children with Hearing Problems," "Feeding the Cerebral Palsied Child," and "Camps for Children with Disabilities."

National Federation of the Blind
1800 Johnson Street, Baltimore, MD 21230
Telephone: 301-659-9314

Membership: $8 annually for families

The largest and oldest self-help organization for the blind and visually impaired, the National Federation of the Blind works for the independence and self-sufficiency of its members. The group lobbies on behalf of the blind, participates in legal actions, and helps to organize local support groups and workshops. A subgroup, Parents of Blind Children, puts members in touch with other families in their area, sponsors local groups where the numbers warrant, and publishes the quarterly magazine *Future Reflections* (available in a print edition or on cassette). A recent issue of the magazine included articles on blind role models for blind children, library services for blind children, and teaching Braille reading, as well as a profile of a noted blind scientist and reviews of several Braille books.

National Information Center for Children and Youth With Handicaps
P.O. Box 1492, Washington, DC 20013
Telephone: 800-999-5599 or 703-893-6061

Membership: no members

A government agency that provides free information to parents, teachers, and caregivers in helping children and youth with disabilities become participating members of the community. A publications list offers resource guides for parents, legal information, fact sheets on various disabilities, and brochures with such titles as "Children with Handicaps: Understanding Sibling Issues" and "Alternatives for Community Living." The center also provides referrals and technical assistance.

Reference Section, National Library Service for the Blind and Physically Handicapped
1291 Taylor Street N.W., Washington, DC 20542
Telephone: 202-707-9286

Membership: no members

A free library service offered to people who are unable to read or use standard printed material because of visual or physical disabilities. Recorded and Braille books and magazines are sent to eligible readers and returned by postage-free mail. Those interested in the program should call or write the Reference Section of the service for an application and for the address of the library serving their area. The same office sends out a number of free publications on disabilities, among them "Parents' Guide to the Development of Pre-School Handicapped Children: Resources and Services," and a number of bibliographies and reference circulars. Write or call for a complete list.

National Multiple Sclerosis Society
Information Resource Center and Library, 205 E. 42nd Street, New York, NY 10017
Telephone: 800-624-8236 or 212-986-3240

Membership: $20 annually

A group founded in 1946 to support research into the cause, prevention, cure, and treatment of multiple sclerosis. The basic membership fee brings a subscription to the newsletter *Inside MS* and voting rights in a local chapter. The society's publications are available to both members and non-members, and a list is sent on request. For the general reader it offers such booklets as "Living with Multiple Sclerosis: A Practical Guide" and more comprehensive books, including *Maximizing Your Health: A Program of Graded Exercise and Meditation for Persons with Multiple Sclerosis*.

Travel

American Youth Hostels, Inc.
National Office, P.O. Box 37613, Washington, DC 20013-7613
Telephone: 202-783-6161

Membership: $20 annually, $10 for youths under 18

American Youth Hostels gives members access to an international network of affordable accommodations. Best known as stopping places for young travelers, many hostels also accommodate adults and families. Members can also order guidebooks, sleeping sacks, and other travel gear from the AYH Travel Store.

Twins and Multiples

Center for Study of Multiple Birth
333 E. Superior Street, Suite 476, Chicago, IL 60611
Telephone: 312-266-9093

Membership: no members

The center will send a catalog of publications to anyone who writes. Included are books like *Make Room for Twins* by Terry Alexander and *The Joy of Twins* by Pamela Novotny, and reports on such subjects as triplet management, twins' private language, and premature twins.

International Twins Association
c/o Lynn Long and Lori Stewart, 6898 Channel Road N.E., Minneapolis, MN 55432
Telephone: 612-571-3022

Membership: $15 annually for adults, $7 for teens, $3 for children

An association organized by and for twins in 1934. It sponsors an annual Twins Convention—held in a different city in the U.S. or Canada each year—and publishes a newsletter. Triplets and other multiples are encouraged to join.

National Organization of Mothers of Twins Clubs
12404 Princess Jeanne N.E., Albuquerque, NM 87112-4640
Telephone: 505-275-0955

Membership: $10 annually for newsletter

A national coordinating body for local clubs across the U.S. The name is a bit misleading, as the group's membership takes in fathers, grandparents, adoptive and foster parents, and parents of triplets, quadruplets, and quintuplets. Those who write for information are mailed an informational brochure, "Your Twins and You," and a referral to the nearest local club or support group. The quarterly newsletter, "Mothers of Twins Clubs Notebook," profiles members, offers practical advice, and reports on club activities.

Parents of Multiple Births Associations of Canada
283 Seventh Avenue S., Lethbridge, AB T1J 1H6, Canada
Telephone: 403-328-9165

Membership: inquire for current rates

The Canadian counterpart to the National Organization of Mothers of Twins Clubs (see above).

The Triplet Connection
P.O. Box 99571, Stockton, CA 95209
Telephone: 209-474-3073 or 0885

Membership: $12 annually (plus $7 for "expectant/new parents information packet")

A nonprofit support organization for families who are expecting larger multiple births (triplets, quadruplets, and quintuplets) or who are raising multiples. Founded in 1983 in response to the lack of solid medical and child-rearing information available to parents of multiples, the group helps members reach out to others to share their experiences and offer help. Members receive a quarterly newsletter, information packets on multiple pregnancy and child care, access to a telephone hot line for answers to questions, and invitations to the biannual conventions.

Twinline
P.O. Box 10066, Berkeley, CA 94709
Telephone: 415-644-0861

Membership: by contribution

A nonprofit agency that provides information and services to parents of twins and larger multiples. Members receive the quarterly newsletter *Twinline Reporter*; access to a telephone "warm line" (the number above) providing information, counseling, and referrals; and an opportunity to shop from the agency's publications catalog. Parents in Northern California can take advantage of classes for new and expectant parents; a resource library; and a respite program that offers emergency child care, donated baby clothing and equipment, and counseling support.

Index

ABC School Supply, 88
AG Industries, 189
A-Plus Products, Inc., 17
ASPO/Lamaze, 217
Abbey Press, 108
Abilities International, 191
Abracadabra Magic Shop, 122
Action BMX Cycle Co., 31
Adoptive Families of America, Inc., 216
Adventures in Learning, 64
Aerie Design, 47
After the Stork, 47, *49*
Afterschool, 191
Aims International Books, Inc., 36
Al-Anon/Alateen, 220
Aladdin Stamps & Celebrations, 180
Alcazar Records, 132
Elaine Aldrich, 47
Alf Magazine, 64
Alice in Wholesale Land, 48
Alive Productions, Ltd., 113
All About Kids, 154
All But Grown-Ups, 95, *96*
All Night Media, 180
All Points Products, 113
Allergy Control Products, 114
Allstar Costume, 79
Alpine Ventures, 178
Ambient Shapes, Inc., 128
American Academy of Pediatrics, 220
American Arts & Graphics, Inc., 161
American Baby Magazine, 145
American Bronzing Co., 17
The American Clockmaker, 118
American College of Obstetricians and Gynecologists, 217
American Council of the Blind, 223
American Family Society, 113
American Foundation for the Blind, 223
American Library Association, 161
American Meteorite Laboratory, 168
American Montessori Consulting, 88

American Mothers, Inc., 113
The American Needlewoman, 118
American Numismatic Association, 222
American Optometric Association, 221
American Red Cross, 221
American Science Center, 168
American Spacemodeling, 222
American Speech-Language-Hearing Association, 223
American Stationery Company, Inc., 108
American Youth Hostels, Inc., 225
Americans for International Aid and Adoption, 216
America's Hobby Center, Inc., 128
Amish Country Collection, 95
Ampersand Press, 102
Anatomical Chart Co., 168
AnCar Enterprises, Inc., 17
Hanna Andersson, 48
Animal Town Game Company, 102
Annabetta, 48
AristoPlay, 102
The Ark Catalog, 191
Armor Products, 209
H. G. Arms Company, 114
Artisans Cooperative, 108
Laura Ashley by Post, 48
Ash's Magic Catalog, 122
Associated Photo Co., 159
Association for Childbirth at Home, International, 217
Association for Childhood Education International, 218
Association for Children and Adults with Learning Disabilities, 223
Association for Retarded Citizens, 223
Astronomy, 71
Atlanta Parent, 152
Atlas Tool Co., 128
Audio-Forum, 88
Audubon Workshop, 168
Austad's, 180
Avalon Hill Game Co., 102

Babe too! Patterns, 124
'BabieKnit' Industries, Inc., 17
Baby & Company, Inc., 18
Baby Biz, 18
Baby Bunz & Co., 18
Baby Care, 18
The Baby Connection, 154
Baby Dreams, 18
Baby Furs by Scandinavian Origins, 18
Baby-Go-To-Sleep Center, 36
Baby Lamb Products, Inc., 174
Baby Name-a-Grams™ Designer Birth Announcements, 33
Baby Safety Specialists, Inc., 114
Baby Table Bumper Productions, 114
Baby Talk, 145
Baby Wrap, Inc., 18
Babygram Service Center, 33
Babymax Angelwear, 49
Baby's Comfort Products, 18
Babysling, Inc., 19
Back to Basics Toys, 192
Badge A Minit, 181
Baltimore's Child, 152
Banbury Cross, 132
Barap Specialties, 209
Barbie Magazine, 64
Barker Enterprises, Inc., 118
Robert Barrow, Furnituremaker, 95
The Bartley Collection, Ltd., 95
The BASIC Teacher, 64
Eddie Bauer, 49
Bay Area Parent, 151
L. L. Bean, Inc., 138
Bear Feet, 178
Bear-in-Mind, 189
Bears to Go, 189
Bobbi Becker—Folk Artist, 161
Beegotten Creations, Inc., 124
Bellerophon Books, 36
Alexander Graham Bell Association for the Deaf, 223
Benny's Express, 192
Berman Leathercraft, 118

Best Products Co., Inc., 13
Best Selection, Inc., 19
Better Beginnings Catalog, 36
Between the Sheets Waterbed Products, Inc., 174
Bicycle USA, 206
Big Apple Parents' Paper, 153
Big Brothers/Big Sisters of America, 215
Big City Kite Co., 121
Big Toys, 184
Bighorn Sheepskin Company, 174
Bike Vermont Tours, 206
Bimbini, 49
Biobottoms, 50
Bird 'n Hand, 168
Birth & Beginnings, 19
Birth Notes, 217
Birth-O-Gram Company, 33
Birth to Three, 154
BirthWrites, 34
Bits & Pieces, 103
Bizzaro, 182
Blacklion Books, 36
Dick Blick Art Materials, 14
Bliss Ridge, 50
Blue Balloon, 50
Blue Sky Cycle Carts, 31
Bluejacket Ship Crafters, 129
Bluestocking Press/Educational Spectrums, 88
The Body Shop, 29
Bohlings, 50
Bruce Bolind, 108
Book Hunters, 37
Books of My Very Own, 45
Books of Wonder, 37
Boomerang Man, 121
Bosom Buddies, 124
Boston & Winthrop, 96
Boston Parents Paper, 153
Boston Proper Mail Order, 50
Boy Scouts of America, 215
Boys' Life, 64
T. E. Breitenbach, 161
Bright Future Futon Co., 174
Brights Creek, 50
Brooklyn Botanic Garden, 107
Brookstone Company, 103
The Brown University Child Behavior and Development Letter, 141
Building Blocks, 145

The Bulletin of the Center for Children's Books, 146
Bumkins International, Inc., 19
Bunting Magnetics Co., 168
W. Atlee Burpee & Co., 107
Button Creations, 50

CARS, 129
C.H.I.L.D., 19
Cabela's, 138
Cabin Creek Quilts, 174
Cabin North, 96
Cahill & Company, 37
California Hot Products, 31
California Pacific Designs, 182
Calvert School, 88
Camp Fire, Inc., 215
Campmor, 138
Canadian Alliance of Home Schoolers, 218
Canadian Rehabilitation Council for the Disabled, 223
Cane & Basket Supply Co., 118
Capri Photo Company, 159
Carrick Knitwear, 51
Caswell-Massey Co. Ltd., 29
CedarWorks, 184
The Center for Early Learning, 88
Center for Environmental Education, 168
Center for Study of Multiple Birth, 225
The Center on Human Policy, 223
Cesarean Prevention Movement, 218
Chad's Newsalog, 89
Channel Island Imports, 51
Chaselle, Inc., 14, 19, 78, 83, 89, 96, 118, 132, 180, 185, 192, 209
Chatelaine's New Mother, 145
Cheatsheet Products, Inc., 76
The Chef's Catalog, 78
Cherry Tree Toys, 210
The Cherubs Collection, 114
Chi Pants, 51
Chicago Parent News Magazine, 152
Chickadee, 62
Child, 141
Child and Youth Care Quarterly, 147
Child Care Action Campaign, 219
Child Care Information Exchange, 147
Child Health Alert, 150
Child Life, 65
Child Life Play Specialties, Inc., 185
Child Safety Catalog, 114

Child Welfare League of America, Inc., 219
Childbirth Alternatives Quarterly, 147
Childbirth Resources, 47
Childcraft, 192
Childhood, The Waldorf Perspective, 147
Childhood Education, 218
The ChildLife Video Collection, 207
Children & Teens Today, 148
Children in Hospitals, 221
Children Today, 148
Children's Album, 65
Children's Book & Music Center, 37, 132, 207
Children's Book News, 146
Children's Choice Book Club, 45
The Children's Collection, 51
The Children's Corner, 19
Children's Defense Fund, 219
Children's Digest, 72
Children's Music House, 132
Children's Playgrounds Inc., 185
Children's Playmate, 65
Children's Reading Institute, 45
Children's Recordings, 133
The Children's Room, 96
The Children's Shop, 51
The Children's Small Press Collection, 38
Children's Video Report, 146
Children's Wear Digest, 51
A Child's Collection, 38
Child's Play, 218
Chinaberry Book Service, 38
Chock Catalog Corporation, 52
Choice, 220
Christian Family Movement, 113
Christian Parenting, 141
Chuckles, 62
Circus Today, 129
Circustamps, 183
Claire's Bears & Collectibles, 189, 192
Classic Toy Trains, 72
Classical Calliope, 65
Classics for Kids, 52
Claus & Crew, 109
Clothcrafters, Inc., 78
Clothkit, 52
Cloud Chart, Inc., 169
Cloud Nine, Futons & Furnishings, 175

Cloudburst Quiltworks, 52
Cobblestone: The History Magazine for Young People, 65
Cohasset Colonials by Hagerty, 97
Collectors Marketing Corp., 118
Color Lab, 159
Colourwheel Designs, 52
Colten Creations, 124
Comfey Carrier, 20
Comfortably Yours, 114
Committee for Children, 219
Committee for Single Adoptive Parents, 216
A Common Reader, 38
Communique, 214
Community Playthings, 89
The Company Store, *138*, 175
Compare and Save Premium Catalogue, 20
The Compassionate Friends, Inc., 213
Compassionate Friends Newsletter, 141
The Compleat Mother, 147
Connecticut Parent, 151
Conran's Mail Order, 97
Constantine's Woodworker's Catalog, 210
Constructive Playthings, 89
Cot'ntot Fashions, 52
Cotton Patch Crafts, 84
Coulicou, 62
Country Cupboard, 84
Country Cycling Tours, 206
Country Spirit Crafts, 178
Country Workshop, 97
The Couple to Couple League, 220
Courier Health Care, Inc., 20
Cozy Baby Products, Inc., 20
Cozy Carrier, 20
Cracker Barrel Old Country Store, 13
Cradle Gram, 34
Crate & Barrel, 97, 193
Crayon Caps, 53
Create-A-Book, 38
Creations by Paula Brown, 125
Creative Kids, 66
Creative Parenting Resources, 38
Creative Playgrounds Ltd., 185, *186*
Creative Publications, 90
Creatively Speaking, 84
Cricket, 66
CritiCard, Inc., 114

Critic's Choice Video, Inc., 207
Cuddledown, 175
Cuddlers Cloth Diapers, 20
Cuisenaire Company of America, Inc., 90
Cumberland Crafters/Kid's Art, 53
Cumberland General Store, 193
Current, 183
Current Consumer & Lifestudies, 72
Custom Cards, 34
Cynthia's Country Store, Inc., 189

Daddy's Tees, 125
Dads Only, 141
Daisy Kingdom, 175
Dallas Alice, 53
Dallas Child, 154
Dance Magazine, 72
Day Care & Early Education, 148
Deaf American, 222
Decent Exposures, 125
John Deere Catalog, 193
Deerskin Trading Post, 20
Delby System, 29
Denver Parent, 151
Depression After Delivery, 113
Design Group, 210
Designer Diapers, 21
Designer Series, 125
Diap-Air, 21
Diaperaps, 21
Didax Educational Resources, 90
Dimensions, 220
Direct-To-You Baby Products, 21
Discovery Corner, 169
Discovery Music, 133
The Disney Catalog, 109
Walt Disney World, 109
Disney's DuckTales Magazine, 66
The Doll Cottage, 84
Doll Domiciles, 81
The Doll Factory, 84
Doll Reader, 72
Doll Repair Parts, Inc., 84
Dollsville Dolls & Bearsville Bears, 85
Dolphin Log, 66
Domestications, 175
The Donnelley Corp., 85
Double A Productions, Inc., 114
Doubleday, 38
Doug's Hobby Shop, 129
Doug's Train World, 129

Dover Publications, Inc., 39
Dover Scientific Co., 169
Dragon Magazine, 72
Walter Drake & Sons, 109
Droll Yankees Inc., 169
Gail Wilson Duggan Designs, Inc., 190
Duncan Toy Co., 194
Duncraft, 169
Durkin Hayes Publishing Ltd., 39
Dy-Dee Service, 21

EBSCO Curriculum Materials, 90
EDC Publishing, 39
E-Z Enterprises, Inc. 21
Early Advantage Programs for Children, 208
Early Learning Centre, 194
Eastman Kodak Co., 159
Eastside Parent, 155
Edmund Scientific, 169
Educational Activities, Inc., 90
Educational Materials Library, 90
Educational Record Center, Inc., 133
Educational Teaching Aids, 91
Elderly Instruments, 133
Electronic Arts Direct Sales, 76
Emily's Toy Box, 194
The Enchanted Child, 21
The Enchanted Doll House, 85
Engineered Knits, 175
English Garden Toys, 186
Engravables, 109
Environments, Inc., 16, 91, *194*, *211*
Epilepsy Foundation of America, 224
Estes Industries, 129
Every Buddies Garden of Puzzles, 103
Exceptional Children, 155
The Exceptional Parent, 155
Exchange, 215
Exploratorium Store, Mail Order Dept., 170
Exposures, 159

F & H Child Safety Co., 115
Fabrications, 161
Faces: The Magazine About People, 66
Families Adopting Children Everywhere, 216
Families Welcome!, 206
Family Clubhouse, 22
Family Communications, 113
Family Day Care Bulletin, 148
Family Day Caring, 148

Family Home Entertainment, IVE, 208
Family Life Products, 115
Family Pastimes, 103
Family Reader, 142
Family Resource Coalition, 113
Family Safety & Health, 150
The Family Software Catalog, 76
Family Travel Times, 156
Family Tree Originals, 109
The Fantasy Den, 190
The Far and Wide Recording Club, 222
Fatherhood Project, 114
Favorites from the Past, 81
Federal Emergency Management Association, 115
Feeling Fine Programs, 208
Fiberarts, 73
Henry Field Seed & Nursery Co., 107
5th Avenue Maternity, 125
Fille, 39
The Fine Tool Shops, 210
FineScale Modeler, 73
First Class, 98
1st Class B*M*X, 31
First Strike, 222
First Teacher, 148
Fisher-Price, 194
The Five Owls, 146
Flap Happy, 53
Florida Playground & Steel Co., 186
Flying Models, 73
Focus on the Family, 142
Henry Ford Museum & Greenfield Village, 170
Jack Ford Science Projects, 170
The Fordham-Scope Catalog, 33
ForParents, 142
Roberta Fortune's Almanac, 109
Foxfire, 73
Free Play, 195
French Creek Sheep & Wool Company, 139
Friends, 54
Frostline Kits, 139
Fun Furniture, 98
The Fun House, 122
Furniture Designs, 98

GCT Inc., 91
Gaines Pet Care Booklets, 158
Gander Mountain, Inc., 139
Gardener's Supply, 107
Garnet Hill, 54

A Gentle Wind, 134
Geode Educational Options, 39
GeorGee's, 54
Gifted Child Quarterly, 148
The Gifted Child Today, 149
Gifted Children Monthly, 142
The Gifted Children's Catalog, 195
Gil & Karen's Toy Box, 195
Girl Scouts of the U.S.A., 216
Gleanings, 40
Gloucester Classics Ltd., Inc., 166
David R. Godine, Publisher, Inc., 40
Golden Press, 46
Golem Computers, 76
Good Apple, 92
Good Gear for Little People, 139
Goods from the Woods, 98
Grand Rapids Parent, 153
Ginny Graves, 15
Great Days Publishing, Inc., 110
Green Tiger Press, 40
Grey Owl Indian Craft Co., 118
Grieger's, 119
Grolier Enterprises Inc., 46
Growing Child, 146
Growing Child, 195
Growing Without Schooling, 149
Growstick Co., 115
Gryphon House, 40
Gurney Seed & Nursery Co., 107
GYM*N*I Playgrounds, Inc., 186

H & F Products, Inc., 34
H.O. Center of the World, 129
H.U.D.D.L.E., 98
Hagenow Laboratories Inc., 170
Hagstrom's Sales Ltd., 196
Hammacher Schlemmer, 196
J. L. Hammett Co., 92
Hancock Toy Shop, 196
Hand in hand, 196
Hangouts, 98
Hansen Planetarium Publications, 162
Harper & Row, Publishers, Inc., 40
Harps of Lorien, 134
H. E. Harris and Co., Inc., 119
Hartline, 166
Abbie Hasse Catalog, 54
Hawk Meadow of New England, 110
Hazelwood, 22
Healthy Alternatives, Inc., 115
Hear You Are, Inc., 115

Heart Thoughts, Inc., 35
Hearthside Quilts, 175
HearthSong, 85, 196
Heartwood Arts, 196
Heath Co., 119
Heir Affair, 110
Herron's Books for Children, 40
Herrschners, Inc., 119
Hibou, 67
High Fly Kite Co., 121
High Q Products, 197
Highlights for Children, 67
Highway 70 Chair Shop, 81
Hill's Dollhouse Workshop, 81
Hobby Shack, 130
Hobby Surplus Sales, 130
Hold Everything, 99
Holst, Inc. 110
John Holt's Book & Music Store, 40, 136
Holy Cow, Inc./Woody Jackson, 54
The Home Business Advocate, 142
Home Education Magazine, 149
Home Education Press, 92
The Homespun Company, 54
Hoover Brothers, Inc., 92
The Horchow Collection, 197
The Horn Book Magazine, 147
Houghton Mifflin Company, 41
The House Ear Institute, 224
House of Oldies, 134
The Huggies Shopper, 22
M. P. Hughes Dulcimer Co., 135
Human Resources Center, 224
Humpty Dumpty, 62
Hyde Bird Feeder Co., 171
Hyperkites, 121

I Love You Drooly, 176
I-Z Industries, 183
Ice Wood Designs, 197
Ident-ify Label Corp., 55
Imagination Toys, 197
Indoor Model Supply, 130
Industrial Arts Supply Co., 119
Indy's Child, 152
Infant Wonders, 22
Information for Private Adoption, 216
John Ingalls Designs, 22
The Initial Place, 55
Initials+ Collection, 110

Institute for Responsive Education, 219
Intensive Caring Unlimited, 156
International Childbirth Education Association, 218
International Concerns Committee for Children, 216
International Twins Association, 226
Interracial Family Alliance, 114
Into the Wind/Kites, 121
Iris Inc., 22
Irish World Imports, 22
Isis Innovations, 187

JMK Enterprises, 115
J & M Designs, Inc., 162
Jack and Jill, 67
Jan's Small World, 82
Jerryco, 171
Johnson & Johnson Child Development Toys, 197
Joy Bee Designs, 35
Judi's Dolls, 85
Judy/Instructo, 104
Julia & Brandon, 198
Julia Toys, 210
Just for Kids!, 198
Just Sew Creative, 162
Justin Discount Boots and Cowboy Outfitters, 55
Juvenile Diabetes Foundation International, 221

KBS Designs, 55
Kansas City Parent, 152
Kaplan School Supply Corp., 92
Karen Studios, 159
Karter News, 73
H. Kauffman & Sons, 158
Kid City, 67
KiddyKube, Inc., 187
The Kids on the Block, Inc., 135
Kids Quarters Inc., 99
Kids Toronto, 155
KidsArt, 15
KidsArt News, 67
Kidstamps, 183
KIDVIDZ, 208
Miles Kimball, 110
Kimbo Educational, 135
Kindred Spirits, 114
King-Aire Products, 115
Klockit, 119
Klutz Flying Apparatus Catalogue, 122

Knit Wits, 176
Koala Club News, 67
KolbeConcepts, Inc., 92
Kotton Koala, 115

L.A. Parent Magazine, 151
L'Ecole des Loisirs Clubs, 46
La Leche League International, 125
Lait-Ette Company, 126
Lambskin, Inc., 176
Lamby, 176
Lamkin, Inc., 190
Lancaster Towne Quilts, 176
Landau, 176
Lands' End, 55
Latin America Parents Association, 217
Laughing Bear, 55
Laura D's Folk Art Furniture, 99
Lauri, Inc., 104
League of American Wheelmen, 222
Learn Me Bookstore, 41
Learning At Home, 92
The Learning Company, 76
The Learning Edge, 149
Learning Gifts, 76
Learning Materials Workshop, 198
Learning Things, Inc., 171
Leclerc Corp., 119
Elizabeth Lee Designs, 126
Hank Lee's Magic Factory, 123
The Left-Handed Complement, 15
LEGO Shop-at-Home Service, 199, 200
Leonard Bear Learning, Ltd., 93
Les Petits, 55
Lick Observatory OP, 171
Lifesaver Charities, 115
Light Impressions, 159
The Lighter Side, 110
Lindsay Publications Inc., 171
Linmar Specialties, Inc., 111
Lionel Service, 130
Little Angels, 135
Little Ears, 135
The Little Fox Factory, 78
Living Skills Press, 41
Living Sound Productions, 135
Livonia, 23
Lollipops, 149
Lopuco Ltd., 126
Lorimar Home Video, 208
Los Angeles Birthing Institute, 41

Val Love, 56
Love-in-Idleness Dolls, 85
Lovely Essentials/Snuggleups Diapers, 23
Loving Fathers, 142
Lucretia's Pieces, 104
The Lyon & Gryphon, 78

MB Enterprises, 41
MJ & Kids, Inc., 116
MMI Corporation, 171
The McGuffey Writer, 67
Archie McPhee & Company, 123
Maher Ventriloquist Studios, 165
Maid of Scandinavia, 157
Maine Baby, 23
Malick's Fossils, 171
Manna Computing Concepts, 77
Manzanita Publications, 15
Marlboro Records, 135
Marriage & Family, 142
Marvelous Toy Works, 200
Mason & Sullivan Co., 119
Massachusetts Society for the Prevention of Cruelty to Animals, 158
Maternity Center Association, 218
Maternity Modes, Inc., 126
Math Games with Manipulatives, 104
M. Matthews Handwoven Fabrics, 176
J. Mavec & Company, Ltd., 23
Merlyn's Pen, 73
Merrell Scientific/World of Science, 172
Metrobaby, 23
The Metropolitan Museum of Art, 23, 111
Mickey Mouse Magazine, 62
Micrograms, 77
Milk & Honey, 85
Milkduds, 56
Mill Pond Farms, 200
The Mind's Eye, 41
The Miniature Shop, 82
Miniatures Showcase, 73
Minnesota Parent, 153
Mobility International U.S.A., 224
Model Aviation, 74
Model Railroader, 74
Modern Homesteader, 56
Mommy's Helper, Inc., 23
Moncour's, 177

The Montgomery Schoolhouse, Inc., 200
Moonflower Birthing Supply, 24
David Morgan, 177
Jim Morris Environmental T-Shirts, 56
William Morrow & Company, 42
Moss & Associates, 162
Mother, 142
Mother & Baby, 145
Mother Nurture Breastfeeding Apparel & Patterns, 126
Motherhood, 126
Mothering, 142
Mothers at Home, 144
Mother's Choice, 143
Mother's Place, 126
Mothers Today, 145
Mother's Wear, 127
Mothers Without Custody, 214
Mothers Work, 127
Motherwear, 127
Mountain Craft Shop, 200
Mountain Shirts, 56
Mountain Toy Makers, 201
Mountaintop Industries, 139
The Mouse Hole Workshop, 86
Movies Unlimited, 208
Mueller-Wood Kraft, Inc., 201
Munchkin Outfitters, 56
Muppet Magazine, 68
Museum Books Mail-Order, 111
Museum of Fine Arts, Boston, 111
The Museum of Modern Art, 111
Museum of the American Indian, 24
Music Box World, 135
Music for Little People, *133*, 135
Music in Motion, 137
My Sister's Shoppe, 82
My Uncle, 82

NAPSAC International, 218
Nantucket Kiteman, 122
Nasco, 15, 93, 172
The National PTA, 220
National Adoption Center, 217
National Association for the Education of Young Children, 219
National Association for Family Day Care, 219
National Association for Parents of the Visually Impaired, 224
National Association of Childbearing Centers, 218

National Association of the Deaf, 224
National Association of Rocketry, 222
National Audubon Society, 172
National Button Society, 222
National Center for Learning Disabilities, 224
National Clearinghouse for Alcohol and Drug Information, 221
National Coalition Against Domestic Violence, 214
National Coalition of Alternative Community Schools, 219
National Committee for Adoption, 217
National Committee for the Prevention of Child Abuse, 214
National Council for Children's Rights, 218
National Council on Child Abuse and Family Violence, 214
National Easter Seal Society, 224
National Federation of the Blind, 225
National Gallery of Art, 162
National Geographic World, 68
National Information Center for Children and Youth With Handicaps, 225
National Institutes of Health, 221
National Library Service for the Blind and Physically Handicapped, 225
The National Mental Health Association, 221
National Multiple Sclerosis Society, 225
National Nanny Newsletter, 149
National Organization of Mothers of Twins Clubs, 226
National Society to Prevent Blindness, 221
National Stuttering Project, 221
The Natural Baby Co., 24, 201
Natural Elements, 24
The Nature Company, 172
Naturepath, 24
Nature's Little Shoes, 178
Nenuco, 30
New Expression, 74
New Moons, 35
New World Trading Co., 58
New York Family Magazine, 153
The New-York Historical Society, *162*, 163
Holly Nicolas Nursing Collection, 127
Noah's Ark, 68
North American Council on Adoptable Children, 217

North Star Toys, 202
Northeastern Scale Models Inc., 82
Northwest Baby, 155
Nurturing Today: For Self and Family Growth, 143
Nutshell News, 74

OURS, 216
Oak Meadow, 93
Odyssey, 68
Old Village Shop, 112
Olsen's Mill Direct, 58
One Step Ahead, 24
Opportunities for Learning, Inc., 77, 93, 172
Orange Cat, 42
Orbix Corporation, 202
Orphan Foundation of America, 216
Orvis, 13
Our Best Friends Pet Catalog, 158
Our Kids, 154
Out of the Woodwork, 100
Over the Moon Handpainted Clothing Co., 58
Over the Shoulder Baby Holder, 25
Owl, 68

Pacific Puzzle Company, 105
Pacifica Crafts Equipment, 120
Pack-O-Fun, 68
Page Boy Maternity, 127, *128*
Palmer Method Handwriting, 93
Pampers Baby Care Catalog, 25
Papa Don's Toys, 202
The Paragon, 112
Parent & Child, 152
The Parent Connection, 146
Parent Express, 155
ParentCare Ltd., 208
Parentguide News, 153
Parenting for Peace & Justice, 143
Parenting Magazine, 143
Parents, 143
Parents Anonymous, 214
Parents' Choice, 147
Parents Magazine, 143
Parents Magazine Newsletter, 46
Parents Magazine Read Aloud Book Club, 46
Parents of Multiple Births Associations of Canada, 226
Parents' Pediatric Report, 150
Parents Plus, 151

Index

Parents Sharing Custody, 218
Parents Without Partners, Inc., 215
Patagonia Mail Order, Inc., 58, *139*
Patterncrafts, 86
The Peaceable Kingdom Press, *35*, *163*
Pediatrics for Parents, 150
Pedigrees, The Pet Catalog, 158
The Pen and Quill, 222
J. C. Penney Company, Inc., 13
Penny Power, 68
Perfect Presents by Suzy, Ltd., 112
Perfectly Safe, 116
H. H. Perkins Co., 120
Persnickety, 163
Personal Stamp Exchange, 183
Personal Statements, 177
Claudia Pesek Designs, 25
Petit Pizzazz, 58
Philip's Foto Co., 160
Picture Book Studio, 42
Pictures—U.S.A., 160
Pieces of Olde, 86
Pierce County Parent, 155
Pillow Pals, Inc., 177
Pine Specialties, 100
Pittsburgh's Child, 154
Placenta Music, Inc., 25
Planned Parenthood Federation of America, Inc., 220
Play-Well Equipment Co., 187
PlayFair Toys, 202
The Playmill, 164
Plays, 69
Pleasant Company, 86
Plow & Hearth, 108
Pockets, 69
Polk's Modelcraft Hobbies, 130
Polywog Press, 42
Portland Family Calendar, 154
Portland Soaker, 25
Portraits, Inc., 164
Posh Impressions, 183
Poster Originals, Ltd., 164
Potomac Children, 152
Potpourri, 112
Pourette Mfg. Co., 120
Practical Parenting, 43
The Prairie Pedlar, 164
Pre-K Today, 149
Prehistoric Times, 69
Premarq, 100

Preschool Perspectives, 149
Preventive Dental Care, Inc., 116
Price Is Rite Shoes, 179
Prince Lionheart, 25, *30*
Printed Personals, 35
Priority Parenting, 143
Prism, 69
Priss Prints, Inc., 164
Pro Libris, 43
Pro-Moms, 25
The Public (Software) Library, 77
Publishers Central Bureau, 43
The Puzzle People, Inc., 105, *106*

Qualex, 160
Quiet Tymes, Inc., 25
Quilts Unlimited, 177

R & G Products, 93
The R. Duck Company, 59
Race Prep, 130
Racing Strollers, Inc., 180
Rail Scene, 59
Railroad Model Craftsman, 74
Rainbow Play Systems, 187
Rainbows & Lollipops, Inc., 26
Ramsey Outdoor, 140
Rand McNally & Co., 124
Random House Media, 77
Ranger Rick, 69
Rascals in Paradise, 207
Read It Again, 43
A Real Doll, 87
Reborn Maternity, 127
Recorded Books, 43
Recreational Equipment, Inc., 140
ReCreations Maternity, 128
Reflections, 70
Relco Industries, 120
Reliance Color Labs, Inc., 160
Renew America Catalog, 202
Research Concepts, 93
Resolve, Inc., 217
Richman Cotton Company, 59
The Ridge Company, Inc., 202
The Right Start Catalog, 26
Robbe Model Sport, Inc., 130
D. Robbins and Co., Inc., 123
The Will Rogers Institute, 222
Rose's Dollhouse Store, 83
Rowhouse Press, 164

Royal Rocking Horses, 167
Rubber Stamps of America, 183
Rubberstampmadness, 70
Ryder Products, 27

S & S Arts & Crafts, 16
S. F. Peninsula Parent, 151
S Gaugian, 74
Saks Fifth Avenue, 59
Salad Days, 43
The Salvation Army, 215
San Diego Family Press, 151
The San Francisco Music Box Co., 137
Santa's Elves, 112
Sara's Ride, Inc., 27
Sax Arts & Crafts, 16
School Products Co., 120
Hans Schumm Woodworks, 203
F. A. O. Schwarz, 203
Scienceland, 63
Scientific Models, Inc., 83
Scott Publishing Co., 120
Sears, Roebuck & Co., 13
Seattle's Child, 155
Seedling Short Story International, 70
SelfCare Catalog, 116
Sensational Beginnings, 203
Sesame Street Magazine, 63
Sesame Street Magazine Parents' Guide, 63
Shaker Workshops, 100
Shar Products Company, 136, 137
The Sheepskin Factory, 177
Shelter Systems, 187
Shibumi Trading, Ltd., 112
Shillcraft, 120
Shining Star, 149
Shoe Tree, 70
Shoes & Socks, Inc., 179
Shofar, 70
Shop Talk, 144
Shoppe Full of Dolls, 87
Sig Manufacturing Co., 131
Signals Catalog, 208
Simply Divine, 59
Single Mothers by Choice, 215
The Single Parent, 144
Skole, 220
Sleeping Fawn Cradleboards, 27
Small Fry Originals, 160
Small Treasures, 27

Small World Toys, 173
Smith & Hawken, Catalog for Gardeners, 108
Smithsonian Institution, 112
Snapi Toy Tethers, 27
Snoopy Magazine, 63
Soft as a Cloud, 60
The Soft Menagerie, 190
Soft Shoes, 179
Soft Star Shoes, 179
Soft Steps, 179
Software-of-the-Month Club, 77
Solutions, 117
Sound Sleep Products, 27
Southern Association of Children Under Six, 220
Southern Emblem Co., 60
Special Delivery, 147
Special Effects Merchandise, 209
Special Ideas, 60
Spencer's, 177
Spinning Fool Top, 204
Spoken Arts, 44
Sports Bookshelf, 180
Sports Illustrated for Kids, 70
Sporty's Preferred Living Catalog, 106
Squiggles and Dots, 100
Stamp Magic, 184
Standard Doll Company, 87
Standard Hobby Supply, 131
A Star Is Born, 35
Stave Puzzles, 106
StenArt, Inc., 164
Stepfamilies & Beyond, 144
Stepfamily Association of America, 215
Stepfamily Bulletin, 144
Stick-Em Up, 33
Stickers Only, 184
Stocking Fillas, 112
Stone Soup: The Magazine by Children, 70
Storage Concepts, Inc., 100
Stork, 63
Stowe Canoe & Snowshoe Co., 140
A. Strader Folk Art, 165
Strauss' Country Ware, 60
Strawberry Kite Collection, 77
Stuf'd 'n Stuff, 190
Sturbridge Yankee Workshop, 100
Suburban Parent, 153

The Sudden Infant Death Syndrome Alliance, 222
Sun Sparks, 204
Surmacz Originals, 160
Surprises: Activities for Today's Kids and Parents, 71
Suzo, 80
Sweet Baby Dreams, 137
Sweet Dreems, Inc., 28
SweetGrass, 101
The Sycamore Tree, 93

T & T Seeds Ltd., 108
TG Magazine, Voices of Today's Generation, 74
T'Owl Productions, 117
Tabor Industries, 101
Tackle-Craft, 140
Taffy's-by-Mail, 80
Tandy Leather Co., 120
Tangoes/Rex Games, 106
Louis Tannen, Inc., 123
Tapestry, 101
Taylor Gifts, 113
Teco, 181
Teddy Bear & Friends, 74
Telegames USA, 77
Telltales, 44
Tendercare Diapers, 28
Terminal Hobby Shop, 131
Texas Child Care Quarterly, 150
Their World, 224
Things of Science, 173
Think Big!, 101
Threads, 75
Three Rivers Amphibian, Inc., 173
3-2-1 Contact, 71
Tiffany & Co., 113
The Timberdoodle, 94
Timbers Woodworking, 210
Toad'ly Kids, 60
Today's Christian Woman, 144
TOPS Learning Systems, 173
Tortellini, 60
Tot Tenders, Inc., 28
Totline, 150
Totline Books, 44
Tots World Company, 117
Towards Life Catalogue, 30
Tower Hobbies, 131
Toy Designs, 211

The Toy Factory, 204
The Toy Works, Inc., 87
Toys and Joys, 211
Toys to Grow On, 204
Toys Unique, 167
Tracy Creations, 128
Trains, 75
Trampoline World, 181
Treasure, 75
Treasured Toys, *167*, 205
Trend Enterprises, 94
Trend-Lines, Inc., Woodworking Supplies, 211
The Triplet Connection, 226
Troll, 205
Tropical Fish Hobbyist, 75
Troubadour, Inc., 137
Tryon Toymakers, 205
Tully Toys, 167
Turtle, 63
Twinline, 226
Twinline Reporter, 226
Twins Magazine, 156

U. B. Cool, 61
U-Bild, 101
U.S. Games Systems, Inc., 106
U.S. Geological Survey, 124
U.S. Government Printing Office, Superintendent of Documents, 44
U.S.A. Toy Library Association, 193, 215
The Ultimate Outlet, 61
Universal Autograph Collectors Club, 222
The University Prints, 165
Unknown Products, Inc., 205

V.I.P. Bronzing Service, 28
Venture, 75
Vermont Bird Company, *28*, 140
The Vermont Country Store, 29
Lillian Vernon, 113
The Victorian Papers, 35
Vida Health Communications, 29
Video F.I.N.D.S., 209
Viking Penguin, Inc., 44
Visual Education Association, 94
Voyageur's, 140

WNY Family Magazine, 153
Waldenvideo, 209

L. T. Walker Co., Inc., 61
Wee Wisdom, 71
Weekly Reader Books, 46
Weekly Reader Software, 77
Welcome Home, 144
Westags, Inc., 117
Westchester Family Magazine, 154
Wet-No-More, 117
Wet Set Gazette, 151
Whittemore-Durgin Glass Co., 120
Wild Bird Supplies, 173
Wild Child, 61
Wilderness Press, 140
Williams-Sonoma, 79
Wilshire Book Co., 158
Wimmer-Ferguson Child Products, 21, 29
Winganna Care Products, 177

Wireless, 113
Wisconsin Wagon Co., 33
Wisconsin Wood Products, 167
Wobkins, 61
Wood Built of Wisconsin, Inc., 188
Woodcraft, 211
The Wooden Soldier, 61
Woodmonger, 167
Woodplay Incorporated, 188
Woodset, Inc., 188
The Woodworkers' Store, 211
Woodworks, 206
Work & Family Life, 144
Working Mothers Network, 215
Working Mother, 145
Workman Publishing, 45
World Wear, 61
World Wide Games, 106

World Wide Sea Shells, 173
The Writewell Co., 35
Dennis Wyatt, 188

Yards of Fun, 188
Ye Olde Huff N Puff, Mfg., 131
Yellow Moon Press, 45
Yield House, 101
York Color Labs, 160
Young Children, 150
Young Rembrandts, 160
Your Big Backyard, 64
Youth View, 152

The Zaadi Company, 117
Zaner-Bloser, 94
The Zany Zoo Gang, 165
Zoobooks, 71

Request to Readers

As time goes by, catalog prices will change, new companies will appear, old ones may close down, and this book will need revision. To make the next edition even better, we will appreciate letters from readers with news of new mail-order suppliers, tips about established ones we may have missed, and comments on your experiences—both good and bad—with the companies we've listed. Because of the volume of our mail, we can't promise to respond to everyone who writes, but we will use your information to make the next edition more comprehensive and informative.

If you write to let us know about a supplier, please include the company's complete address, and, if you know it, the catalog price and a telephone number. Tell us what you like about the business, whether it be a particular product or simply the efficient service.

We also welcome news about sources of hard-to-find items. Just after we'd set the type and finished the layouts for this edition, for example, we learned that cot-size sheets (the size needed for bunk beds) are virtually impossible to find except through the J. C. Penney and Sears Roebuck & Co. catalogs. We'd like to put more of that kind of information in the next edition.

Please address your correspondence to us at:

Steam Press
15 Warwick Road
Watertown, MA 02172

About the Author

Hal Morgan was born in 1954 and graduated from Hampshire College. He is the author of *Symbols of America* and *The Mail Order Gardener* and co-author of *Amazing 3-D* and *Prairie Fires and Paper Moons: The American Photographic Postcard, 1900–1920*. He and his wife, Kerry Tucker, have collaborated to write *Rumor!*, *More Rumor!*, *The Shower Songbook*, *The Kids' Bathtub Songbook*, and *The Kids' Bathtub Rhyme Book*. Together they run Steam Press, a book production company in Watertown, Massachusetts.

When not writing books, Morgan works as a graphic designer, digs in his garden, and cares for his son Wesley, born in January 1988 (seen modeling a T-shirt from Over the Moon Handpainted Clothing Co. on page 58).

Notes

Notes

Notes